HEART REGENERATION
Stem Cells and Beyond

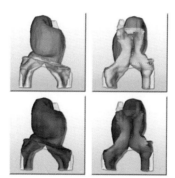

HEART REGENERATION

Stem Cells and Beyond

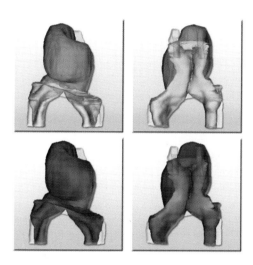

edited by

Felix B Engel

Max Planck Institute for Heart & Lung Research, Germany

 **World Scientific**

NEW JERSEY · LONDON · SINGAPORE · BEIJING · SHANGHAI · HONG KONG · TAIPEI · CHENNAI

Published by

World Scientific Publishing Co. Pte. Ltd.

5 Toh Tuck Link, Singapore 596224

USA office: 27 Warren Street, Suite 401-402, Hackensack, NJ 07601

UK office: 57 Shelton Street, Covent Garden, London WC2H 9HE

British Library Cataloguing-in-Publication Data
A catalogue record for this book is available from the British Library.

HEART REGENERATION
Stem Cells and Beyond

ISBN-13 978-981-4299-80-0
ISBN-10 981-4299-80-4

Typeset by Stallion Press
Email: enquiries@stallionpress.com

Printed in Singapore by Mainland Press Pte Ltd.

Contents

Contributors

Erin Bohula, M.D., D.Phil.
Cardiovascular Division
Brigham and Women's Hospital
75 Francis Street
Boston, MA 02115, USA
Email: EBOHULA11@PARTNERS.ORG

Thilo Borchardt, Ph.D.
Dept. I — Cardiac Development and Remodelling
Max Planck Institute for Heart and Lung Research
Ludwigstrasse 43
61231 Bad Nauheim, Germany
Email: thilo.borchardt@mpi-bn.mpg.de

Thomas Braun, M.D., Ph.D.
Dept. I — Cardiac Development and Remodelling
Max Planck Institute for Heart and Lung Research
Ludwigstrasse 43
61231 Bad Nauheim, Germany
Email: thomas.braun@mpi-bn.mpg.de

Kenneth R. Chien, M.D., Ph.D.
MGH Cardiovascular Research Center
Department of Stem Cell and Regenerative Biology
Harvard Stem Cell Institute, Harvard University
Holyoke Center, Suite 727W
1350 Massachusetts Avenue
Cambridge, MA 02138, USA
Email: KRCHIEN@PARTNERS.ORG

William C. Claycomb, Ph.D.
Department of Biochemistry and Molecular Biology
LSU Health Sciences Center
1901 Perdido Street
New Orleans, LA 70112, USA
E-mail: wclayc@lsuhsc.edu

Michael E. Davis, Ph.D.
Wallace H. Coulter Department of Biomedical Engineering
Emory University and Georgia Institute of Technology
101 Woodruff Circle
Atlanta, GA 30322, USA
Email: michael.davis@bme.gatech.edu

Stefanie Dimmeler, Ph.D.
Institute for Cardiovascular Regeneration
Centre of Molecular Medicine
Goethe University
Theodor Stern-Kai 7
60590 Frankfurt, Germany
Email: dimmeler@em.uni-frankfurt.de

Victor J. Dzau, M.D.
Chancellor for Health Affairs, Duke Medicine
Department of Medicine
Duke University Medical Center
Box 3701 DUMC
Durham, NC 27710, USA
Email: victor.dzau@duke.edu

Felix B. Engel, Ph.D.
Sofja Kovalevskaja Research Group
Dept. I — Cardiac Development and Remodelling
Max Planck Institute for Heart and Lung Research
Parkstrasse 1
61231 Bad Nauheim, Germany
Email: felix.engel@mpi-bn.mpg.de

Thomas Eschenhagen, M.D.
Department of Experimental Pharmacology and Toxicology
University Medical Center Hamburg Eppendorf
Martinistr. 52
20246 Hamburg, Germany
Email: t.eschenhagen@uke.de

Massimiliano Gnecchi, M.D., Ph.D.
Department of Molecular Medicine, University of Pavia
Laboratory of Experimental Cardiology for Cell
  and Molecular Therapy
Intensive Coronary Care Unit
Department of Cardiology
Fondazione IRCCS Policlinico San Matteo di Pavia
Viale Golgi 18
27100 Pavia, Italy
Email: m.gnecchi@unipv.it

Patricia A. Kelly, M.D.
Department of Cardiology
Missoula Heart Institute
2827 Fort Missoula Rd
Missoula, MT 59804, USA
Email: patrkelly@aol.com

Masamichi Koyanagi, M.D., Ph.D.
Department of Cardiovascular
  Respiratory, and Geriatric Medicine
Kyushu University Beppu Hospital
Kyushu University
4546, Tsurumihara, Tsurumi
Beppu, 874-0838, Japan
Email: m_koya@beppu.kyushu-u.ac.jp

Michael A. Laflamme, M.D., Ph.D.
Department of Pathology
Center for Cardiovascular Biology
Institute for Stem Cell and Regenerative Medicine
University of Washington
815 Mercer Street
Brotman Building Room 442
Mail Stop 358050
Seattle, WA 98109, USA
Email: laflamme@u.washington.edu

Jason T. Lam, Ph.D.
Klinikum rechts der lsar und Deutsches Herzzentrum
Technische Universität München
I. Medizinische Klinik — Molekulare Kardiologie
Ismaninger Straße 22
81675 Munchen, Germany
Email: jtlam@med1.med.tum.de

Karl-Ludwig Laugwitz, M.D.
Klinikum rechts der Isar und Deutsches Herzzentrum
Technische Universität München
I. Medizinische Klinik — Molekulare Kardiologie
Ismaninger Straße 22
81675 Munchen, Germany
Email: klaugwitz@med1.med.tum.de

Richard T. Lee, M.D.
Professor of Medicine, Harvard Medical School
Harvard Stem Cell Institute
Cardiovascular Division
Brigham and Women's Hospital
Partners Research Facility
65 Landsdowne St., Room 280
Cambridge, MA 02139, USA
Email: RLEE@PARTNERS.ORG

May Liang-Lam, Ph.D.
LSU Health Sciences Center
1901 Perdido Street
New Orleans, LA 70112, USA
Email: MLam@lsuhsc.edu

JoAnn Lindenfeld, M.D.
Professor of Medicine, Division of Cardiology
Director, Heart Transplantation Program
University of Colorado Denver
1635 Aurora Court, F749
Aurora, CO 80045, USA
Email: JoAnn.Lindenfeld@ucdenver.edu

Scott D. Lundy, Ph.D.
Department of Bioengineering
Institute for Stem Cell and Regenerative Medicine
University of Washington
815 Mercer Street
Brotman Building Room 442
Mail Stop 358050
Seattle, WA 98109, USA
Email: slundy@u.washington.edu

Robert J. Major, Ph.D.
Department of Biology
Indiana University of Pennsylvania
1011 South Drive
Indiana, PA 15705, USA
Email: rmajor@iup.edu

Maria Mirotsou, Ph.D.
Department of Medicine
Duke University Medical Center
Box 3701 DUMC
Durham, NC 27710, USA
Email: maria.mirotsou@duke.edu

Antoon F. M. Moorman, Ph.D.
Professor of Embryology
Department of Anatomy
    Embryology, and Physiology
University of Amsterdam
Meibergdreef 15
1105AZ Amsterdam, The Netherlands
Email: a.f.moorman@amc.uva.nl

Alessandra Moretti, Ph.D.
Klinikum rechts der lsar und Deutsches Herzzentrum
Technische Universität München
I. Medizinische Klinik — Molekulare Kardiologie
Ismaninger Straße 22
81675 München, Germany
Email: amoretti@med1.med.tum.de

Kara White Moyes
Department of Pathology
Center for Cardiovascular Biology
Institute for Stem Cell and Regenerative Medicine
University of Washington
815 Mercer Street
Brotman Building Room 442
Mail Stop 358050
Seattle, WA 98109, USA
Email: karaw2@u.washington.edu

Tatyana Novoyatleva, Ph.D.
Sofja Kovalevskaja Research Group
Max Planck Institute for Heart and Lung Research
Parkstrasse 1
61231 Bad Nauheim, Germany
Email: tatyana.novoyatleva@mpi-bn.mpg.de

Kenneth D. Poss, Ph.D.
Howard Hughes Medical Institute
Department of Cell Biology
Duke University Medical Center
Durham, NC 27710, USA
Email: kenneth.poss@duke.edu

Kathryn M. Rafferty, Ph.D.
Wallace H. Coulter Department of Biomedical Engineering
Emory University and Georgia Institute of Technology
101 Woodruff Circle
Atlanta, GA 30322, USA
Email: Kathryn.Rafferty@cchmc.org

Kai C. Wollert, M.D.
Department of Molecular and Translational Cardiology
Hans-Borst-Center for Heart and Stem Cell Research
Department of Cardiology and Angiology
Hannover Medical School
Carl-Neuberg-Str. 1
30625 Hannover, Germany
Email: wollert.kai@mh-hannover.de

Machteld J. van Amerongen, Ph.D.
Sofja Kovalevskaja Research Group
Max Planck Institute for Heart and Lung Research
Parkstrasse 1
61231 Bad Nauheim, Germany
Email: machteld.van-Amerongen@mpi-bn.mpg.de

Gert van den Berg, Ph.D.
Department of Anatomy, Embryology, and Physiology
Heart Failure Research Center
Academic Medical Center
Meibergdreef 15
1105AZ Amsterdam, The Netherlands
Email: G.vandenBerg@amc.uva.nl

Maurice J. B. van den Hoff, Ph.D.
Department of Anatomy, Embryology, and Physiology
Heart Failure Research Center
Academic Medical Center
Meibergdreef 15
1105AZ Amsterdam, The Netherlands
E-mail m.j.vandenhoff@amc.uva.nl

Bram van Wijk, M.D., Ph.D.
Department of Cardio-thoracic Surgery
Academic Medical Center
University of Amsterdam
PO Box 22660
1100 DD Amsterdam, The Netherlands
Email: a.vanwijk@amc.uva.nl

Wolfram-Hubertus Zimmermann, M.D., Ph.D.
Department of Pharmacology
Heart Research Center Göttingen (HRCG)
University Medical Center Göttingen
Georg-August-University Göttingen
Robert-Koch-Str. 40
37075 Göttingen, Germany
Email: w.zimmermann@med.uni-goettingen.de

Foreword
The Art of Reversing Heart Disease

"Art: A skill acquired by experience, study, or observation; the conscious use of skill and creative imagination especially in the production of aesthetic objects...."

—Merriam Webster's Dictionary

In the field of regenerative cardiovascular medicine, science is art. Careful observation of newts, zebrafish, and, most recently, mice, have uncovered unsuspected capabilities for regenerating entire regions of heart muscle following injury. The study of cardiogenesis has identified a family of multipotent heart progenitors that not only have challenged conventional thinking about heart formation but also suggested new approaches for cardiac regeneration. Skillful use of genetically engineered mouse models that will allow genetic tracking of the ancestors that form specific cardiac, smooth muscle, conduction system, and endothelial lineages have provided a potential map for generating these specific differentiated cells from pluripotent stem cell lines from specific patients. Creating functional heart muscle from pluripotent stem cells has led to the skillful integration of tissue engineering and stem cell biology towards the generation of heart parts. A brilliant example of the scientific imagination is provided by studies of human heart cell turnover based on carbon radioisotope pulse labeling from nuclear testing in the 60's, leading to a seminal translational advance in our thinking about human heart muscle regeneration. New insight from clinical experience has suggested both the challenges, as well as the opportunities in cell based therapy. In short, we are entering a new era of cardiovascular science and medicine, where the fundamental tools of stem cell biology are positioning our field to enter a new era where we actually can envision the reversal of heart disease.

In this regard, the book by Dr. Engel and colleagues brings together many of the young thought leaders in this new field, where each of the topics above is covered in detail. The text covers work in fundamental aspects of regenerative biology in lower vertebrates, and moves on to highly topical areas of tissue engineering, driving heart muscle cell replication by a variety of approaches, new insights into heart progenitors and approaches to unlock their potential therapeutic value, delivery devices, and a summary of recent clinical studies. In my view, this compilation is timely, focused, and a must read for anyone interested in the "art" of cardiovascular regenerative biology and medicine.

Few would argue that reversing heart disease will be one of the most daunting challenges for the field of cardiovascular medicine in the coming decades. Aging populations, improvements in post MI care, exponential growth of cardiovascular disease in rapidly developing nations such as China and India, epidemics of diabetes as a growing driver of disease progression all point to chronic cardiovascular disease in general, and heart failure in particular. Taken together, there has never been a more compelling case for rethinking our approach to chronic degenerative heart diseases. This text imagines a future based not only on slowing down disease progression, but on reversing heart disease. Perhaps in the coming decades, Art Basel will spring an offshoot of CardioBasel, and we will look back at this early work as forming the basis for the next generation of Rothko's in this rapidly growing field. Got art?

Kenneth R. Chien, MD, PhD
MGH Cardiovascular Research Center
and the Department of Stem Cell and Regenerative Biology
Harvard University
Cambridge, MA 02138, USA

I. Introduction to Cardiac Disease

Erin Bohula and Richard T. Lee

1. CARDIOVASCULAR DISEASE: THE LEADING CAUSE OF DEATH

Cardiovascular disease (CVD) is the leading cause of death world-wide, accounting for 17 million or 30% of deaths in 2004.[1] In middle- and high-income countries, the proportionate toll of cardiovascular disease is even higher. Cardiovascular disease was responsible for 36% of deaths in the United States in 2005, representing 2400 deaths daily or one death every 37 seconds. CVD claims as many lives each year as cancer, chronic lower respiratory diseases, accidents and diabetes combined. The estimated cost of CVD in 2009 was US$475.3 billion in the US alone.[2] The future burden of heart disease is expected to increase in high-income countries as the population ages and in low-income countries as infectious diseases diminish and the incidence of cardiac risk factors such as hypertension, obesity, and diabetes increases. Given all of this, it is critically important for the medical community to develop improved treatments to prevent, treat and reverse heart disease.

Ischemic heart disease is the leading cause of acute and chronic heart failure. Ischemia is usually caused by coronary artery obstruction and reduced blood flow, and acute reductions in blood flow can lead to myocardial infarction and potentially massive cardiomyocyte loss. Despite attempts by the remaining cardiac tissue to compensate via hypertrophy, loss of cardiomyocytes can progress to heart failure. While atherosclerosis is the most common cause of heart failure, numerous other etiologies exist, including valvular disease, toxins such as ethanol, pressure overload as seen with chronic hypertension, myocarditis and genetic cardiomyopathies. Regardless of the mechanism by which heart failure develops, the final common pathway in many cases is the loss of cardiomyocytes. Our current therapies attempt to prevent or minimize cardiomyocyte death, but to-date, we have limited options to effectively replace damaged or lost cardiac tissue other than organ transplantation. And, unfortunately, as will be discussed below, the therapies available are limited either by their longevity or poor availability.

2. CONVENTIONAL TREATMENT STRATEGIES

Much of the focus over the last several decades in cardiovascular medicine has been on prevention of coronary disease and heart failure and minimizing damage when it occurs. We have identified modifiable risk factors, like cholesterol, that when uncontrolled contribute to development of coronary artery disease. As a result there has been a major focus on preventative cardiovascular medicine, with targeting of lipid, diabetic and blood pressure goals and the addition of antiplatelet therapy. In addition to prevention, we have made major advances in treatment of acute myocardial infarction. Since the introduction of reperfusion therapy with thrombolysis, the mortality from transmural myocardial infarctions has decreased from around 11% in the mid-1980s to 7% or less in recent trials.[3,4] There has been an even further reduction in morbidity and mortality since the advent of acute percutaneous coronary intervention as compared to fibrinolysis and with the addition of antiplatelet agents, such as glycoprotein IIb/IIIa inhibitors.[5]

Despite these advances, heart failure has increased in prevalence in the last several decades.[6] This may be in part a product of our own successes in treating acute MI, where mortality has decreased significantly. Some patients who would have succumbed to a large MI are now surviving and progressing to heart failure. Other possible explanations may include the increasing number of people at risk as the population ages and with the development of cardiac risk factors at a younger age. Regardless of the cause, we are seeing a rise in the burden of heart failure in the US and worldwide.[6,7]

In addition to advances made in the areas of prevention and acute ischemia, there have also been improvements in the treatment of acute and chronic heart failure. Optimal medical management of systolic heart failure now includes angiotensin converting enzyme inhibitors or angiotensin receptor blockers, beta-blockers, and aldosterone antagonists, each of which has demonstrated a significant mortality benefit.[8-10] Furthermore, the use of implantable automatic cardiac defibrillators further reduced mortality by about 7% over five years in the SCD-HeFT trials.[11] Despite these additions to therapy for acute and chronic heart failure, mortality rates are still significant, estimated at ~4% in-hospital mortality, ~10% mortality at 60 days post discharge and as much as 30% mortality after one year.[12-14]

A possible explanation for the persistently high mortality seen with heart failure is that our current therapies slow progression but do not reverse the underlying process of cardiomyocyte loss. As mentioned previously, there are currently only two therapies that effectively replace missing cardiac tissue: heart transplantation and mechanical ventricular assist devices. Cardiac transplantation continues to be the best option for treatment of many patients with end-stage heart failure refractory to medical therapy, with survival rates of more than 85% at one year and 50% at ten years.[15] Unfortunately, the widespread use of this therapy is limited by a demand that massively outstrips donor supply.[16] Ventricular assists devices (VAD) have been used successfully as a bridge to transplantation and more recently as destination therapy. The Randomized Evaluation of Mechanical Assistance for the Treatment of Congestive Heart Failure (REMATCH) trial demonstrated improved overall survival and improved quality of life in the LVAD arm as compared to optimal medical management. However, poor survival two years after

LVAD implantation and frequent device-associated adverse events has led to limited use of destination implants until simpler, more durable devices become available.[17,18] Therefore, although these treatments options can be effective, they are limited by availability in the case of transplant and by long-term efficacy for LVADs.

3. MYOCARDIAL REGENERATION

The relatively high morbidity and mortality of heart failure, combined with several recent advancements in the field of developmental biology, has led to significant excitement about the possibility of myocardial regeneration as a treatment for cardiovascular disease. For many decades, we have thought that the heart was a terminally differentiated organ without any ability to replace damaged myocardium. This dogma has been challenged in recent years by findings in both non-mammalian and mammalian systems. Studies in non-mammalian model systems, including newts and zebrafish, have demonstrated surprising regenerative potential after injury. Newts regenerate cardiac tissue through a pathway of dedifferentiation of existing cardiomyocytes.[19,20] While prior studies suggested that zebrafish recruited undifferentiated precursor cells to repopulate injured myocardium, more recent studies suggest that zebrafish hearts, like newts, regenerate cardiac tissue primarily through proliferation and limited dedifferentiation of existing cardiomyocytes.[21–24] These findings raise two interesting questions: do mammalian hearts have similar regenerative mechanisms in place and, if so, is there a way to exploit these endogenous mechanisms as a means of treatment? While the answer to this question is far from complete, we are accumulating data that suggests that cardiac regeneration does in fact occur in mammalian hearts. Multiple studies using varied techniques suggest that the adult mammal has some degree of cardiomyocyte turnover. For example, exhaustive microscopic studies using immunohistochemistry markers of cellular division or direct visualization of mitotic figures showed low level turnover.[25–27] Chimeric studies of male transplant recipients of female donor hearts found Y chromosome positive cardiomyocytes on autopsy, suggesting post transplant cardiomyocyte birth from the recipient progenitor pool.[28] A mouse genetic fate mapping model demonstrated formation of cardiomyocytes from an immature myocyte or progenitor pool following pressure overload and infarction.[29] A study using retrospective birth dating of cells, a technique that takes advantage of a brief increase in atmospheric radioactive carbon during above ground nuclear bomb testing, demonstrated a low level of cardiomyocyte turnover (~1% annually).[30]

4. SUMMARY AND CONCLUDING REMARKS

As the data accumulates, most now agree that the traditional view of the heart as post mitotic is incorrect. This then brings us to the question of how to best harness the regenerative potential of the heart for therapeutics. In particular, the following questions arise: 1) what cells and tissues should be targeted for regeneration, 2) what is the most efficacious method for encouraging tissue regeneration and finally, 3) what is the optimal timing for cardiac regenerative therapies? Because the underlying pathology of heart failure is the irreversible loss of cardiomyocytes, presumably, successful regeneration strategies will involve delivery or generation of new cardiomyocytes. However, benefit could be derived from neovascularization, or generation of new vascular tissue, both for existing inadequately perfused tissue and for newly introduced or generated tissue. In fact, it is believed that in several trials of cell-based therapy, the improvement in cardiac function may have been through vasculogenesis rather than cardiomyocyte regeneration.[31,32]

Much of the research in the field has focused on the second question: how to best facilitate tissue regeneration. The basic approaches have fallen primarily into two categories: cell-based therapies and paracrine stimulation of endogenous mechanisms. Cell-based therapies involve exogenous introduction of differentiated cardiomyocytes or stem cell populations. Several different types of stem cell and progenitor cells have been shown to improve cardiac function through various mechanisms, including formation of new myocytes or vasculature and paracrine effects (reviewed in Ref. 33). The numerous candidate cells types, including embryonic stem cells, bone marrow-derived mononuclear cells, circulating hematopoietic stem cells, mesenchymal stem cells, skeletal myoblasts, endothelial progenitor cells and inducible progenitor stem cells, will be discussed in detail in Chapter IV. An alternative or adjunct to exogenous cell therapy is to stimulate endogenous mechanisms of cardiac repair and regeneration in a paracrine fashion. In fact, it has been proposed that the functional improvement seen after stem cell transfer in animal models may be secondary to secretion of soluble growth factors and cytokines that induce regeneration, promote vasculogenesis and discourage mal-adaptive remodeling after injury.[34] This theory is supported by experiments showing that administration of conditioned media from adult stem cells was able to match improvements in ventricular function seen with stem cell therapy post injury in adult rat hearts.[35] It remains to be seen whether cell-based approaches, paracrine strategies or a combination will yield the most successful cardiac regenerative therapies.

Finally, there is an important question of whether the optimal timing of regenerative therapies will be shortly after acute myocardial infarction or at the point of chronic heart failure. An argument for early post MI therapy is that fibrosis and ventricular remodeling have not yet occurred and pro-survival and angiogenic paracrine mediators might be of most benefit in this period. However, the intense inflammatory and hypoxic environment could limit survival of delivered or recruited stem cells. Advantages of treatment in the chronic stage are the ability to deliver therapy in a less time-sensitive manner, to a larger patient population. At this point, the preferred timing of regenerative therapies remains to be determined.

In the setting of continued morbidity and mortality from cardio-vascular disease, there is increasing interest in therapeutic myocardial regeneration. The following chapters will outline both the advances in the field, as well as the hurdles and questions still present. This text outlines our understanding of normal cardiac development and how that informs translation to clinically viable therapies. It addresses the evidence for cardiac regeneration in mammalian and non-mammalian systems during normal growth and after injury. It summarizes the current state of cardiac stem cell therapy, including a discussion of the various candidate progenitor cell populations and the paracrine effect of stem cells. And, finally, it describes efforts to create an artificial heart *ex vivo*, via tissue engineering and mechanical approaches.

REFERENCES

1. (2008) *The Global Burden of Disease: 2004 Update.* World Health Organization, Geneva.
2. Lloyd-Jones D, Adams R, Carnethon M, De Simone G, *et al.* (2009) Heart disease and stroke statistics — 2009 update: a report from the American Heart Association Statistics Committee and Stroke Statistics Subcommittee. *Circulation* 119: 480–486.
3. (1994) Indications for fibrinolytic therapy in suspected acute myocardial infarction: collaborative overview of early mortality and major morbidity results from all randomised trials of more than 1000 patients. Fibrinolytic Therapy Trialists' (FTT) Collaborative Group. *Lancet* 343: 311–322.
4. Antman EM, Morrow DA, McCabe CH, Murphy SA, *et al.* (2006) Enoxaparin versus unfractionated heparin with fibrinolysis for ST-elevation myocardial infarction. *N Engl J Med* 354: 1477–1488.
5. Schomig A, Kastrati A, Dirschinger J, Mehilli J, *et al.* (2000) Coronary stenting plus platelet glycoprotein IIb/IIIa blockade compared with tissue plasminogen activator in acute myocardial infarction. Stent versus Thrombolysis for Occluded Coronary Arteries in Patients with Acute Myocardial Infarction Study Investigators. *N Engl J Med* 343: 385–391.
6. Kannel WB (2000) Incidence and epidemiology of heart failure. *Heart Fail Rev* 5: 167–173.
7. Mendez GF, Cowie MR (2001) The epidemiological features of heart failure in developing countries: a review of the literature. *Int J Cardiol* 80: 213–219.
8. Pitt B, Zannad F, Remme WJ, Cody R, *et al.* (1999) The effect of spironolactone on morbidity and mortality in patients with severe heart failure. Randomized Aldactone Evaluation Study Investigators. *N Engl J Med* 341: 709–717.

9. Granger CB, McMurray JJ, Yusuf S, Held P, *et al.* (2003) Effects of candesartan in patients with chronic heart failure and reduced left-ventricular systolic function intolerant to angiotensin-converting-enzyme inhibitors: the CHARM-Alternative trial. *Lancet* 362: 772–776.

10. (1987) Effects of enalapril on mortality in severe congestive heart failure. Results of the Cooperative North Scandinavian Enalapril Survival Study (CONSENSUS). The CONSENSUS Trial Study Group. *N Engl J Med* 316: 1429–1435.

11. Bardy GH, Lee KL, Mark DB, Poole JE, *et al.* (2005) Amiodarone or an implantable cardioverter-defibrillator for congestive heart failure. *N Engl J Med* 352: 225–237.

12. Gheorghiade M, Abraham WT, Albert NM, Greenberg BH, *et al.* (2006) Systolic blood pressure at admission, clinical characteristics, and outcomes in patients hospitalized with acute heart failure. *JAMA* 296: 2217–2226.

13. Jong P, Vowinckel E, Liu PP, Gong Y, *et al.* (2002) Prognosis and determinants of survival in patients newly hospitalized for heart failure: a population-based study. *Arch Intern Med* 162: 1689–1694.

14. Yancy CW, Lopatin M, Stevenson LW, De Marco T, *et al.* (2006) Clinical presentation, management, and in-hospital outcomes of patients admitted with acute decompensated heart failure with preserved systolic function: a report from the Acute Decompensated Heart Failure National Registry (ADHERE) Database. *J Am Coll Cardiol* 47: 76–84.

15. Taylor DO, Edwards LB, Aurora P, Christie JD, *et al.* (2008) Registry of the International Society for Heart and Lung Transplantation: twenty-fifth official adult heart transplant report — 2008. *J Heart Lung Transplant* 27: 943–956.

16. Langone AJ, Helderman JH (2003) Disparity between solid-organ supply and demand. *N Engl J Med* 349: 704–706.

17. Rose EA, Gelijns AC, Moskowitz AJ, Heitjan DF, *et al.* (2001) Long-term mechanical left ventricular assistance for end-stage heart failure. *N Engl J Med* 345: 1435–1443.

18. Park SJ, Tector A, Piccioni W, Raines E, *et al.* (2005) Left ventricular assist devices as destination therapy: a new look at survival. *J Thorac Cardiovasc Surg* 129: 9–17.

19. Oberpriller JO, Oberpriller JC, Arefyeva AM, Mitashov VI, *et al.* (1988) Nuclear characteristics of cardiac myocytes following the proliferative response to mincing of the myocardium in the adult newt, Notophthalmus viridescens. *Cell Tissue Res* 253: 619–624.

20. Laube F, Heister M, Scholz C, Borchardt T, *et al.* (2006) Reprogramming of newt cardiomyocytes is induced by tissue regeneration. *J Cell Sci* 119: 4719–4729.

21. Jopling C, Sleep E, Raya M, Marti M, *et al.* (2010) Zebrafish heart regeneration occurs by cardiomyocyte dedifferentiation and proliferation. *Nature* 464: 606–609.

22. Kikuchi K, Holdway JE, Werdich AA, Anderson RM, *et al.* (2010) Primary contribution to zebrafish heart regeneration by gata4(+) cardiomyocytes. *Nature* 464: 601–605.

23. Lepilina A, Coon AN, Kikuchi K, Holdway JE, *et al.* (2006) A dynamic epicardial injury response supports progenitor cell activity during zebrafish heart regeneration. *Cell* 127: 607–619.

24. Poss KD, Wilson LG, Keating MT (2002) Heart regeneration in zebrafish. *Science* 298: 2188–2190.

25. Urbanek K, Quaini F, Tasca G, Torella D, *et al.* (2003) Intense myocyte formation from cardiac stem cells in human cardiac hypertrophy. *Proc Natl Acad Sci USA* 100: 10440–10445.

26. Kajstura J, Leri A, Finato N, Di Loreto C, *et al.* (1998) Myocyte proliferation in end-stage cardiac failure in humans. *Proc Natl Acad Sci USA* 95: 8801–8805.

27. Beltrami CA, Di Loreto C, Finato N, Rocco M, *et al.* (1997) Proliferating cell nuclear antigen (PCNA), DNA synthesis and mitosis in myocytes following cardiac transplantation in man. *J Mol Cell Cardiol* 29: 2789–2802.

28. Quaini F, Urbanek K, Beltrami AP, Finato N, *et al.* (2002) Chimerism of the transplanted heart. *N Engl J Med* 346: 5–15.

29. Hsieh PC, Segers VF, Davis ME, Macgillivray C, *et al.* (2007) Evidence from a genetic fate-mapping study that stem cells refresh adult mammalian cardiomyocytes after injury. *Nat Med* 13: 970–974.

30. Bergmann O, Bhardwaj RD, Bernard S, Zdunek S, *et al.* (2009) Evidence for cardiomyocyte renewal in humans. *Science* 324: 98–102.

31. Erbs S, Linke A, Schachinger V, Assmus B, *et al.* (2007) Restoration of microvascular function in the infarct-related artery by intracoronary transplantation of bone marrow progenitor cells in patients with acute myocardial infarction: the Doppler Substudy of the Reinfusion of Enriched Progenitor Cells and Infarct Remodeling in Acute Myocardial Infarction (REPAIR-AMI) trial. *Circulation* 116: 366–374.

32. Kawamoto A, Tkebuchava T, Yamaguchi J, Nishimura H, *et al.* (2003) Intramyocardial transplantation of autologous endothelial progenitor cells for therapeutic neovascularization of myocardial ischemia. *Circulation* 107: 461–468.

33. Segers VF, Lee RT (2008) Stem-cell therapy for cardiac disease. *Nature* 451: 937–942.

34. Gnecchi M, Zhang Z, Ni A, Dzau VJ (2008) Paracrine mechanisms in adult stem cell signaling and therapy. *Circ Res* 103: 1204–1219.

35. Gnecchi M, He H, Noiseux N, Liang OD, *et al.* (2006) Evidence supporting paracrine hypothesis for Akt-modified mesenchymal stem cell-mediated cardiac protection and functional improvement. *FASEB J* 20: 661–669.

II. Growth and Differentiation of the Developing Heart

Bram van Wijk, Gert van den Berg,
Maurice J. B. van den Hoff and Antoon F. M. Moorman

1. INTRODUCTION

Due to the limited regenerative capacity of the adult heart, loss of cardiomyocytes leads to irreversible cardiac damage. For instance, heart muscle that is lost after a myocardial infarction is replaced by fibroblasts, which will form a fibrotic scar. This scar warrants the structural integrity of the damaged cardiac wall, but also predisposes for arrhythmias[1] and does not contribute to the cardiac output. Subsequent hypertrophy and dilation of the heart progresses into congestive cardiac failure. To resolve this problem, many strategies to stimulate cardiac regeneration are under investigation. A challenge of current cardiac developmental biology is to offer inroads into the development of new strategies to stimulate myocardial growth and/or differentiation and to repair the damaged heart.

Many processes that occur during cardiac development, if reactivated, could be of great importance to the repair of the adult failing heart. These processes comprise the differentiation of mesodermal cells into myocardium,[2] the reinitiation of proliferation of the forming myocardial chambers,[3] and the diversification of cardiomyocytes into the components of the conduction system and the chambers.[4] Another important process is the separation of the distinct cardiac lineages, such as cardiomyocytes and cardiac fibroblasts, from the same precursor pool.[5,6]

The non-myocardial cells that reside in the heart might be interesting with respect to the development of new regenerative strategies. Although cardiomyocytes make up the bulk of the myocardial volume, the non-myocytes are the most numerous cells. In the adult heart 30% of the cells are cardiomyocytes, while the remaining 70% of the cells are non-myocardial.[7] These non-myocardial cells comprise the endocardium, the endothelium and the smooth muscle cells of the coronary vessels, the epicardium, cells present in the valves, and the cardiac fibroblasts.[8] Except for the endocardial cells, these cells have not always been present in the heart, but were added, during development, via the pro-epicardium.[9–12]

The pro-epicardium is a villous structure that is formed at the inflow of the heart, at 9.5 days of development in mice and 22 days in humans (Carnegie Stage 10). Until this stage, the heart tube lays "naked" within the pericardial cavity (Figure 1). The pro-epicardium protrudes into the pericardial cavity and attaches to the tubular heart, from where the pro-epicardial cells spread over the myocardium, to cover the heart with an epicardium. From this stage onwards the heart thus contains an endocardial inner layer, an epicardial outer

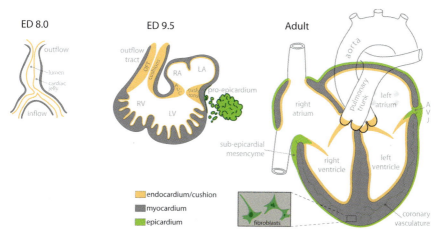

Figure 1.   Development of the endocardial, myocardial and epicardial lineage.
At eight days of development in mouse, the primitive heart tube encompasses a
myocardial outer layer and an endocardial inner layer, separated by cardiac jelly. The
primitive heart tube still lies naked within the pericardial cavity. At 9.5 days of development, epicardial cells are added to the developing heart via a structure called the
pro-epicardium, which develops from splanchnic mesoderm at the inflow of the
heart. Epicardial cells then migrate over the myocardium and cover the myocardium.
In the adult heart, endocardium-derived cells have contributed to the atrio-ventricu-
lar valves, part of the semilunar valves, and still cover the lumen of the cardiac
chambers. Epicardium-derived cells have contributed to the coronary vasculature, the
subepicardial mesenchyme, a minor part of the atrioventriclar valves, and the cardiac
fibroblasts.

layer and a myocardial layer in between (Figure 1). During further
development, these three layers will contribute to almost all cell types
that are present in the adult heart. Besides the cells derived from these
layers some additional cell types will be added to the heart. Firstly,
mesenchymal cells at the systemic inflow and around the pulmonary
veins will differentiate into myocardium.[13,14] Secondly, neural crest-
derived cells will populate the heart at various sites.[15] And finally a
small contribution of hematopoietic cells to the heart has been
reported.[16]

Nonetheless, most cells of the adult four-chambered heart are
derivatives of the myocardial, endocardial, and epicardial cells, present

at embryonic day 10 in mouse and day 24 in human. These three cell types are intimately associated with one another and their interactions result in a highly coordinated pattern of proliferation and differentiation of the developing heart. Insights into these processes undoubtedly will lead directly, or indirectly, to novel hypotheses on how to approach cardiac regeneration. In this chapter we shortly describe the proliferation and differentiation of the embryonic myocardium, starting at early heart field stages and ending at the four-chambered heart. We will give an overview of processes and factors that are involved in the development of the myocardium, endocardium and epicardium and the cells that are derivatives of these lineages. For more detailed information regarding the formation of the heart the reader is referred to other textbooks and reviews.[4,9,17–19]

2. MYOCARDIUM

2.1. Formation of the Early Heart Tube

After gastrulation, the flat embryo consists of three germ layers. These are, going from ventral to dorsal, the endoderm, mesoderm, and ectoderm. The mesoderm separates into a somatic and a splanchnic layer by the formation of the coelomic cavity. It is the initially flat splanchnic mesoderm, which faces the endoderm that forms the primitive heart tube by several morphological transitions and differentiation steps (Figure 2). Firstly, the splanchnic mesoderm starts to form the bilateral vitelline veins, which bulge out from the splanchnic part of the coelomic wall into the coelomic cavity. Then, by the process of folding, the cranial and the lateral aspects of the embryonic disc bend inwards. This forms the foregut as a pocket in the endoderm, and brings the left and right vitelline veins in the midline, where they fuse to form the primary heart tube. The mesodermal walls of the fused vessels have started to express sarcomeric proteins[20] and will, shortly after, initiate slow waves of peristaltic contractions, always starting at the venous pole of the heart.[21] Progressive folding and fusion of the vitelline veins lengthens the primary heart tube in

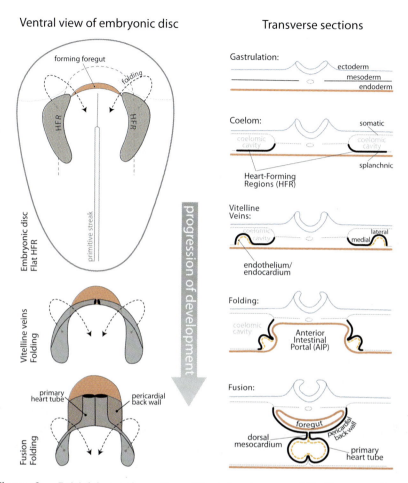

Figure 2. Initial heart formation. The left column spatially illustrates, from a ventral view, early heart formation in chicken. The bilateral heart-forming regions adjoin in midline by folding of the embryonic disc (hatched arrows). Progressive folding lengthens the heart tube in caudal direction, forms the foregut, and establishes the pericardial back wall. The right column illustrates the same process in transverse sections. The flat embryonic disc contains three germ layers (ecto-, meso-, and endoderm). The mesoderm separates into a somatic and splanchnic layer by the formation of the coelomic cavity. It is the splanchnic mesoderm that will form the heart. Firstly, lateral splanchnic mesoderm forms the lumina of the primitive vitelline veins. Then, by folding, these veins swing towards the midline, where they fuse to form the heart tube. Medial splanchnic mesoderm remains to overlie the endoderm of the foregut, so forming the pericardial back wall. The pericardial back wall and the primary heart tube are connected via the dorsal mesocardium.

the direction of its venous pole, while, at the same time, cells are also added to the arterial pole of the heart (reviewed in Ref. 22).

The myocardium of the early heart tube was previously reported to be slow-proliferating.[23,24] Recent work from our laboratory showed the myocardium of the early chicken heart to have a cell-cycle time of 5.5 days; therefore, this early cardiac tube should be considered a non-proliferating structure.[25] The number of cardiomyocytes within this early heart tube, however, was shown to increase rapidly during the same developmental time-frame.[24] These observations indicate that the early heart grows by addition of precursor cells to the myocardial lineage. The precursors that form the heart are currently attributed to two distinct developmental fields.[26] The so-called first heart field is considered to give rise to the linear heart tube and to contain the precursors for the left ventricle and atrioventricular canal, while a second heart field contains the precursors for the rest of the heart, added to both the venous and arterial pole of the heart. Whether the two fields are distinct lineages is debated.[27]

For instance, the hallmarking gene of the second heart field, *Islet1*,[28] was recently found to be already expressed in the precursors of the primary heart tube.[29] Also, classic transplantation studies of the precardiac mesoderm argue against an early specification of the fate of subsets of this mesoderm.[30] The concept of a first and second heart field may therefore better be considered to be a working model, rather than a concept of two discrete embryological entities.[31] Our own studies of cardiac growth in chicken further underline this concept. Within the splanchnic mesoderm, outside the non-proliferating early heart tube, a single center of rapidly dividing cells was observed caudally and medially to the inflow of the heart tube (Figure 3).[25] This growth center dispatches cells laterally and medially. The lateral cells were added to the inflow pole of the heart, while the medial cells moved, via the pericardial back wall, to the arterial pole of the heart.[25,32] Recent observations further showed that cells from the second heart field also contribute to extra-cardiac structures, such as mesenchyme within the aortic arches.[33,34] This indicates that the cells within this field do not have an exclusive cardiac potential, and harmonizes with the observation that the cells that lie even more medial

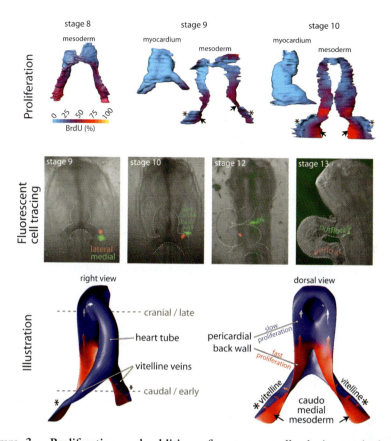

Figure 3. Proliferation and addition of precursor cells during early heart formation. The top row shows proliferation in the splanchnic mesoderm and the myocardium of the forming chicken heart, as measured by the fraction of cells labeled with BromodeoxyUridine (BrdU) after one hour of exposure. At Hamburger and Hamilton stage 8, the splanchnic mesoderm is fusing cranially. At the point of fusion, the cells display slow proliferation, while caudal mesoderm proliferates rapidly. At stage 9, fusion of mesoderm progressed in caudal direction, and also some mesoderm differentiated into myocardium. The newly-formed myocardium displays a virtual absence of proliferation, as also applies to the caudal and lateral mesoderm (*). Caudal and medial mesoderm, on the other hand, shows rapid proliferation (arrowheads). At stage 10, development progressed, as shown by the lengthening and looping of the heart tube. The patterns of proliferation are similar to those at stage 9. The bottom row shows the fate of the splanchnic mesoderm, traced with fluorescent dyes. The lateral, slow proliferating, mesoderm is incorporated into the inflow of the heart with folding (red dye). The medial, rapidly proliferating, mesoderm moves upward via the pericardial back-wall into the outflow of the heart tube (green dye).

to the proliferating growth center contribute to the pharyngeal arch mesoderm.[32,35,36]

Altogether, these observations strongly point towards a model in which the second heart field should be considered to be a spatio-temporal separation of a greater heart-forming region within the splanchnic mesoderm. The lateral cells of this region are the first to exit the cell-cycle and will form the primary heart tube with folding, while the medial mesoderm continues to proliferate and gradually adds cells to both the inflow and outflow of the heart. Be that as it may, the notion that an extra-cardiac, heart-forming region provides the precursor cells for the forming heart could be of great clinical interest from the stance of cardiac regeneration, that is, if the factors that orchestrate this embryonic formation of heart muscle from an extra-cardiac source of cells, could also be employed in the damaged adult heart.

2.2. Formation of the Cardiac Chambers and the Cardiac Conduction System

In the heart, different types of myocardium exist; most notably the working myocardium of the chambers, and the myocardium of the conduction system, such as the nodes and the bundles. For regenerative therapies it is of great importance to introduce the proper myocardial cell type at the right place in the heart. For instance, myocardial conduction cells are needed to repair the sinus node, while the same cells might lead to life-threatening arrhythmias when placed in the ventricles, trying to improve the pump function of the chambers after a myocardial infarction. Below we will discuss the formation of the chamber and conduction myocardium during the development of the heart.

The tubular heart is a relatively undifferentiated structure, containing no chambers and propelling blood from its venous to its arterial pole in a peristaltic fashion. The phenotype of this myocardium is dubbed "primary".[4] Primary myocardium displays low contractility, contains few sarcomeres, and expresses low-conductance gap-junctions composed of Connexin (Cx) 45 and Cx30.3,[37] resulting in slow conduction of the wave of depolarization. The phenotype of the myocardial cells of the tubular heart resembles that of the cells of the adult, so-called central conduction system, which comprises the sinus- and atrioventricular node.

With progression of development, the straight heart tube will loop towards the right. At the so-created outer curvatures of the looped heart, cardiomyocytes locally reinitiate proliferation, resulting in the expansion of the future chambers.[24] The myocardium of the forming chambers not only starts to proliferate, but will also differentiate, as it increases its number of sarcomeres, starts to express high-conductance gap-junction subunits like Cx40 and Cx43,[37] and its sarcoplasmic reticulum becomes functional.[38] The phenotype of this myocardium is called "working" myocardium.[4] The concept of local differentiation and expansion of the cardiac chambers at the outer curvatures of the looped heart tube, and local maintenance of nodal-like myocardium at the inner curvature, atrioventricular canal, and outflow tract is dubbed the ballooning model of chamber formation (Figure 4).[4] The maintenance of the primary phenotype and the slow rate of proliferation are achieved by the action of the transcription factors Tbx2 and Tbx3.[39]

Tbx3 is selectively expressed in the developing and mature conduction system.[40] Indeed, mature sinuatrial node cells resemble embryonic "primary" myocytes.[4,41] Intriguingly, deficiency of Tbx3 in the myocardium results in expansion of the expression of working myocardial genes (i.e. *Cx40, Cx43, Nppa and Scna5a*) into the sinus node domain.[39] Forced expression of Tbx3, on the other hand, results in ectopic development of functional pacemaker tissue and extended cushion formation.[39] Not all cells that initially expressed Tbx2 or Tbx3 become conduction system. Tbx2-positive atrioventricular myocardial cells of mice labeled between E8 and E9.5 not only give rise to the definitive atrioventricular node, but also to the adjacent working myocardium of the left ventricular free wall.[42]

The restriction of Tbx2, and Tbx3, to the atrioventricular canal and the concomitant demarcation of the domains of primary and working myocardium is mediated by BMP-signaling. BMP2 is sufficient to activate Tbx2 and Tbx3.[43] The absence of Tbx2 in the

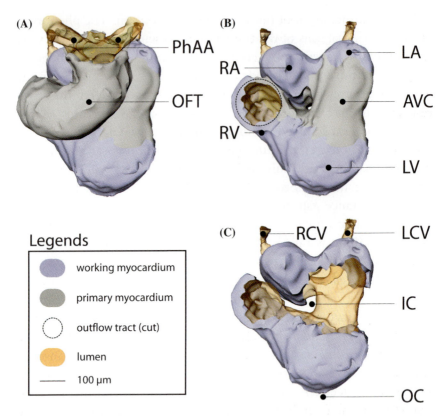

Figure 4.   The ballooning model. This figure shows a ventral view of a reconstruction of expression of Connexin 40, a marker of working myocardium, in a mouse heart of 9.5 days of development. Note how Connexin40 is only expressed in myocardium at the outer curvatures, while the primary myocardium at the inner curvatures remains devoid of expression. **(A)** The entire reconstruction; **(B)** and **(C)** OFT is removed. AVC: atrioventricular canal, IC: inner curvature, LA: left atrium, LCV: left caval vein, LV: left ventricle, OC: outer curvature, OFT: outflow tract, PhAA: pharyngeal arch arteries, RA: right atrium, RCV: right caval vein, RV: right ventricle.

working myocardium is regulated by another T-box transcription factor, Tbx20. Tbx20 is required for the formation of the heart tube and the chambers.[44] Deficiency of Tbx20 leads to widespread ectopic expression of Tbx2 in the entire heart tube.[45] It has been shown that Tbx20 directly binds to Smad, thereby inhibiting the inductive signal downstream of BMP on the promoter of Tbx2.[46]

The ballooning model of chamber formation shows how chambers and conduction myocardium develop from the primary heart tube by local transcriptional repression. Among others, this concept provided valuable insights into the molecular control of sinus node formation that may very well be translated to new techniques with respect to the development of bio-artificial pacemakers.

So far we discussed the growth and differentiation of the myocardium in the developing heart. Below we will describe the differentiation and addition of the other cell types that are present in the adult four-chambered heart.

3. ENDOCARDIUM

3.1. Separation of Endocardial and Myocardial Cells from the Precardiac Mesoderm

The first cardiac specification during development is the formation of precardiac mesoderm from mesoderm. This precardiac mesoderm then separates into the myocardial and endocardial lineage, occurring at four days of development in mice.[47] The differentiation of precardiac mesoderm into these two lineages starts cranially,[47] and induction of this cardiac gene program is regulated by paracrine signals from the endoderm and the ectoderm, like activins,[48] bone morphogenetic proteins (BMPs),[49,50] fibroblast growth factors (FGFs),[47,51] Wnts,[52–54] and inhibitors of Wnts, like Dickkopf1.[55,56] Members of the transforming growth factor β (TGF β) family and vascular endothelial growth factors (VEGFs) induce the expression of endocardial genes.[47]

The signals necessary for differentiation of mesoderm into precardiac mesoderm, and subsequent separation into endocardium and myocardium, are utilized to direct embryonic stem cells into the myocardial and endocardial lineages.[2] The stimulating paracrine effect of endodermal cells during myocardial development is applied in stem cell biology, where endoderm-like cells are used in co-culture with human embryonic stem cells to stimulate myocardial differentiation.[57,58]

3.2. Development of Cushion Mesenchyme

In the linear heart tube the endocardial and myocardial layer are separated by cardiac jelly. Cardiac jelly is an extracellular matrix secreted by the cardiomyocytes. With ongoing development, the cardiac jelly further disappears in the regions where the atria and ventricles will develop. This allows for a direct contact between the endocardial and myocardial cells, leading to a fundamental different process in the atrioventricular canal and outflow tract, compared to the working myocardium of the chambers. Within the atrioventricular canal and the outflow tract, the cardiac jelly expands and becomes populated by mesenchymal cells.[59-62]

The understanding of the contribution of specific cell lineages to the developing heart has benefited from the use of genetic labeling strategies in mice (Figure 5).[63] In this labeling strategy two mouse lines are interbred. In one mouse, the expression of a gene called *Cre* (*Causes Recombination Event*) is driven by a tissue-specific promoter. The second mouse contains a reporter, i.e. β-galactosidase, driven by a ubiquitous promoter. The transcription of this reporter, however, is blocked by the presence of a stop-site, flanked by LoxP sites. Upon expression of *Cre*, the two LoxP sites recombine, resulting in a permanent removal of the stop-site. As a result the ubiquitous promoter now drives expression of the reporter gene. This genomic change is irreversibly inherited, and, consequently, all daughter cells will continue to express the reporter, irrespective of the continuation of the expression of *Cre*. Subsequently, the lineage of recombined cells can be visualized by the expression of the reporter protein (Figure 5). The results obtained in these lineage tracing analyses depends on the promoter used to drive *Cre*, and on the sensitivity of detection of the reporter. A detailed discussion of the potential pitfalls of this system, however, falls beyond the scope of this chapter.

The contribution of endocardium-derived cells to the heart has been investigated using *Cre* driven by the *Tie2* promoter, which is specifically activated in endocardium.[64,65] Endocardium-derived cells were found to materially contribute to almost all cells of the atrioventricular valves, the

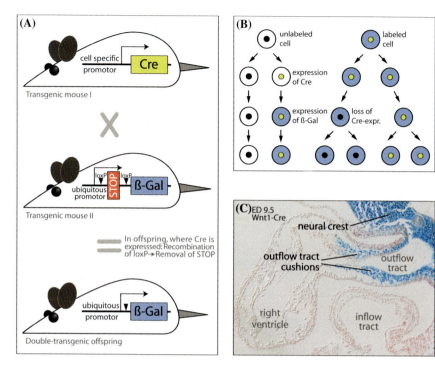

Figure 5. Cre-Lox lineage analysis. **(A)** Two transgenic mice that are crossed in a *Cre*-Lox lineage analysis. In Mouse I, *Cre* is targeted into an endogenous gene of interest, or *Cre* is driven by a tissue specific promoter. Mouse II carries a ubiquitous promoter that does not drive expression of a reporter gene (here βGal) because of a stop sequence that is flanked by loxP sites. **(B)** and **(C)** After breeding of mouse I and II, in their progeny the sequence between the loxP sites recombines wherever *Cre* is expressed. This deletes the stop sequence that prevents βGal expression. As a result, βGal is active in all cells in which *Cre* is, or once has been, expressed. βGal can then be visualized using a specific stain to show the lineage of the cells that expressed the Cre-coupled gene of interest.

chordae tendineae, and to some cells of the semilunar valves.[66] The mesenchymal cells that populate the atrioventricular canal cushions are formed from the overlaying endocardium by a process called epithelial-to-mesenchymal-transformation (EMT). During EMT, endocardial cells loosen their epithelial context and migrate into the cardiac jelly, thereby forming the cardiac cushion mesenchyme. BMP2, secreted by the

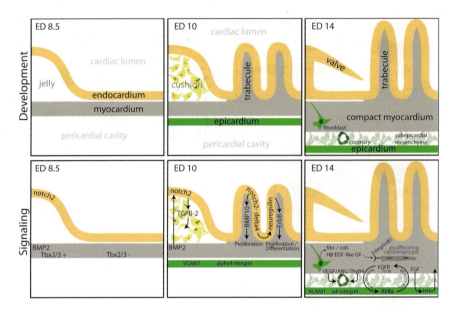

Figure 6. Intercellular signaling of endocardial, myocardial and epicardial cells. At day 8.5 in mice, the cardiac jelly has disappeared in developing chambers of the heart. In the atrioventricular canal myocardium (gray), the transcriptional repressors Tbx2 and Tbx3 are expressed, which is stimulated by the action of the growth factor BMP2. At ten days of development, the cardiac jelly of the cushions of the atrioventricular canal become populated by mesenchymal cells derived from Notch2-positive endocardial cells (orange). In the cardiac chambers, the differentiation and proliferation of myocardial cells in the trabecules is stimulated via an interaction between endocardial and myocardial cells via Notch2-Delta4 signaling in the endocardium, which is stimulated via the receptor ErbB and the growth factor BMP10. The mesenchyme of the cushions of the atrioventricular canal has transformed into valves. At 14 days of development, the epicardium (green) has covered the myocardium and has formed epicardial derived cells (EPDCs). Paracrine signals from the myocardium stimulate EPDCs to differentiate into cells of the coronary vessels and cardiac fibroblast. Via an intercellular interaction, cardiac fibroblasts stimulate cardiomyocytes of the compact zone to proliferate.

atrioventricular canal myocardium, induces a Notch2-positive endocardial subpopulation to undergo EMT.[67] TGFβ2 regulates the loosening of this population from their epithelial context and their concurrent migration into the cardiac jelly (Figure 6).[68–70]

In the outflow tract cushions, initially, the majority of the mesenchymal cells are cardiac neural crest-derived.[66,71–73] Endocardium-derived mesenchymal cells have only a small contribution to the proximal part of the outflow tract cushions.[66,74] In the proximal portion of the cushions, the neural crest-derived mesenchymal cells become dispersed and disappear by apoptosis. Concomitant with the observed apoptosis, the proximal outlet septum becomes populated by cardiomyocytes, both by migration of flanking myocardial cells, and by differentiation of mesenchymal cells.[75–78] In the distal part of the outflow septum, the neural crest-derived cells remain and contribute to the intra-pericardial part of the aorta and the pulmonary trunk.[15]

The capacity of the adult endocardium to initiate EMT and produce mesenchymal cells was recently described. Upon pressure overload, induced by trans-aortic banding, the adult endocardium (re)initiated EMT, resulting in the formation of cardiac-fibroblasts, and leading to cardiac fibrosis.[79] This process could be inhibited by BMP7, and stimulated by TGFβ1. Insight into this mechanism might be of importance for the development of a therapy for cardiac fibrosis, a major problem in cardiac disease.

3.3. Interplay between Endocardial and Myocardial Cells

Due to the absence of cardiac jelly in the atria and ventricles, a direct contact exists between endocardium and myocardium. This contact allows a cellular interaction, which is important in the initiation of differentiation of working myocardium, and, amongst others, is manifested by the formation of trabecules. In the endocardium, signaling via Delta4 and Notch1 is necessary.[80] In the endocardium signaling via Delta4 and Notch1 induces the expression of Neuregulin. Neuregulin stimulates, via ErbB, differentiation and proliferation of myocardial cells, resulting in formation of the trabecules. In this process BMP10 is necessary for the proliferation of the myocardial cells that populate the trabecules (Figure 6).[67]

These signaling pathways form important inroads to develop strategies to reinitiate proliferation in the adult cardiomyocytes. Recently, it has been shown, in mice, that adult cardiomyocytes can be induced to proliferate by stimulation of neuregulin1/ErbB4 signaling.[81]

4. EPICARDIUM

4.1. Separation of the Epicardial Lineage from the Precardiac Mesoderm

Although the pro-epicardium contributes to the majority of the non-myocardial cells of the adult heart, (pro-)epicardial cells, like myocardial cells, are derivatives of the precardiac mesoderm. Using the Cre-lox system in mice,[5,82] and DiI labeling in chicken[6] it was shown that pro-epicardial cells separate from the myocardial lineage.

(Pro-)epicardial and myocardial cells of the inflow are formed from T-box transcription factor (Tbx18)-positive cells directly adjacent to the heart.[6] Induction of Tbx18 and Wilms Tumor 1 (Wt1), another transcription factor that marks the epicardium, in the precardiac mesoderm is initiated by signals derived from the liver at stage 11 in chicken embryos.[83] Transplantation of the liver bud to another location in the mesoderm induced ectopic expression of Tbx18 or Wt1 in the adjacent mesoderm, pointing to the hepatocytes of the forming liver as a potential source of the inducing factors.[83] The separation of myocardial and pro-epicardial cells from this common precursor is regulated by an intracellular interaction between BMP- and FGF-signaling pathways.[6] In this process several EMT-related genes, like Wt1 and Snail are activated.[84,85] Cellular polarity also plays a crucial role in the development of the pro-epicardium, as deletion of the mammalian homolog of the *Caenorhabditis elegans* polarity proteins, PAR3, results in defective epicardial development.[86]

In the pro-epicardium, at least two different cell types can be detected: mesothelial cells that cover the pro-epicardial villi, and, along with accumulations of extracellular matrix, mesenchymal cells within the villi. The epithelial cells contact the myocardium and, subsequently, migrate over the "naked" heart tube to cover it entirely.

Retinoic acid, its receptor RXRα,[87–90] and genes involved in cell-cell interactions, like VCAM1 and α4-integrin are indispensable for the contact between the epicardium and myocardium,[91,92] as well as for the migration of epicardial cells over the myocardium.[93,94] The transcription factors GATA4 and Wt1 were shown to be important for the formation of the epicardium, as disruption of these genes resulted in aberrant formation of the pro-epicardium.[95]

4.2. Development of the Subepicardial Mesenchyme

From 11.5 days of development in mouse, and 33 days in human, when the myocardium is largely covered with epicardial cells, a subset of epicardial cells undergoes EMT, leading to the formation of the so-called epicardium-derived cells (EPDCs). EMT in the epicardium starts at the base of the ventricles and proceeds towards the apex.[96] Like in endocardial cells, FGFs and TGFs stimulate EMT in the epicardium.[12] In mice deficient for the zinc-finger transcription factor, friend of GATA 2 (FOG2), the epicardium does not undergo EMT. This process is rescued by transgenic expression of Fog2 in the myocardium, suggesting that signals from the myocardium are important in the regulation of epicardial EMT.[97,98]

EPDCs either reside within the subepicardial space or migrate into the myocardium. The fibroblast growth factor receptor type 1 (FGFR1) is indispensable for EPDC migration into the myocardium.[99] EPDCs contribute to the interstitial fibroblasts, the coronary vessels, and may have a contribution to the atrioventricular valves (Figure 1).[66,100,101] In the adult situation, the accumulation of subepicardial mesenchyme has largely disappeared, except in the atrioventricular junction, thereby forming the annulus fibrosis, which is of major importance in the electrical insulation of the atria and ventricles.

From genetic lineage tracing analyses, using Tbx18 or Wt1, it was recently concluded that, during development, the epicardium significantly contributes to the myocardial lineage. Seven to ten percent of the cardiomyocytes, located within the interventricular septum and dispersed throughout the ventricular wall, were indicated to originate from the epidardium.[82,102] Tbx18, however, is not only expressed in the epicardium, but also in the cardiomyocytes of the ventricular septum itself. Therefore, the concluded presence of epicardium-derived cardiomyocytes within the septum, traced using *Cre* under control of the Tbx18-promoter,[102] was disproven.[103] Nevertheless, lineage-positive cardiomyocytes were also detected within other areas of the heart, like the ventricular free wall and the ventricular wall adjacent to the atrioventricular junction. Intriguingly, lineage tracing of the epicardium, using an inducible Wt1 Cre, also identified lineage-positive myocardial cells throughout the myocardium. Thus far, myocardial Wt1 expression has not been reported, suggesting an epicardial contribution to the myocardial component of the heart. If confirmed, these studies indicate that the (embryonic) epicardium contributes for a minor part to the myocardial component of the heart. This capacity would make the (embryonic) epicardium a highly interesting cell type for the development of new regenerative therapies.

4.3. Development of the Coronary Vasculature

Between day 11.5 and 13.5 in mice, and day 33 and 44 in human, the majority of the cells of the coronary plexus are formed from EPDCs within the subepicardial space.[10,101] This plexus later is remodeled into the mature coronary vessels.[95,104,105] Hypoxia is supposed to be an important initiating factor in the formation of the coronary tree.[106] FGFs secreted from adjacent myocardium are important players in the development of the coronary vasculature. These FGFs initiate a base to apical wave of Hedgehog signaling within the epicardium that covers the ventricles, thereby inducing the expression of pro-angiogenic factors, like vascular endothelial growth factors A, B and angiotensin.[107] Wt1 is also involved in the vascularization of the heart by trans-activating the TrkB neutrophin receptor (Figure 6).[108]

As already discussed, EPDCs contribute to the majority of cells of the coronary plexus. The origin of the endothelial cells of the coronary vessels, however, is still controversial. Chicken quail chimeras showed coronary endothelial cells being derived from the pro-epicardium, suggesting *de novo* formation of endothelial cells (vasculo-genesis).[11,101,109] In line with these observations, genetic lineage analyses using Wt1 as a driver of *Cre*, also showed positive coronary endothelial cells, while the comparable analysis using Tbx18-Cre did not.[82,102] The absence of positive endothelial cells in the Tbx18 line-age was explained by the presence of Tbx18 negative, Flk1 positive endothelial precursor cells within the pro-epicardium. However, other groups have reported formation of the coronary vessels by a process of sprouting (angiogenesis) from the systemic venous veins.[104,105] A recent study underscored these conclusions and, showed that the endothelial cells of the coronaries are derived by sprouting from the endothelium of the common systemic venous veins, called by the authors sinus venosus. Unfortunately, no distinc-tion had been made between the development of endothelial cells of the coronary arteries and the coronary veins.[110] Subsequent to dedif-ferentiation, the endothelial cells of the common systemic veins were shown to migrate via the connection of this region with the dorsal side of heart formed by the (pro-)epicardium, and to redifferentiate into endothelial cells of the coronaries in the myocardium.

These above-mentioned discrepancies in the explanation of the formation of the coronary vessels might be resolved by recent obser-vations by Perez-Pomares and co-workers, as yet unpublished, indicating distinct differences in the development of coronary arteries as compared to coronary veins. These analyses indicate that the coro-nary arterial capillary plexus forms by vasculogenesis in the subepicardial mesenchyme, while the coronary veins primarily develop by angiogenesis of the inflow endocardium into the subepi-cardial space.

Important processes for the development of the coronary arter-ies are used in the development of regenerative therapies. For example, activation of Hedgehog signaling in adult mice has been used to stimulate coronary neovasculogenesis to improve the oxygenation of ischemic myocardium.[107] The adult epicardium has also been shown to be a source for vascular progenitors. Neovascularization of the ischemic heart by epicardium-derived cells was shown to be enhanced by stimulation with thymosin ß4.[111]

These observations support the idea that reactivation of the adult epicardium to neovascularize the heart after ischemic disease is a realistic and promising option.

4.4. Interplay Between Epicardial and Myocardial Cells

Signals from the endocardium induce the development of tra-becules,[67,80] whereas signals from the epicardium stimulate the formation of the compact myocardium (Figure 6).[67] Removal of the (pro-)epicardium in chicken results in a thin myocardial wall of the chambers, and an arrest of coronary development.[96] Retinoic acid, its receptor RXRalpha, and Epo are important players in the mitogenic interaction between epicardium and myocardium.[89,112–114] Mice deficient for RXRalpha or Raldh2, which are important for the intracellular production of retinoic acid, die early during development, showing ventricular hypoplasia.[115,116] Epicardial-derived FGF 9, 16 and 20 signal to the myocardium via the FGFR receptors type 1c and 2c, regulating myocardial proliferation. FGF9 expression in the epicardium is induced by retinoic acid (Figure 6).[95,96]

Besides paracrine signals directly from the epicardium, it was recently shown that cardiac fibroblasts, which are derived from the epicardium, also influence the proliferation of myocardial cells in the compact zone. Myocardial cells of the compact zone are stimulated to proliferate by direct cellular interaction via the ß1-integrin receptor with fibronectin, collagen, and heparin binding EGF-like growth factor (HBEGF) produced by the cardiac fibroblasts (Figure 6).[117]

5. CARDIAC REGENERATION

To compensate for myocardial loss caused by ischemia or other damage, the heart needs to regenerate cardiomyocytes. Most approaches to improve cardiac function after loss of cardiomyocytes have focused on the regulation of myocardial differentiation of stem cells of various sources, which are injected into the failing heart. These approaches did transiently improve cardiac function,[118] but the formation of new cardiomyocytes from these cells has hardly been observed, most probably due to the limited integration and survival of these cells in the myocardium.[119] An alternative approach would be to recruit myocardial cells from non-myocardial lineages in the adult heart or to stimulate myocardial cells to proliferate. With the lessons learned from the analysis of the development of the heart it might become possible to manipulate the non-myocardial cells to differentiate into myocardial cells or stimulate myocardial proliferation after damage, thereby possibly inducing a regenerative response from endogenous populations of cells, already present at the location where they are needed.

Although the prevailing dogma poses that the heart is a terminally differentiated organ, recent reports have shown that the human heart is able to form new cardiomyocytes during life. In a very elegant study of Bergmann and co-workers, a virtual pulse-chase experiment was used to calculate the turnover of adult cardiomyocytes.[120] The incorporation of carbon-14, generated by nuclear bomb tests during the Cold War, into the genomic DNA was used to determine the turnover of human cardiomyocytes. A turnover of 1% per year at the age of 25 was observed, decreasing to 0.45% per year at the age of 75. These observations imply that during a normal life span approximately 50% of the cardiomyocytes are renewed. These results are not in line with the assertion that the cardiomyocytes that are present in our heart are those that we were born with. These results also show that cardiomyocytes do not proliferate at a high rate.[121] It should be noted that the study of Bergmann and co-workers does not necessarily imply that the newly formed cardiomyocytes are derived by proliferation of existing cardiomyocytes. This newly formed muscle might be derived from the non-myocardial component of the heart, or even from an extra-cardiac source.

Taken together, stimulation of the (very limited) capacity of the adult heart to regenerate cardiomyocytes seems a prerequisite for cardiac rejuvenation. A first attempt to do so requires that the source of these newly formed cardiomyocytes should be assessed. A genetic fate-mapping study, in mice, showed that after myocardial injury, newly formed cardiomyocytes are not recruited from the myocardial component of the heart.[122] This indicates that proliferation of existing, α-myosin heavy chain expressing cardiomyocytes is not the source of the newly formed cardiomyocytes after injury. It also has been shown that the endocardial lineage does not contribute to the myocardial component in adult hearts after myocardial stress.[79] Endocardium-derived cells were, however, found to contribute fibroblasts, resulting in cardiac fibrosis. From a developmental point of view, the epicardium might be a potential source of myocardial cells. As discussed above, epicardial-derived cells contribute to the myocardial lineage during development.[82]

This notion is further underscored by the finding that the adult epicardium is reactivated after cardiac damage in zebrafish, a species that is able to regenerate their heart.[123] Along with epicardial activation, induction of proliferation of existing cardiomyocytes is crucial for cardiac regeneration in zebrafish.[124,125] Upon injury, expression of GATA4 is triggered in a subpopulation of cardiomyocytes located in the sub-epicardial layer and starts to proliferate.[125] This activation parallels the epicardial activation observed after myocardial damage and suggests an important relation between these cell populations in the regenerative response of the zebrafish heart. We do note, however, that fish, like amphibians and reptiles, have low pressure hearts, and, consequently, lack a compact myocardium, which may permit a higher regenerative capacity.

Whether the epicardium, like in zebrafish, also plays a role during cardiac regeneration in mammals, has not been shown yet, albeit some studies suggest that the epicardium plays a role in mammalian cardiac regeneration.[126–129] For instance, the locations of niches of cardiac progenitor cells found in the adult heart might suggest a role for the adult epicardium in the regenerative response of the heart. In the atria, atrioventricular junction, and around the apex of the heart, more cardiac progenitor cells were found compared to other areas of the heart.[130] These sites are also the locations where a myocardial contribution of epicardial-derived cells to myocardial cells was found in the Tbx18 and Wt1 lineage studies discussed above.[82,102] Our own studies, as yet unpublished, show that the epicardium overlying these parts of the adult heart, still expresses embryonic genes like Wt1 and Tbx18.

Unraveling the source of the limited endogenous regenerative capacity of the mammalian heart, as well as the development of approaches to stimulate this process, will be a great challenge for the coming decades. Undoubtedly, understanding of the mechanisms that drive cardiac development will be instrumental in this search.

REFERENCES

1. van Dessel PF, de Bakker JM, van Hemel NM, Linnenbank AC, *et al.* (2001) Pace mapping of postinfarction scar to detect ventricular tachycardia exit sites and zones of slow conduction. *J Cardiovasc Electrophysiol* 12: 662–670.

2. Yang L, Soonpaa MH, Adler ED, Roepke TK, *et al.* (2008) Human cardiovascular progenitor cells develop from a KDR+ embryonic-stem-cell-derived population. *Nature* 453: 524–528.

3. van Amerongen MJ, Engel FB (2008) Features of cardiomyocyte proliferation and its potential for cardiac regeneration. *J Cell Mol Med* 12: 2233–2244.

4. Moorman AF, Christoffels VM (2003) Cardiac chamber formation: development, genes, and evolution. *Physiol Rev* 83: 1223–1267.

5. Zhou B, von Gise A, Ma Q, Rivera-Feliciano J, *et al.* (2008) Nkx2-5- and Isl1-expressing cardiac progenitors contribute to proepicardium. *Biochem Biophys Res Commun* 375: 450–453.

6. van Wijk B, van den Berg G, Abu-Issa R, Barnett P, *et al.* (2009) Epicardium and myocardium separate from a common precursor pool by crosstalk between bone morphogenetic protein- and fibroblast growth factor-signaling pathways. *Circ Res* 105: 431–441.

7. Vliegen HW, van der Laarse A, Cornelisse CJ, Eulderink F (1991) Myocardial changes in pressure overload-induced left ventricular hypertrophy. A study on tissue composition, polyploidization and multinucleation. *Eur Heart J* 12: 488–494.

8. Nag AC, Crandell TF, Cheng M (1981) Competence of embryonic mammalian heart cells in culture: DNA synthesis, mitosis and differentiation. *Cytobios* 30: 189–208.

9. van den Hoff MJ, Kruithof BP, Moorman AF (2004) Making more heart muscle. *Bioessays* 26: 248–261.

10. Munoz-Chapuli R, Macias D, Gonzalez-Iriarte M, Carmona R, *et al.* (2002) [The epicardium and epicardial-derived cells: multiple functions in cardiac development.]. *Rev Esp Cardiol* 55: 1070–1082.

11. Wessels A, Perez-Pomares JM (2004) The epicardium and epicardially derived cells (EPDCs) as cardiac stem cells. *Anat Rec A Discov Mol Cell Evol Biol* 276: 43–57.

12. Winter EM, Gittenberger-de Groot AC (2007) Epicardium-derived cells in cardiogenesis and cardiac regeneration. *Cell Mol Life Sci* 64: 692–703.

13. Christoffels VM, Mommersteeg MT, Trowe MO, Prall OW, *et al.* (2006) Formation of the venous pole of the heart from an Nkx2-5-negative precursor population requires Tbx18. *Circ Res* 98: 1555–1563.

14. Mommersteeg MT, Brown NA, Prall OW, de Gier-de Vries C, *et al.* (2007) Pitx2c and Nkx2-5 are required for the formation and identity of the pulmonary myocardium. *Circ Res* 101: 902–909.

15. Kirby ML, Gale TF, Stewart DE (1983) Neural crest cells contribute to normal aorticopulmonary septation. *Science* 220: 1059–1061.

16. Zhang Z, Huynh T, Baldini A (2006) Mesodermal expression of Tbx1 is necessary and sufficient for pharyngeal arch and cardiac outflow tract development. *Development* 133: 3587–3595.

17. Moorman AFM, van den Berg G, Anderson RH, Christoffels VM (2009) Growth and differentiation of the developing heart. In: Harvey RP, Rosenthal N (eds.), *Heart Development and Regeneration.* Elsevier, Philadelphia, pp. 1–29.

18. Abu-Issa R, Kirby ML (2007) Heart field: from mesoderm to heart tube. *Annu Rev Cell Dev Biol* 23: 45–68.

19. Brand T (2003) Heart development: molecular insights into cardiac specification and early morphogenesis. *Dev Biol* 258: 1–19.

20. Tokuyasu KT, Maher PA (1987) Immunocytochemical studies of cardiac myofibrillogenesis in early chick embryos. I. Presence of immunofluorescent titin spots in premyofibril stages. *J Cell Biol* 105: 2781–2793.

21. Patten BM, Kramer TC (1933) The initiation of contraction in the embryonic chicken heart. *Am J Anat* 53: 349–375.

22. van den Berg G, Moorman AF (2009) Concepts of cardiac development in retrospect. *Pediatr Cardiol* 30: 580–587.

23. Sissman NJ (1966) Cell multiplication rates during development of the primitive cardiac tube in the chick embryo. *Nature* 210: 504–507.

24. Soufan AT, van den Berg G, Ruijter JM, de Boer PA, *et al.* (2006) Regionalized sequence of myocardial cell growth and proliferation characterizes early chamber formation. *Circ Res* 99: 545–552.

25. van den Berg G, Abu-Issa R, de Boer BA, Hutson MR, *et al.* (2009) A caudal proliferating growth center contributes to both poles of the forming heart tube. *Circ Res* 104: 179–188.

26. Buckingham M, Meilhac S, Zaffran S (2005) Building the mammalian heart from two sources of myocardial cells. *Nat Rev Genet* 6: 826–835.

27. Moorman AF, Christoffels VM, Anderson RH, van den Hoff MJ (2007) The heart-forming fields: one or multiple? *Philos Trans R Soc Lond B Biol Sci* 362: 1257–1265.

28. Cai CL, Liang X, Shi Y, Chu PH, *et al.* (2003) Isl1 identifies a cardiac progenitor population that proliferates prior to differentiation and contributes a majority of cells to the heart. *Dev Cell* 5: 877–889.

29. Prall OW, Menon MK, Solloway MJ, Watanabe Y, *et al.* (2007) An Nkx2-5/Bmp2/Smad1 negative feedback loop controls heart progenitor specification and proliferation. *Cell* 128: 947–959.

30. de Haan RL, Fujii S, Satin J (1990) Cell interactions in cardiac development. *Dev Growth Differ* 32: 233–241.

31. Rochais F, Mesbah K, Kelly RG (2009) Signaling pathways controlling second heart field development. *Circ Res* 104: 933–942.

32. Abu-Issa R, Kirby ML (2008) Patterning of the heart field in the chick. *Dev Biol* 319: 223–233.

33. Durland JL, Sferlazzo M, Logan M, Burke AC (2008) Visualizing the lateral somitic frontier in the Prx1Cre transgenic mouse. *J Anat* 212: 590–602.

34. Waldo KL, Hutson MR, Ward CC, Zdanowicz M, *et al.* (2005) Secondary heart field contributes myocardium and smooth muscle to the arterial pole of the developing heart. *Dev Biol* 281: 78–90.

35. Nathan E, Monovich A, Tirosh-Finkel L, Harrelson Z, *et al.* (2008) The contribution of Islet1-expressing splanchnic mesoderm cells to distinct branchiomeric muscles reveals significant heterogeneity in head muscle development. *Development* 135: 647–657.

36. Tzahor E (2009) Heart and craniofacial muscle development: a new developmental theme of distinct myogenic fields. *Dev Biol* 327: 273–279.
37. Kreuzberg MM, Willecke K, Bukauskas FF (2006) Connexin-mediated cardiac impulse propagation: connexin 30.2 slows atrioventricular conduction in mouse heart. *Trends Cardiovasc Med* 16: 266–272.
38. Moorman AF, Schumacher CA, de Boer PA, Hagoort J, *et al.* (2000) Presence of functional sarcoplasmic reticulum in the developing heart and its confinement to chamber myocardium. *Dev Biol* 223: 279–290.
39. Hoogaars WM, Engel A, Brons JF, Verkerk AO, *et al.* (2007) Tbx3 controls the sinoatrial node gene program and imposes pacemaker function on the atria. *Genes Dev* 21: 1098–1112.
40. Mommersteeg MT, Hoogaars WM, Prall OW, de Gier-de Vries C, *et al.* (2007) Molecular pathway for the localized formation of the sinoatrial node. *Circ Res* 100: 354–362.
41. de Haan RL (1961) Differentiation of the atrioventricular conducting system of the heart. *Circulation* 24: 458–470.
42. Aanhaanen WT, Brons JF, Dominguez JN, Rana MS, *et al.* (2009) The Tbx2+ primary myocardium of the atrioventricular canal forms the atrioventricular node and the base of the left ventricle. *Circ Res* 104: 1267–1274.
43. Yamada M, Revelli JP, Eichele G, Barron M, *et al.* (2000) Expression of chick Tbx-2, Tbx-3, and Tbx-5 genes during early heart development: evidence for BMP2 induction of Tbx2. *Dev Biol* 228: 95–105.
44. Stennard FA, Costa MW, Elliott DA, Rankin S, *et al.* (2003) Cardiac T-box factor Tbx20 directly interacts with Nkx2-5, GATA4, and GATA5 in regulation of gene expression in the developing heart. *Dev Biol* 262: 206–224.
45. Singh MK, Christoffels VM, Dias JM, Trowe MO, *et al.* (2005) Tbx20 is essential for cardiac chamber differentiation and repression of Tbx2. *Development* 132: 2697–2707.
46. Singh R, Horsthuis T, Farin HF, Grieskamp T, *et al.* (2009) Tbx20 interacts with smads to confine tbx2 expression to the atrioventricular canal. *Circ Res* 105: 442–452.
47. Lough J, Sugi Y (2000) Endoderm and heart development. *Dev Dyn* 217: 327–342.

48. Mangiacapra FJ, Fransen ME, Lemanski LF (1995) Activin A and transforming growth factor-beta stimulate heart formation in axolotls but not rescue cardiac lethal mutants. *Cell Tissue Res* 282: 227–236.

49. Ladd AN, Yatskievych TA, Antin PB (1998) Regulation of avian cardiac myogenesis by activin/TGFbeta and bone morphogenetic proteins. *Dev Biol* 204: 407–419.

50. Schultheiss TM, Burch JB, Lassar AB (1997) A role for bone morphogenetic proteins in the induction of cardiac myogenesis. *Genes Dev* 11: 451–462.

51. Alsan BH, Schultheiss TM (2002) Regulation of avian cardiogenesis by Fgf8 signaling. *Development* 129: 1935–1943.

52. Marvin MJ, Di Rocco G, Gardiner A, Bush SM, *et al.* (2001) Inhibition of Wnt activity induces heart formation from posterior mesoderm. *Genes Dev* 15: 316–327.

53. Nakamura T, Sano M, Songyang Z, Schneider MD (2003) A Wnt- and beta-catenin-dependent pathway for mammalian cardiac myogenesis. *Proc Natl Acad Sci USA* 100: 5834–5839.

54. Schneider VA, Mercola M (2001) Wnt antagonism initiates cardiogenesis in *Xenopus laevis*. *Genes Dev* 15: 304–315.

55. Monaghan AP, Kioschis P, Wu W, Zuniga A, *et al.* (1999) Dickkopf genes are co-ordinately expressed in mesodermal lineages. *Mech Dev* 87: 45–56.

56. Pfeffer PL, De Robertis EM, Izpisua-Belmonte JC (1997) Crescent, a novel chick gene encoding a Frizzled-like cysteine-rich domain, is expressed in anterior regions during early embryogenesis. *Int J Dev Biol* 41: 449–458.

57. Beqqali A, Kloots J, Ward-van Oostwaard D, Mummery C, *et al.* (2006) Genome-wide transcriptional profiling of human embryonic stem cells differentiating to cardiomyocytes. *Stem Cells* 24: 1956–1967.

58. Mummery C, Ward-van Oostwaard D, Doevendans P, Spijker R, *et al.* (2003) Differentiation of human embryonic stem cells to cardiomyocytes: role of coculture with visceral endoderm-like cells. *Circulation* 107: 2733–2740.

59. Eisenberg LM, Markwald RR (1995) Molecular regulation of atrioventricular valvuloseptal morphogenesis. *Circ Res* 77: 1–6.

60. Kolditz DP, Wijffels MC, Blom NA, van der Laarse A, *et al.* (2008) Epicardium-derived cells in development of annulus fibrosis and persistence of accessory pathways. *Circulation* 117: 1508–1517.

61. Markwald R, Eisenberg C, Eisenberg L, Trusk T, *et al.* (1996) Epithelial-mesenchymal transformations in early avian heart development. *Acta Anat* 156: 173–186.

62. Ramsdell AF, Markwald RR (1997) Induction of endocardial cushion tissue in the avian heart is regulated, in part, by TGFbeta-3-mediated autocrine signaling. *Dev Biol* 188: 64–74.

63. Soriano P (1999) Generalized lacZ expression with the ROSA26 Cre reporter strain. *Nat Genet* 21: 70–71.

64. Constien R, Forde A, Liliensiek B, Grone HJ, *et al.* (2001) Characterization of a novel EGFP reporter mouse to monitor Cre recombination as demonstrated by a Tie2 Cre mouse line. *Genesis* 30: 36–44.

65. Kisanuki YY, Hammer RE, Miyazaki J, Williams SC, *et al.* (2001) Tie2-Cre transgenic mice: a new model for endothelial cell-lineage analysis *in vivo*. *Dev Biol* 230: 230–242.

66. de Lange FJ, Moorman AF, Anderson RH, Manner J, *et al.* (2004) Lineage and morphogenetic analysis of the cardiac valves. *Circ Res* 95: 645–654.

67. Wagner M, Siddiqui MA (2007) Signal transduction in early heart development (II): ventricular chamber specification, trabeculation, and heart valve formation. *Exp Biol Med* 232: 866–880.

68. Niessen K, Karsan A (2008) Notch signaling in cardiac development. *Circ Res* 102: 1169–1181.

69. Snarr BS, Kern CB, Wessels A (2008) Origin and fate of cardiac mesenchyme. *Dev Dyn* 237: 2804–2819.

70. van Wijk B, Moorman AF, van den Hoff MJ (2007) Role of bone morphogenetic proteins in cardiac differentiation. *Cardiovasc Res* 74: 244–255.

71. Hutson MR, Kirby ML (2003) Neural crest and cardiovascular development: a 20-year perspective. *Birth Defects Res C Embryo Today* 69: 2–13.

72. Hutson MR, Kirby ML (2007) Model systems for the study of heart development and disease. Cardiac neural crest and conotruncal malformations. *Semin Cell Dev Biol* 18: 101–110.

73. Jiang X, Rowitch DH, Soriano P, McMahon AP, *et al.* (2000) Fate of the mammalian cardiac neural crest. *Development* 127: 1607–1616.

74. Ya J, van den Hoff MJ, de Boer PA, Tesink-Taekema S, *et al.* (1998) Normal development of the outflow tract in the rat. *Circ Res* 82: 464–472.

75. Kruithof BP, van den Hoff MJ, Tesink-Taekema S, Moorman AF (2003) Recruitment of intra- and extracardiac cells into the myocardial lineage during mouse development. *Anat Rec A Discov Mol Cell Evol Biol* 271: 303–314.

76. Moralez I, Phelps A, Riley B, Raines M, *et al.* (2006) Muscularizing tissues in the endocardial cushions of the avian heart are characterized by the expression of h1-calponin. *Dev Dyn* 235: 1648–1658.

77. van den Hoff MJ, Kruithof BP, Moorman AF, Markwald RR, *et al.* (2001) Formation of myocardium after the initial development of the linear heart tube. *Dev Biol* 240: 61–76.

78. van den Hoff MJ, Moorman AF, Ruijter JM, Lamers WH, *et al.* (1999) Myocardialization of the cardiac outflow tract. *Dev Biol* 212: 477–490.

79. Zeisberg EM, Tarnavski O, Zeisberg M, Dorfman AL, *et al.* (2007) Endothelial-to-mesenchymal transition contributes to cardiac fibrosis. *Nat Med* 13: 952–961.

80. Grego-Bessa J, Luna-Zurita L, del Monte G, Bolos V, *et al.* (2007) Notch signaling is essential for ventricular chamber development. *Dev Cell* 12: 415–429.

81. Bersell K, Arab S, Haring B, Kuhn B (2009) Neuregulin1/ErbB4 signaling induces cardiomyocyte proliferation and repair of heart injury. *Cell* 138: 257–270.

82. Zhou B, Ma Q, Rajagopal S, Wu SM, *et al.* (2008) Epicardial progenitors contribute to the cardiomyocyte lineage in the developing heart. *Nature* 454: 109–113.

83. Ishii Y, Langberg JD, Hurtado R, Lee S, *et al.* (2007) Induction of proepicardial marker gene expression by the liver bud. *Development* 134: 3627–3637.

84. Carmona R, Gonzalez-Iriarte M, Macias D, Perez-Pomares JM, *et al.* (2000) Immunolocalization of the transcription factor Slug in the developing avian heart. *Anat Embryol* 201: 103–109.

85. Carmona R, Gonzalez-Iriarte M, Perez-Pomares JM, Munoz-Chapuli R (2001) Localization of the Wilm's tumour protein WT1 in avian embryos. *Cell Tissue Res* 303: 173–186.

86. Hirose T, Karasawa M, Sugitani Y, Fujisawa M, *et al.* (2006) PAR3 is essential for cyst-mediated epicardial development by establishing apical cortical domains. *Development* 133: 1389–1398.

87. Kastner P, Messaddeq N, Mark M, Wendling O, *et al.* (1997) Vitamin A deficiency and mutations of RXRalpha, RXRbeta and RARalpha lead to early differentiation of embryonic ventricular cardiomyocytes. *Development* 124: 4749–4758.

88. Kubalak SW, Hutson DR, Scott KK, Shannon RA (2002) Elevated transforming growth factor beta2 enhances apoptosis and contributes to abnormal outflow tract and aortic sac development in retinoic X receptor alpha knockout embryos. *Development* 129: 733–746.

89. Sucov HM, Dyson E, Gumeringer CL, Price J, *et al.* (1994) RXR alpha mutant mice establish a genetic basis for vitamin A signaling in heart morphogenesis. *Genes Dev* 8: 1007–1018.

90. Xavier-Neto J, Shapiro MD, Houghton L, Rosenthal N (2000) Sequential programs of retinoic acid synthesis in the myocardial and epicardial layers of the developing avian heart. *Dev Biol* 219: 129–141.

91. Gurtner GC, Davis V, Li H, McCoy MJ, *et al.* (1995) Targeted disruption of the murine VCAM1 gene: essential role of VCAM-1 in chorioallantoic fusion and placentation. *Genes Dev* 9: 1–14.

92. Yang JT, Rayburn H, Hynes RO (1995) Cell adhesion events mediated by alpha 4 integrins are essential in placental and cardiac development. *Development* 121: 549–560.

93. Pae SH, Dokic D, Dettman RW (2008) Communication between integrin receptors facilitates epicardial cell adhesion and matrix organization. *Dev Dyn* 237: 962–978.

94. Sengbusch JK, He W, Pinco KA, Yang JT (2002) Dual functions of [alpha]4[beta]1 integrin in epicardial development: initial migration and long-term attachment. *J Cell Biol* 157: 873–882.

95. Lavine KJ, Ornitz DM (2009) Shared circuitry: developmental signaling cascades regulate both embryonic and adult coronary vasculature. *Circ Res* 104: 159–169.

96. Lavine KJ, Ornitz DM (2007) Rebuilding the coronary vasculature: hedgehog as a new candidate for pharmacologic revascularization. *Trends Cardiovasc Med* 17: 77–83.

97. Crispino JD, Lodish MB, Thurberg BL, Litovsky SH, *et al.* (2001) Proper coronary vascular development and heart morphogenesis depend on interaction of GATA-4 with FOG cofactors. *Genes Dev* 15: 839–844.

98. Tevosian SG, Deconinck AE, Tanaka M, Schinke M, *et al.* (2000) FOG-2, a cofactor for GATA transcription factors, is essential for heart morphogenesis and development of coronary vessels from epicardium. *Cell* 101: 729–739.

99. Pennisi DJ, Mikawa T (2009) FGFR-1 is required by epicardium-derived cells for myocardial invasion and correct coronary vascular lineage differentiation. *Dev Biol* 328: 148–159.

100. Gittenberger-de Groot AC, Vrancken Peeters MP, Mentink MM, Gourdie RG, *et al.* (1998) Epicardium-derived cells contribute a novel population to the myocardial wall and the atrioventricular cushions. *Circ Res* 82: 1043–1052.

101. Manner J (1999) Does the subepicardial mesenchyme contribute myocardioblasts to the myocardium of the chick embryo heart? A quail-chick chimera study tracing the fate of the epicardial primordium. *Anat Rec* 255: 212–226.

102. Cai CL, Martin JC, Sun Y, Cui L, *et al.* (2008) A myocardial lineage derives from Tbx18 epicardial cells. *Nature* 454: 104–108.

103. Christoffels VM, Grieskamp T, Norden J, Mommersteeg MT, *et al.* (2009) Tbx18 and the fate of epicardial progenitors. *Nature* 458: E8–9; discussion E9–10.

104. Vrancken Peeters MP, Gittenberger-de Groot AC, Mentink MM, Hungerford JE, *et al.* (1997) Differences in development of coronary arteries and veins. *Cardiovasc Res* 36: 101–110.

105. Vrancken Peeters MP, Gittenberger-de Groot AC, Mentink MM, Hungerford JE, *et al.* (1997) The development of the coronary vessels and their differentiation into arteries and veins in the embryonic quail heart. *Dev Dyn* 208: 338–348.

106. Krishnan J, Ahuja P, Bodenmann S, Knapik D, *et al.* (2008) Essential role of developmentally activated hypoxia-inducible factor 1alpha for cardiac morphogenesis and function. *Circ Res* 103: 1139–1146.

107. Lavine KJ, White AC, Park C, Smith CS, *et al.* (2006) Fibroblast growth factor signals regulate a wave of Hedgehog activation that is essential for coronary vascular development. *Genes Dev* 20: 1651–1666.

108. Wagner N, Wagner KD, Theres H, Englert C, *et al.* (2005) Coronary vessel development requires activation of the TrkB neurotrophin receptor by the Wilms' tumor transcription factor Wt1. *Genes Dev* 19: 2631–2642.

109. Perez-Pomares JM, Macias D, Garcia-Garrido L, Munoz-Chapuli R (1997) Contribution of the primitive epicardium to the subepicardial mesenchyme in hamster and chick embryos. *Dev Dyn* 210: 96–105.

110. Red-Horse K, Ueno H, Weissman IL, Krasnow MA (2010) Coronary arteries form by developmental reprogramming of venous cells. *Nature* 464: 549–553.

111. Smart N, Risebro CA, Melville AA, Moses K, *et al.* (2007) Thymosin beta4 induces adult epicardial progenitor mobilization and neovascularization. *Nature* 445: 177–182.

112. Chen TH, Chang TC, Kang JO, Choudhary B, *et al.* (2002) Epicardial induction of fetal cardiomyocyte proliferation via a retinoic acid-inducible trophic factor. *Dev Biol* 250: 198–207.

113. Merki E, Zamora M, Raya A, Kawakami Y, *et al.* (2005) Epicardial retinoid X receptor alpha is required for myocardial growth and coronary artery formation. *Proc Natl Acad Sci USA* 102: 18455–18460.

114. Stuckmann I, Evans S, Lassar AB (2003) Erythropoietin and retinoic acid, secreted from the epicardium, are required for cardiac myocyte proliferation. *Dev Biol* 255: 334–349.

115. Niederreither K, Subbarayan V, Dolle P, Chambon P (1999) Embryonic retinoic acid synthesis is essential for early mouse post-implantation development. *Nat Genet* 21: 444–448.

116. Niederreither K, Vermot J, Messaddeq N, Schuhbaur B, *et al.* (2001) Embryonic retinoic acid synthesis is essential for heart morphogenesis in the mouse. *Development* 128: 1019–1031.

117. Ieda M, Tsuchihashi T, Ivey KN, Ross RS, *et al.* (2009) Cardiac fibroblasts regulate myocardial proliferation through beta1 integrin signaling. *Dev Cell* 16: 233–244.

118. Schachinger V, Assmus B, Erbs S, Elsasser A, *et al.* (2009) Intracoronary infusion of bone marrow-derived mononuclear cells

abrogates adverse left ventricular remodelling post-acute myocardial infarction: insights from the reinfusion of enriched progenitor cells and infarct remodelling in acute myocardial infarction (REPAIR-AMI) trial. *Eur J Heart Fail* 11: 973–979.

119. Freund C, Mummery CL (2009) Prospects for pluripotent stem cell-derived cardiomyocytes in cardiac cell therapy and as disease models. *J Cell Biochem* 107: 592–599.

120. Bergmann O, Bhardwaj RD, Bernard S, Zdunek S, *et al.* (2009) Evidence for cardiomyocyte renewal in humans. *Science* 324: 98–102.

121. Beltrami AP, Urbanek K, Kajstura J, Yan SM, *et al.* (2001) Evidence that human cardiac myocytes divide after myocardial infarction. *N Engl J Med* 344: 1750–1757.

122. Hsieh PC, Segers VF, Davis ME, MacGillivray C, *et al.* (2007) Evidence from a genetic fate-mapping study that stem cells refresh adult mammalian cardiomyocytes after injury. *Nat Med* 13: 970–974.

123. Lepilina A, Coon AN, Kikuchi K, Holdway JE, *et al.* (2006) A dynamic epicardial injury response supports progenitor cell activity during zebrafish heart regeneration. *Cell* 127: 607–619.

124. Jopling C, Sleep E, Raya M, Marti M, *et al.* (2010) Zebrafish heart regeneration occurs by cardiomyocyte dedifferentiation and proliferation. *Nature* 464: 606–609.

125. Kikuchi K, Holdway JE, Werdich AA, Anderson RM, *et al.* (2010) Primary contribution to zebrafish heart regeneration by gata4(+) cardiomyocytes. *Nature* 464: 601–605.

126. Di Meglio F, Castaldo C, Nurzynska D, Romano V, *et al.* (2010) Epicardial cells are missing from the surface of hearts with ischemic cardiomyopathy: a useful clue about the self-renewal potential of the adult human heart? *Int J Cardiol* 145: e44–46.

127. Gherghiceanu M, Popescu LM (2009) Human epicardium: ultrastructural ancestry of mesothelium and mesenchymal cells. *J Cell Mol Med* 13: 2949–2951.

128. Limana F, Zacheo A, Mocini D, Mangoni A, *et al.* (2007) Identification of myocardial and vascular precursor cells in human and mouse epicardium. *Circ Res* 101: 1255–1265.

129. Winter EM, Grauss RW, Hogers B, van Tuyn J, *et al.* (2007) Preservation of left ventricular function and attenuation of remodeling

after transplantation of human epicardium-derived cells into the infarcted mouse heart. *Circulation* 116: 917–927.

130. Urbanek K, Cesselli D, Rota M, Nascimbene A, *et al.* (2006) Stem cell niches in the adult mouse heart. *Proc Natl Acad Sci USA* 103: 9226–9231.

III. Does Heart Regeneration Occur?

A. Cardiac Regeneration in the Newt

Thilo Borchardt and Thomas Braun

1. INTRODUCTION

Loss of cardiomyocytes and functional impairment of the myocardium results from numerous different causes. Dying cardiac cells trigger a sequence of events that ultimately might lead to irreversible scarring and severely reduced functionality of the heart in species that lack comprehensive regenerative capabilities. Numerous attempts have been made to avoid such fatal consequences. The induction of cell cycle re-entry in cardiomyocytes is one of the most obvious approaches to replace lost myocardium.[1,2] Another option is the use of cells from other organs, such as skeletal myoblasts[3,4] or bone marrow-derived cells,[5,6] which might substitute for lost cardiomyocytes or exert positive effects on the remodeling myocardium. Although positive results from preclinical studies have led to several clinical trials performed in patients with acute myocardial infarction or end-stage heart failure, major improvements of cardiac function in these patients were not achieved.[4,7,8] In recent years, the identification of a number of cardiac resident progenitor cell populations has raised hope about alternative routes to regenerate damaged myocardium.[8-10] The potential of such cardiac resident stem cells *in vivo* remains to be determined. Yet, the inability for spontaneous reversion of severe cardiac injuries in patients demonstrates only a limited potential of endogenous progenitors to regenerate the heart in mammals under pathophysiological conditions. Animal species, which efficiently regenerate hearts, might lead a way out of this dilemma once the mechanisms that are employed to repair their hearts are understood.

Urodele amphibians are known for more than 200 years to own exceptional regenerative capabilities. Adult newts of the genus *Notophthalmus* are able to regenerate entire limbs and tails after attacks by natural predator and parts of the central nervous system or ocular tissues, like lens and retina.[11,12] Cardiac regeneration in the newt *Notophthalmus viridescens* can occur after ventricular apex amputation[13,14] or after mechanical disruption of more than 50% of the ventricular myocardium within three to four months.[15,16] A detailed understanding of the molecular processes underlying efficient cardiac regeneration in newts, may lead to novel therapeutic strategies that can efficiently stimulate cardiac regeneration in mammals.

2. MORPHOLOGICAL CHARACTERISTICS OF CARDIAC REGENERATION IN NEWTS

2.1. Cardiac Regeneration After Ventricular Apex Amputation

The cardiac ventricle of the newt resembles embryonic mammalian hearts in its trabeculated structure, in which a single layer of epicardial-like cells envelops myocytes, fibroblasts and nerve fibers. First hints of regeneration of adult newt hearts came from ventricular apex amputation studies.[14] Amputation of the ventricular apex leads to a rapid wound closure by blood clotting. This blood clot is dissolved after one week and is followed by invasion of lymphocytes at around ten days post amputation at the site of the initial margin of the clot. This period marks first mitotic cells in the trabeculae adjacent to the wound. Only three to four days later macrophage activity can be observed, removing remnants of necrotic cardiomyocytes. At the same time a layer of newly formed epicardium seals the inner surface of the initial blood clot. Soon after, connective tissue fibers appear within the wound area. An increasing mitotic activity co-localizes within trabecular structures near the wound. At the endpoint of observation, four weeks after amputation, scattered myocytes, smaller than those found in uninjured hearts, reappear within the wound area that is now mainly composed of connective tissue. The authors therefore concluded, that regeneration in the newt heart does not occur completely and hence might result in semi-functional tissue containing a mixture of contractile and residual connective tissue.

Follow-up experiments set out to increase the reactive area of the wound after ventricular apex amputation. Removed pieces of the ventricle were minced and placed onto the amputated plane.[13,17,18] By extending the observation period to 70 days the authors were able to observe a proliferative phase of partially dedifferentiated cardiomyocytes during the second and third week after injury. Finally, the grafts reorganized into functional mini ventricles consisting primarily of cardiac muscle. Seventy-day-old grafts contained myocytes resembling those of uninjured ventricles by carrying numerous myofibrillae, mitochondria, glycogen and intercellular junctions demonstrated that newt hearts can effectively generate organized cardiac muscle even in adult individuals.

2.2. Cardiac Regeneration After Mechanical Injury of Ventricular Myocardium

We have reinvestigated heart regeneration in the newt by mechanical disruption of one half of the newt ventricle through repeated squeezing with fine forceps. This model disorganizes the ventricular myocardium at the wound site, maintaining the epicardial seal of the newt ventricle.[15,16] Within 24 hours after injury, the expression of the cardiac markers alpha-myosin heavy chain and cardiac troponin T in remaining cardiomyocytes largely disappeared. Re-expression of mature cardiac markers in the wound area only reappeared from around three weeks after mechanical injury. We observed a complete functional restoration of the trabeculated ventricular myocardium within three to four months (Figure 1). We therefore conclude that under situations mimicking ischemic loss of cardiomyocytes, the newt heart is able to fully regenerate the loss of contractile tissue.

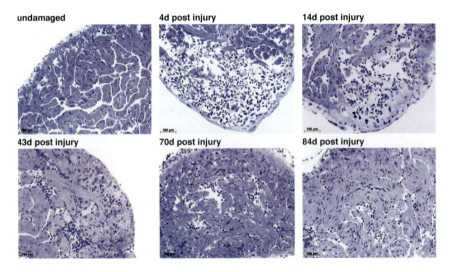

Figure 1.   Morphological characteristics of newt heart regeneration. At four days after mechanical injury, a massive invasion of lymphocytes in the damaged area occurs (4d post injury). Ten days later, newly formed trabeculae containing mitotic figures appear (14d post injury). After 43 days, the original wound area is almost filled with newly formed myocardium (43d post injury), with only small areas not containing reorganized myofibrils, that further decrease at 70 days after injury (70d post injury). After around three months (84d post injury), the ventricular myocardium is completely rebuilt and resembles uninjured newt hearts (undamaged). Scale bars 100 μm.

3. GENERATION OF NEWLY FORMED CARDIAC TISSUE BY PARTIAL DEDIFFERENTIATION OF MATURE CELLS

Findings from appendage and lens studies[12,19] favor the view that regeneration in the newt is mostly accomplished by the generation of new progenitors through dedifferentiation of mature cells. This concept has been refined recently in a study[20] where the authors have utilized transgenic axolotls (*Ambystoma mexicanum*), with a genetically encoded GFP transgene. Using this tool the authors were able to track the fate of the major limb tissues after transplantation during regeneration. To their surprise and in contrast to what has been proposed before[21] each tissue produced progenitor cells with only restricted differentiation potential. The authors concluded that the blastema is not a homogenous pool of pluripotent progenitor cells. These findings elegantly underlined that regeneration of complex tissues does not necessarily need a niche of resident stem cells for each individual tissue and thus makes it unlikely that newt tissues maintain a reservoir of different progenitor populations, which are predetermined to differentiate into mature progeny.

Additional evidence for the concept of "limited deprogramming" comes from studies of fluorescently labeled cardiomyocytes that were heterotopically transplanted into intact and regenerating limb buds of the newt.[16] Transplanted cardiomyocytes did not react through dedifferentiation and cell cycle re-entry when implanted into intact limbs. However, cardiomyocytes rapidly downregulated expression of mature cardiac maker genes and activated the expression of genes characteristic for undifferentiated blastema cells when implanted into a limb blastema five days after amputation. The expression of blastema markers disappeared at later stages and cardiomyocyte-derived cells started to express markers of mature skeletal muscle. Even more, the transplanted cells contributed to small muscle fibers re-established in regenerating limb stumps. Transplanted cells did not only transdifferentiate robustly into muscle lineages of the recipient organ but also expressed phosphorylated histone H3, indicating active contribution to limb regeneration by proliferation. These findings demonstrated a remarkable plasticity of cardiomyocytes and favor the model that cardiac regeneration in newts is based on partial dedifferentiation and not a complete reprogramming into pluripotency. Although skeletal and heart muscle lineages segregate during embryonic development several heart muscle genes are also expressed in skeletal muscle and vice versa, which makes it likely that regulatory molecules shared by both cell types provided the necessary cues for cell type-specific redifferentiation.

Only recently, reprogramming of somatic cells into pluripotent progenitors for therapeutic approaches has become a more widely recognized issue. The ability of induced pluripotent stem cells to generate various cell types and to contribute to germ line competent chimeric animals[22–24] has raised hopes for new therapeutic approaches. However, most chimeras and their progeny derived from induced pluripotent cells engender tumors in the presence,[22] or absence of the proto-oncogene c-myc.[25] Thus, major concerns about the safety of such iPS cells remain. Interestingly, newts seem to be largely resistant to chemically induced tumor formation[26] and only few studies have reported spontaneous tumor formation in newts.[27,28] A thorough understanding of the molecular processes, which lead to restricted dedifferentiation of differentiated cells without tumor formation, will help to control adverse effects and thus might be of major therapeutic relevance. The newt will definitely serve as a reference model organism to understand and develop therapeutic strategies based on dedifferentiation and tissue reprogramming.

4. PROLIFERATIVE POTENTIAL OF NEWT CARDIOMYOCYTES *IN VIVO* AND *IN VITRO*

4.1. Proliferative Potential *in Vivo*

It is not clear whether dedifferentiation of cardiomyocytes is an essential precondition to allow cell cycle re-entry *in vivo* although cardiomyocytes of the uninjured heart are virtually non-proliferative.[18] Only upon injury they respond with a massive increase in DNA synthesis and mitosis.[13,14,17,18] Also atrial myocytes can respond to ventricular damage with proliferation[29] or after direct atrial damage.[30]

Within the wounded area up to 30% of cardiomyocytes are able to initiate DNA synthesis and roughly 10% of wound-site cardiomyocytes enter the mitotic phase. During progression of cardiac regeneration, cardiomyocytes represent more than 80% of all mitotically active cells in the wound area.[13,14,17]

In contrast to mammalian hearts,[31] adult cardiomyocytes in uninjured newt ventricles are mostly (98%) mononucleated and diploid. Although approximately 50% of cardiomyocytes have undergone one round of DNA synthesis during cardiac regeneration, nine out of ten remain mononucleated and diploid[18] indicating that most cardiomyocytes complete mitosis and cytokinesis after DNA synthesis, which enables the newt heart to repair even extensive damages.

4.2. Proliferative Potential *in Vitro*

Cultured newt cardiomyocytes from uninjured hearts possess a limited proliferative potential without a prominent loss of sarcomeric proteins,[16,32,33] but dedifferentiation of cardiomyocytes enhances the proliferation potential of cardiomyocyte-derived cells dramatically (Borchardt *et al.*, unpublished observations). *Notophthalmus viridescens* is therefore the only adult vertebrate model system that allows growth of actively proliferating cardiomyocytes *in vitro*. In organ cultures from newt heart[34,35] as well as in dissociated primary cultures,[36] up to 9% of all cardiomyocytes are mitotically active. Comparable to the *in vivo*, situation, 80% of mononucleated cardiomyocytes produce mononucleated daughter cells in culture. The level of binucleation and multinucleation increases only with increasing culture time,[37] indicating that the process of bi- and multinucleation commences only after more than one round of cell replication. Most mononucleated cardiomyocytes might replicate only once but initiate a second round of DNA synthesis under culture conditions. Stimulation of DNA synthesis by more than two-fold can be induced by conditioned media of older cardiomyocyte cultures or from conditioned media of non-cardiomyocytes[38] indicating the presence of stimulatory factors, which are actively secreted from cardiac cells. Several molecules have been confirmed to stimulate cardiomyocyte DNA synthesis *in vitro*, like PDGF,[39] TPA, retinoic acid, FGF-1, ET-1, PDGF, bombesin and FGF-2 while other factors such as thyroxin, TGF-beta and heparin inhibited DNA synthesis.[38]

The potential of individual cardiomyocytes to undergo more than one round of cell division *in vitro* was examined by time lapse imaging in dissociated cultures.[33] More than 75% of individually tracked cardiomyocytes that synthesized DNA also underwent mitosis. Almost one third of all imaged cells successfully completed one or more rounds of cell division giving rise to beating daughter cells. Interestingly, even 20% of all bi- or multinucleated cardiomyocytes successfully completed cell division suggesting that bi- or multinucleation does not constitute an absolute block for cell proliferation.

The differences in the proliferative potential of cultured cardiomyocytes and undamaged cardiomyocytes *in vivo* is mainly mediated by mammalian serum components.[33] Increasing concentrations in fetal bovine serum stimulated DNA synthesis in up to 30% of cardiomyocytes in culture reaching a maximum at 10% FBS. It was therefore hypothesized that serum activated pathways control the phosphorylation of Rb-protein and might thus enable cell cycle progression similar to the control of skeletal newt myotubes.[40] Indeed, inhibition of Rb-phosphorylation through p16[Ink4] expression in cardiomyocytes caused a 13-fold inhibition of S-phase re-entry of cardiomyocytes compared to uninjected cells.[33] A better understanding of the key pathways responding to these factors will provide insight into the process of regeneration and may help to understand, how cell cycle re-entry of cardiomyocytes can be initiated.

5. FUTURE PERSPECTIVES AND CONCLUSIONS

Complex tissue regeneration is a naturally occurring process in urodele amphibians. Hence, newts provide a powerful model system to gain basic information about pathways that direct heart regeneration. A detailed understanding of the molecular circuits that control heart regeneration in newts might also open the way for heart regeneration in mammals.

At present, the newt is not a well-established model system such as the zebrafish and the mouse despite a rich history as an experimental model, which resulted in the assignment of the Nobel prize for the discovery of the "organizer" to Hans Spemann. Undoubtedly, rapid advances in DNA sequencing techniques and high throughput proteomics have made the newt amendable for advanced molecular studies, which will help to unravel the mechanisms underlying cardiac regeneration. Recent sequencing approaches on regenerating salamander limb and tails[41,42] and the establishment of large EST datasets for newt heart regeneration[43] will help to disclose differences between mere repair and bona fide regeneration. A newly developed high throughput method to monitor expressional changes of thousands of proteins by SILAC-based *in vivo* labeling of newts[44] will greatly facilitate regeneration research in newts, helping to identify yet unknown newt proteins involved in cardiac regeneration. Techniques to efficiently manipulate gene expression in newts exist.[45,46] The availability of transgenic newt models[47,48] will further help to elucidate the molecular mechanisms of organ regeneration. Although major differences in the physiology of mammalian and amphibian hearts exist we believe that the molecular pathways, which control tissue regeneration in newts, might still be present in mammals and might eventually be employed for the development of innovative approaches to improve pathological conditions in humans.

REFERENCES

1. Hassink RJ, Pasumarthi KB, Nakajima H, Rubart M, *et al.* (2008) Cardiomyocyte cell cycle activation improves cardiac function after myocardial infarction. *Cardiovasc Res* 78: 18–25.
2. Laflamme MA, Murry CE (2005) Regenerating the heart. *Nat Biotechnol* 23: 845–856.
3. Menasche P, Hagege AA, Scorsin M, Pouzet B, *et al.* (2001) Myoblast transplantation for heart failure. *Lancet* 357: 279–280.
4. Menasche P (2009) Cell-based therapy for heart disease: a clinically oriented perspective. *Mol Ther* 17: 758–766.

5. Ang KL, Shenje LT, Srinivasan L, Galinanes M (2006) Repair of the damaged heart by bone marrow cells: from experimental evidence to clinical hope. *Ann Thorac Surg* 82: 1549–1558.
6. Rubart M, Field LJ (2006) Cardiac regeneration: repopulating the heart. *Annu Rev Physiol* 68: 29–49.
7. Dimmeler S, Burchfield J, Zeiher AM (2008) Cell-based therapy of myocardial infarction. *Arterioscler Thromb Vasc Biol* 28: 208–216.
8. Stamm C, Choi YH, Nasseri B, Hetzer R (2009) A heart full of stem cells: the spectrum of myocardial progenitor cells in the postnatal heart. *Ther Adv Cardiovasc Dis* 3: 215–229.
9. Lam JT, Moretti A, Laugwitz KL (2009) Multipotent progenitor cells in regenerative cardiovascular medicine. *Pediatr Cardiol* 30: 690–698.
10. Segers VF, Lee RT (2008) Stem-cell therapy for cardiac disease. *Nature* 451: 937–942.
11. Alvarado AS, Tsonis PA (2006) Bridging the regeneration gap: genetic insights from diverse animal models. *Nat Rev Genet* 7: 873–884.
12. Brockes JP, Kumar A (2002) Plasticity and reprogramming of differentiated cells in amphibian regeneration. *Nat Rev Mol Cell Biol* 3: 566–574.
13. Bader D, Oberpriller JO (1978) Repair and reorganization of minced cardiac muscle in the adult newt (*Notophthalmus viridescens*). *J Morphol* 155: 349–357.
14. Oberpriller JO, Oberpriller JC (1974) Response of the adult newt ventricle to injury. *J Exp Zool* 187: 249–253.
15. Borchardt T, Braun T (2007) Cardiovascular regeneration in non-mammalian model systems: what are the differences between newts and man? *Thromb Haemost* 98: 311–318.
16. Laube F, Heister M, Scholz C, Borchardt T, *et al.* (2006) Reprogramming of newt cardiomyocytes is induced by tissue regeneration. *J Cell Sci* 119: 4719–4729.
17. Bader D, Oberpriller J (1979) Autoradiographic and electron microscopic studies of minced cardiac muscle regeneration in the adult newt, *Notophthalmus viridescens*. *J Exp Zool* 208: 177–193.
18. Oberpriller JO, Oberpriller JC, Arefyeva AM, Mitashov VI, *et al.* (1988) Nuclear characteristics of cardiac myocytes following the proliferative response to mincing of the myocardium in the adult newt, *Notophthalmus viridescens*. *Cell Tissue Res* 253: 619–624.

19. Tsonis PA (2006) How to build and rebuild a lens. *J Anat* 209: 433–437.

20. Kragl M, Knapp D, Nacu E, Khattak S, *et al.* (2009) Cells keep a memory of their tissue origin during axolotl limb regeneration. *Nature* 460: 60–65.

21. Echeverri K, Tanaka EM (2002) Ectoderm to mesoderm lineage switching during axolotl tail regeneration. *Science* 298: 1993–1996.

22. Okita K, Ichisaka T, Yamanaka S (2007) Generation of germline-competent induced pluripotent stem cells. *Nature* 448: 313–317.

23. Kang L, Wang J, Zhang Y, Kou Z, *et al.* (2009) iPS cells can support full-term development of tetraploid blastocyst-complemented embryos. *Cell Stem Cell* 5: 135–138.

24. Kim JB, Sebastiano V, Wu G, Arauzo-Bravo MJ, *et al.* (2009) Oct4-induced pluripotency in adult neural stem cells. *Cell* 136: 411–419.

25. Miura K, Okada Y, Aoi T, Okada A, *et al.* (2009) Variation in the safety of induced pluripotent stem cell lines. *Nat Biotechnol* 27: 743–745.

26. Tsonis PA, Eguchi G (1981) Carcinogens on regeneration. Effects of N-methyl-N'-nitro-N-nitrosoguanidine and 4-nitroquinoline-1-oxide on limb regeneration in adult newts. *Differentiation* 20: 52–60.

27. Bryant SV (1973) Spontaneous epidermal tumor in an adult newt, *Cynops pyrrhogaster*. *Cancer Res* 33: 623–625.

28. Zilakos NP, Tsonis PA (1991) A spontaneous melanoma-like tumor in the adult newt Triturus cristatus. *Tumour Biol* 12: 120–124.

29. McDonnell TJ, Oberpriller JO (1983) The atrial proliferative response following partial ventricular amputation in the heart of the adult newt. A light and electron microscopic autoradiographic study. *Tissue Cell* 15: 351–363.

30. McDonnell TJ, Oberpriller JO (1984) The response of the atrium to direct mechanical wounding in the adult heart of the newt, *Notophthalmus viridescens*. An electron-microscopic and autoradiographic study. *Cell Tissue Res* 235: 583–592.

31. Rumyantsev PP (1991) Replicative behavior of different types of cardiomyocytes in terms of experimental conditions, age and systematic position of animals. In: Oberpriller JO, Oberpriller JC, Mauro A (eds.), *The Development and Regenerative Potential of Cardiac Muscle.* Harwood Academic Publishers, London, pp. 81–92.

32. Tate JM, Oberpriller JO, Oberpriller JC (1989) Analysis of DNA synthesis in cell cultures of the adult newt cardiac myocyte. *Tissue Cell* 21: 335–342.
33. Bettencourt-Dias M, Mittnacht S, Brockes JP (2003) Heterogeneous proliferative potential in regenerative adult newt cardiomyocytes. *J Cell Sci* 116: 4001–4009.
34. Nag AC, Cheng M, Healy CJ (1980) Studies of adult amphibian heart cells *in vitro*: DNA synthesis and mitosis. *Tissue Cell* 12: 125–139.
35. Nag AC, Healy CJ, Cheng M (1979) Organ culture of adult amphibian heart: a fine structural analysis. *Tissue Cell* 11: 231–248.
36. Tate JM, Oberpriller JO (1989) Primary cell culture and morphological characterization of ventricular myocytes from the adult newt, *Notophthalmus viridescens*. *Anat Rec* 224: 29–42.
37. Matz DG, Oberpriller JO, Oberpriller JC (1998) Comparison of mitosis in binucleated and mononucleated newt cardiac myocytes. *Anat Rec* 251: 245–255.
38. Soonpaa MH, Oberpriller JO, Oberpriller JC (1994) Factors altering DNA synthesis in the cardiac myocyte of the adult newt, *Notophthalmus viridescens*. *Cell Tissue Res* 275: 377–382.
39. Soonpaa MH, Oberpriller JO, Oberpriller JC (1992) Stimulation of DNA synthesis by PDGF in the newt cardiac myocyte. *J Mol Cell Cardiol* 24: 1039–1046.
40. Tanaka EM, Gann AA, Gates PB, Brockes JP (1997) Newt myotubes reenter the cell cycle by phosphorylation of the retinoblastoma protein. *J Cell Biol* 136: 155–165.
41. Habermann B, Bebin AG, Herklotz S, Volkmer M, *et al.* (2004) An *Ambystoma mexicanum* EST sequencing project: analysis of 17,352 expressed sequence tags from embryonic and regenerating blastema cDNA libraries. *Genome Biol* 5: R67.
42. Putta S, Smith JJ, Walker JA, Rondet M, *et al.* (2004) From biomedicine to natural history research: EST resources for ambystomatid salamanders. *BMC Genomics* 5: 54.
43. Borchardt T, Looso M, Bruckskotten M, Weis P, *et al.* (2010) Analysis of newly established EST databases reveals similarities between heart regeneration in newt and fish. *BMC Genomics* 11: 4.

44. Looso M, Borchardt T, Krueger M, Braun T (2010) Advanced identification of proteins in uncharacterized proteomes by pulsed *in vivo* stable isotope labeling-based mass spectrometry. *Mol Cell Proteomics*, 9: 1157–1166.

45. Schnapp E, Tanaka EM (2005) Quantitative evaluation of morpholino-mediated protein knockdown of GFP, MSX1, and PAX7 during tail regeneration in *Ambystoma mexicanum. Dev Dyn* 232: 162–170.

46. Morrison JI, Loof S, He P, Alestrom P, *et al.* (2007) Targeted gene delivery to differentiated skeletal muscle: a tool to study dedifferentiation. *Dev Dyn* 236: 481–488.

47. Ueda Y, Kondoh H, Mizuno N (2005) Generation of transgenic newt Cynops pyrrhogaster for *regeneration study. Genesis* 41: 87–98.

48. Sobkow L, Epperlein HH, Herklotz S, Straube WL, *et al.* (2006) A germline GFP transgenic axolotl and its use to track cell fate: dual origin of the fin mesenchyme during development and the fate of blood cells during regeneration. *Dev Biol* 290: 386–397.

III. Does Heart Regeneration Occur?

B. Zebrafish Heart Regeneration

Robert J. Major and Kenneth D. Poss

1. INTRODUCTION

Acute myocardial infarction is a leading cause of death throughout the world. Loss of oxygen and nutrient delivery, most commonly due to occlusion of a coronary artery, can quickly kill cardiac muscle. Those fortunate to survive a heart attack replace necrotic tissue with scar. Scar tissue deposition can be detrimental to cardiac function, potentially resulting in future pathology and tissue damage.

Therefore, the development of therapies for aiding cardiac muscle survival or regeneration is of utmost medical and societal importance.

We are currently in the midst of an exciting resurgence in the field of tissue regeneration. Scientists are employing modern genetic and molecular tools to unlock the mysteries of regeneration in adult animals. Although one area of immense interest is the isolation and transplantation of mammalian stem cells, we must also harness information from non-mammalian vertebrate organisms with robust capacities to naturally regenerate adult tissues.

Unlike mammalian vertebrates, zebrafish are capable of mounting a significant regenerative response after loss of heart tissue. Instead of scarring, the initial fibrin clot is replaced with a new myocardial wall within one to two months. This chapter navigates what is known at this early stage in investigations of heart regeneration in adult zebrafish. It will highlight the genetic and molecular tools available to zebrafish researchers, and discuss the bright future for this organism in informing mechanisms of natural cardiac regeneration.

2. ZEBRAFISH: A UNIQUE VERTEBRATE COMBINATION OF REGENERATION AND MODERN TOOLS

Although the concept of regeneration, at least on a mythological and observational basis, has fascinated people for thousands of years, documented scientific presentations and reports of regeneration in crayfish appendages and hydra first emerged in the early 18th century.[1] Many invertebrate organisms, including starfish, hydra, and planarians, have been on stage through the years for their surprising capacities to regenerate an entire animal from small regions. Indeed, prior to his pioneering genetic research in *Drosophila*, Thomas Hunt Morgan studied regeneration for many years in animals from planaria to goldfish.[2] Today, scientists have begun to use modern molecular tools like RNA interference to reinvestigate the regenerative phenomena identified by Morgan and others in invertebrate animals.

We have known for some time that many vertebrates have a profound ability to regenerate body parts, first demonstrated in the 18th century by Lazarus Spallanzani who reported limb regeneration in adult salamanders.[3] Indeed, amputation of a newt limb results in the full regeneration of missing structures with normal function and pattern. Salamanders regenerate a variety of tissues including spinal cord, brain tissue, retina, lens, tail, and limb, giving these animals center stage as the champions of regeneration among vertebrates.

Mammalian tissues also possess significant regenerative capacity, with the liver as a classic example. Following partial hepatectomy, the remaining tissue undergoes compensatory hyperplasia to restore lost mass.[4,5] Such regenerative capacity is not surprising, considering the regular assault on the liver by toxins. Regenerative prowess is also apparent in gut and skin epithelia, as well as in the hematopoietic system, which undergo regular natural replenishment in the absence of overt injury.[5–7] However, unlike salamanders, mammals show little ability to regenerate after major tissue loss from brain, limb, or heart.

Several attributes have made 2–3 cm zebrafish a favored model system for studying patterning and organogenesis within the vertebrate embryo. These include small size and transparency of embryos, speed of development outside the mother, low cost of maintenance, and amenability to forward and reverse genetic manipulations.[8,9] It has also become clear that zebrafish exhibit a robust regenerative capacity analogous to salamanders, as they regenerate a variety of complex structures following injury or severe tissue loss, including fins, retina, spinal cord, and heart.[10–15]

Zebrafish have five types of fins, each capable of regenerating following amputation, with most research attention directed to the caudal fin. Fins consist of segmented bony rays comprised of concave, facing hemirays that are coated with bone-depositing scleroblasts and surround mesenchymal tissue, nerves, and blood vessels. Following amputation of up to 95% of the fin, the lost distal structures regenerate fully within ten to 14 days. This process involves formation of a blastema, a proliferative mass of undifferentiated mesenchymal cells that is maintained as a zone of progenitor tissue for new structures.[10]

Remarkable regenerative responses are also observed within the central nervous system. Following complete trans-section of the spinal cord, many of the axons are coaxed into recovering, traversing the lesion, and re-establishing functional connections.[11] In addition, acute injury or phototoxic ablation of *teleost* retinas results in regeneration along with production of new rod and cone cells.[12,15] It is understood that regeneration of new photoreceptor neurons is driven, at least in part, by stem cells residing within the inner nuclear layer of retinal cells. Neuronal regeneration studies have been extended to the zebrafish lateral line, a stretch of mechanoreceptive hair cells that runs along the anteroposterior axis and detects environmental movement or vibration.[16] As these hair cells are sensitive to aminoglycoside antibiotics, they can be ablated with a short treatment of neomycin.[17,18] Remarkably, within 24–48 hours of treatment, new sensory hairs regenerate along the flank of zebrafish larvae. Sensory hair cells in the mammalian ear fail to regenerate after injury; thus, zebrafish represent a promising model to shed light on congenital hearing disorders. The molecular details garnered from these types of studies will allow for a better understanding of how to enhance neural survival and repair.

In particular, while molecular genetic approaches are difficult to apply to classic regenerative model systems like lizard and salamanders, those who work with zebrafish have access to key tools. Because regeneration is generally studied in adult animals, these approaches most often must involve an element of experimental control that restricts application of the tool during embryonic development. Indeed, adult regenerative events are known to employ some of the same genes and pathways that act during ontogenetic development.

For the study of adult regenerative events, there are two powerful functional approaches available in zebrafish. First, there are reverse genetic approaches. Halloran and co-workers demonstrated that genes driven by the *hsp70* promoter are experimentally activated in transgenic zebrafish through heat-shock induction.[19] Thus, one can use this important technology to 1) withhold ectopic expression of a candidate gene until adulthood; and 2) experimentally induce expression of that gene to various extents to test the effects of that factor on regeneration in adults. A second powerful approach possible in zebrafish is forward genetics. Johnson and Weston were the first to recognize that mutational analysis could be applied to regeneration in vertebrates.[20] They performed a screen for mutations that block regeneration of amputated adult zebrafish caudal fins in a temperature-sensitive (ts) manner. From a similar screen, groups identified four genes required for regeneration through positional cloning: 1) the mitotic checkpoint kinase *mps1*;[21] 2) *sly1*, encoding a protein involved in vesicle trafficking and protein secretion;[22] 3) the heat-shock factor *hsp60*;[23] and 4) *fgf20a*, an Fgf ligand that is released after injury and essential for blastema formation.[24] These mutant lines have not only been informative about fin regeneration, but, as conditional mutations, they have helped investigations of heart regeneration,[13] retinal regeneration,[14] germ cell aneuploidy,[25] and fin homeostasis.[26]

Zebrafish have also enabled a variety of studies examining the effects of small molecule compounds on embryogenesis.[27] Although the logistics of working with the adult animal can limit these types of screens with respect to regeneration, initial pilot screens have been performed to look at inhibitors of larval fin fold regeneration.[28,29] These inhibitors have then been followed up to examine effects in the adult system. Investigators have also been able to dissect gene function during adult fin regeneration by electroporation of antisense morpholinos into the regenerate, in attempts to inhibit specific pathways or microRNAs.[30–32]

In summary, zebrafish present an excellent model system for addressing questions of tissue regeneration and maintenance. The system offers multiple injury models in a variety of tissue types and at different developmental stages. Most importantly, its amenability to forward genetic and transgenic approaches have set this species apart from traditional vertebrate and invertebrate model systems of regeneration.

3. HEART REGENERATION IN ZEBRAFISH

While it had been known for years that *teleost* fish regenerate the spinal cord, retina, and fins, the first report of heart regeneration in *teleosts* was very recent, in 2002.[13] Because of the small size of adult zebrafish, traditional ischemic infarction models involving occlusion of a coronary artery, routine in mouse or rat studies, are virtually impossible. However, one can gain easy access to the zebrafish heart with a ventral incision into the pericardial cavity in an anesthetized animal. Then, iridectomy scissors are used to surgically remove up to 20% of the ventricular apex. It is possible that this type of mechanical injury tests the animal's response under the strongest possible stimuli for regeneration. Because the ventricular lumen is penetrated by the injury, there is a large amount of initial blood loss prior to clotting at the injury site.

While remarkable that the animal survives such an intense injury, what happens during the next two months is even more fascinating. Unlike in mammalian hearts, the initial fibrin clot formed within the first few days is not replaced with scar tissue. In fact, collagen staining reveals very little scar deposition by one to two months post injury.[13] By this time, a new cardiac wall has formed at the injury site, restoring approximate chamber size (Figures 1A and 1B). The zebrafish ventricle is comprised of two main types of cardiac muscle: 1) inner myofibers organized into long, elaborate trabeculae separated from the lumen by the endocardial layer; and 2) a thin wall of compact muscle enveloped by the epicardial layer. The regenerated muscle returns largely in the form of a thickened compact myocardial wall.[13]

In order to call this process "regeneration," one must see evidence that new cardiomyocytes are generated during muscle restoration. There is little or no proliferation by mammalian cardiomyocytes after cardiac injury, and instead the existing cardiomyocytes undergo hypertrophy in response to injury or disease. By contrast, adult zebrafish cardiomyocytes show robust indices of proliferation, beginning at the end of the first week after injury.[13] Cardiomyocyte proliferation is highest within the first two weeks of resection, but can continue for weeks beyond this (Figure 1C). Pulse-chase experiments with the nucleotide analog BrdU have indicated that early peripheral proliferating cardiomyocytes are displaced inward by newer proliferating cells as regeneration proceeds. Thus, unlike in mammals, the injured zebrafish heart heals by the creation of new, proliferative cardiomyocytes, restoring a new myocardial wall after partial ventricular resection.

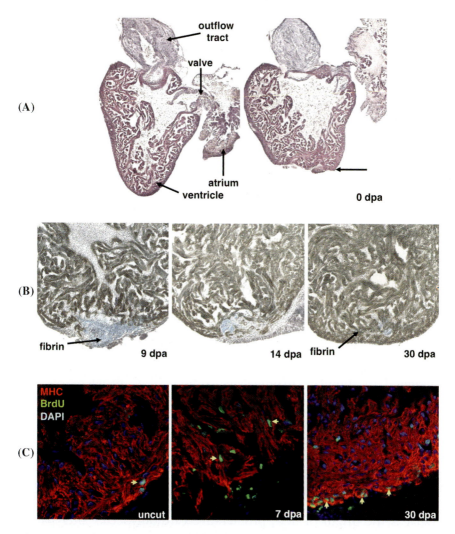

Figure 1.   Heart regeneration in zebrafish. (**A**) Hematoxylin and eosin stain of the adult zebrafish heart before and after 20% ventricular resection. (**B**) Ventricular sections stained for myosin heavy chain for cardiac muscle (brown) and aniline blue for fibrin (blue). A clot filled with mature fibrin seals the wound by 7 dpa, and is gradually replaced by cardiac muscle. The wall is typically restored by 30 dpa. (**C**) BrdU incorporation, a marker of DNA synthesis and cellular proliferation, is rare in the uninjured ventricle but visible in many cardiomyocytes at the injury site by 7 dpa. Cardiomyocyte proliferation is still detectable by 30 dpa. Figure adapted from Ref. 13.

3.1. A Balance of Fibrosis and Proliferation

As discussed above, a genetic screen for defects in zebrafish fin regeneration identified a strain carrying a temperature-sensitive mutation in the cell cycle regulator, *mps1*.[21] Under restrictive temperatures, *mps1* mutants show proliferative defects in the fin blastema, blocking regenerative outgrowth. Interestingly, *mps1* mutants also fail to regenerate heart muscle after ventricular resection. Instead of filling the wound with cardiac muscle, injured mutants healed injuries with large, collagen-rich scars.[13] Similar results were obtained when cardiac regeneration was inhibited by transgenic blockade of Fgf receptors (see below).[33] Thus, zebrafish have the capacity to repair cardiac injuries by scarring, but this mechanism is typically overridden by a robust regenerative process. These results have supported the hypothesis that muscle regeneration and scarring are competing processes after cardiac injury, and that evolutionary differences in the balance between them will determine the mechanism of healing. That there appear to be lower numbers of collagen-depositing fibroblasts within the adult zebrafish heart might contribute to a more regenerative environment.[34] An exciting implication of these results is that therapeutic advances designed to stimulate mammalian cardiomyocyte proliferation may have additional anti-scarring benefits. Indeed, recent attempts to induce cardiomyocyte proliferation after myocardial infarction in mouse and rat has appeared to also impact scarring.[35,36]

3.2. The Origin of Proliferating Cardiomyocytes

What is the origin of new, proliferative cardiomyocytes after partial resection of the zebrafish ventricle? While mitotic cells expressing contractile genes emerge during regeneration, there are multiple mechanisms that could account for their ultimate origin, separating into two main possibilities. First, there could be an undifferentiated stem or progenitor cell that is activated by injury to differentiate into a cardiomyocyte that proliferates and builds the new muscle. This is essentially what occurs during embryonic heart development: undifferentiated mesoderm cells adopt a cardiac fate, turn on contractile genes, and differentiate into cardiomyocytes that proliferate and build the heart. The second possible mechanism is one of simple duplication. In this case, existing cardiomyocytes spared by injury, or perhaps a subpopulation(s) of cardiomyocytes, are activated by injury to divide and directly build new muscle. Such a mechanism might be inclusive of dedifferentiation, or the transient reduction of contractile gene programs, to facilitate myocyte proliferation.

The first experiments directly addressing this question used a unique developmental timing assay.[33] Adult zebrafish were established with two transgenes; *Enhanced Green Fluorescent Protein* (*EGFP*) and *nuclear-localized DsRed2* (a red fluorescent protein) each under the control of the *cardiac myosin light chain 2* (*cmlc2*) promoter. *cmlc2* is a robust marker of the first embryonic cardiomyocytes, and is highly expressed in adult cardiomyocytes.[37] The use of its promoter provides an excellent readout of nascent cardiac muscle tissue, particularly within the context of these two transgenes. EGFP has a faster folding time than nuc-DsRed2, so that green fluorescence will be observed approximately one day prior to red fluorescence within newly formed cardiomyocytes. Ventricular resection of the adult zebrafish heart revealed a zone of EGFP[pos]/nuc-DsRed2[neg] cardiomyocytes at the apical regenerating edge, suggesting that cardiomyocytes are maturing freshly from an undifferentiated state. To uncover possible cardiac progenitors, analysis of transcription factors expressed during embryonic heart development was performed. The markers *tbx20*, *nkx2.5*, and *hand2* were expressed at enhanced levels in cells surrounding the injury site and appear to coincide with newly regenerated tissue. Taken together, the data suggested that the adult injured zebrafish heart employs an undifferentiated intermediate during regeneration, analogous to a blastema.

More recently, different reagents and technologies have identified a new source of regenerated cardiomyocytes.[38] These studies employed a transgenic zebrafish in which regulatory sequences of the transcription factor *gata4* were used to drive EGFP expression. Interestingly, partial ventricular resection induced expression of the reporter in cardiomyocytes of the outer compact layer within seven days of injury, before fluorescence distributed to regenerating muscle

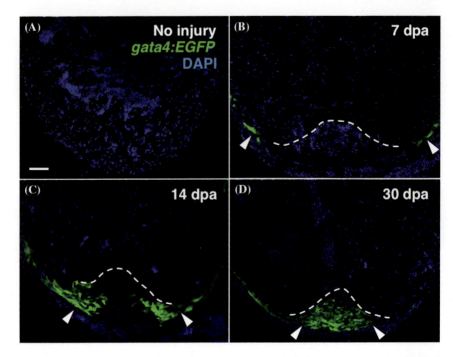

Figure 2.   Activation and contribution of cardiomyocytes to regenerated muscle. *gata4*:EGFP expression (arrowheads) in **(A)** uninjured and **(B–D)** regenerating ventricles. Fluorescence is induced in subepicardial myocytes by 7 dpa and indicates new muscle as regeneration progresses. Dotted line indicates approximate plane of resection. Figure adapted from Ref. 38.

surrounding and within the injury site by 14 and 30 days post amputation (dpa) (Figure 2). To further explore a role for *gata4*-positive cells in building the regenerate, several new transgenic lines were designed to facilitate lineage-tracing analysis. One of these lines had the same *gata4* regulatory sequences driving a Cre recombinase, in this case a fusion protein that could be induced in the presence of the drug 4-hydroxytamoxifen to translocate to the nucleus and catalyze the excision of DNA sequences between *loxp* recognition sites. This line was crossed to a transgenic indicator line permitting the visualization of cardiomyocyte EGFP fluorescence only after excision of *loxP*-flanked transcriptional stop sequences. When 4-hydroxytamoxifen was injected into the animals just prior to evident regeneration, EGFP fluorescence was seen in the subepicardial population at an early timepoint (9 dpa), and then at later timepoints (14–30 dpa) in many cardiac muscle cells within the regenerate. A second line had the Cre fusion cassette driven by *cmlc2* regulatory sequences. This line was also crossed to the indicator line, and 4-hydroxytamoxifen was used to label the majority of cardiac muscle prior to any injury. After injury and regeneration, a similar portion of regenerated cardiac muscle displayed the EGFP marker, indicating that existing cardiomyocytes expressing *cmlc2* give rise to most or all of the muscle within the regenerate.

A second study used similar reagents and approaches to indicate that regenerated adult muscle cells derive from myocytes genetically labeled via *cmlc2* expression at two days post fertilization.[39] Together, these lineage-tracing studies indicate that as part of the injury response, a portion of cardiomyocytes within the subepicardial layer induces *gata4* expression, proliferates, and contributes new cardiomyocytes to the regenerate. It is possible that previous developmental timing assays visualized dedifferentiation events from existing muscle and/or the participation of cells with low activity of the *cmlc2* promoter. Indeed both recent studies found that regenerating cardiomyocytes display a reduced contractile phenotype as compared to non-regenerating muscle cells. Future studies of interest will attempt to confirm and study dedifferentiation, or to assay other possible contributors to regeneration by lineage-tracing.

3.3. The Role of Non-Myocardial Tissues During Regeneration

As mentioned earlier, inner trabecular muscle fibers are protected from the luminal blood by an inner endothelium, the endocardium, while the compact muscle is covered with a thin epicardial layer. During embryogenesis, cells originating from the pro-epicardial organ migrate to and envelope the developing heart after it has completed looping, forming the epicardium.[40] A subset of epicardial cells undergoes an epithelial-to-mesenchymal transition (EMT), dives deep within the cardiac muscle, and differentiate into vascular support cells that make up the coronary vasculature, as well as into cardiac fibroblasts.[41,42] Functional studies mainly in mice have highlighted the importance of the epicardium in the growth of the embryonic heart, primarily through its role as vascular progenitor tissue.

One of the most exciting products of recent studies on the epicardium is the response of the adult zebrafish heart to a focal injury.[33] Expression analyses of two embryonic epicardial markers, *raldh2* and *tbx18*, during adult heart regeneration, revealed a striking "activation" of the epicardium as early as six hours after resection of the ventricular apex. Expression initiates in epicardial cells of the outflow tract and atrium most distant from the wound, and then spreads over the entire organ (Figure 3A). There is an expansion of the epicardial population evident by proliferation assays, and epicardial cells arrive at the injury site and begin to cover the wound by approximately seven days post amputation. Cells positive for *tbx18* are incorporated into the wound several cell layers deep by 14 and 30 dpa, in a process that resembles embryonic epicardial EMT. Use of transgenic reporters showed that a dense population of blood vessels forms within the regenerating tissue at these stages (Figure 3B). Together, these studies suggest that, as during embryogenesis, the adult epicardium is called upon to aid new muscle regeneration, in part by its role in forming new vasculature.

3.4. Fibroblast Growth Factor During Heart Regeneration

The key strength of the zebrafish heart regeneration model is its potential to act as a conduit for discovering factors or manipulations that either block or enhance natural heart regeneration. These factors and manipulations in turn provide excellent candidate strategies to alter the regenerative capacity of the injured mammalian heart. Therefore, revealing important molecular signals is of utmost interest to the field. One key component is the Fibroblast growth factor (Fgf) signaling pathway, represented by a large number of ligand members,

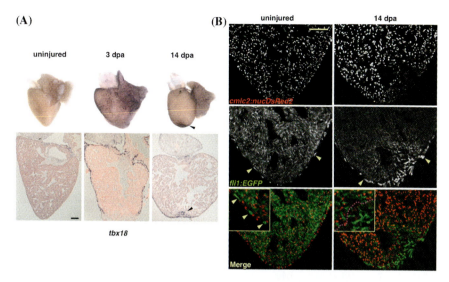

Figure 3.   Epicardial activation and vascularization. (A) Whole-mount (top) and section (bottom) *in situ* hybridization of uninjured and 3 and 14 dpa adult zebrafish hearts for *tbx18*, a marker of the embryonic epicardium. Expression of *tbx18* is low or undetectable in the uninjured heart, but induced in epicardial tissue surrounding the atrium and ventricle. By 14 dpa, *tbx18* expression is localized to cells that have penetrated the apical regenerate (arrowhead). **(B)** *cmlc2:nuc-DsRed2; fli1:EGFP* ventricles display red fluorescence in cardiomyocyte nuclei (top) and EGFP in endocardial and vascular endothelial cells (middle). More intense fluorescence is reported in endothelial cells of the coronary vasculature (arrowheads). By 14 dpa, regenerating myocardium acquires a dense, new coronary plexus. These events occur concomitantly with invasion of *tbx18*-positive cells. Figure adapted from Ref. 33.

and several receptor tyrosine kinases. While Fgfs have long been investigated for roles in tissue regeneration,[43–45] a direct role in cardiac regeneration was indicated from the effects of Fgf receptor inhibition on the regenerative process. The key tool was a transgenic zebrafish line enabling heat-induced expression of a dominant-negative form of the Fgf receptor.[33] As a result of Fgfr inhibition, muscle regeneration arrested at ~14 dpa, displaying reduced integration of epicardial cells into the injury site, and reduced neovascularization. Because the *fgf17b* ligand was shown to be present in regenerating muscle, and the

Fgf receptors Fgfr2 and Fgfr4 within epicardial-derived cells, the authors proposed a mechanism in which Fgf ligands like Fgf17b recruit epicardial cells to regenerating muscle, facilitating neovascularization.[33] Indeed, it is know that Fgf signaling is important for epicardial EMT and coronary vascular development during embryonic heart development.[46–48]

4. HOMEOSTATIC CARDIAC GROWTH AND MAINTENANCE IN ZEBRAFISH

In contrast to mammals, a variety of non-mammalian vertebrates display indeterminate growth, and have the capacity to rapidly increase adult mass in response to environmental changes in nutrition and population density. Increases in organ size accompany and support such growth. In order to address the molecular aspects of adult cardiac homeostasis in the absence of a mechanical injury, aquarium conditions can be established be zebrafish that control the pace of animal growth.[49] Animals placed in conditions that stimulated rapid growth (low aquarium density) showed a tripling of animal mass within 14 days when compared to the slow growth (high aquarium density) population. During these two weeks, the heart displayed a corresponding increase under rapid growth conditions. Cell proliferation assays indicated that hearts from the rapid growth group displayed approximately eight times the number of proliferating cardiomyocytes when compared to the slow growth group (Figures 4A–4C). Therefore, cardiac growth and homeostasis in the adult zebrafish is primarily due to a hyperplastic program, and distinguishing it from the dominant mammalian response to injury.

In addition to myocyte hyperplasia, other mechanisms are employed similarly during cardiac regeneration and homeostatic maintenance in zebrafish. For example, rapid growth conditions induce activation of the epicardium, increasing *raldh2* and *tbx18* expression organ-wide. A pulse-chase study after labeling the epicardial layer with a fluorescent dye indicated recurring contributions of epicardial cells to the underlying compact myocardial wall (Figure 4D).[49] Investigators also blocked Fgf signaling during

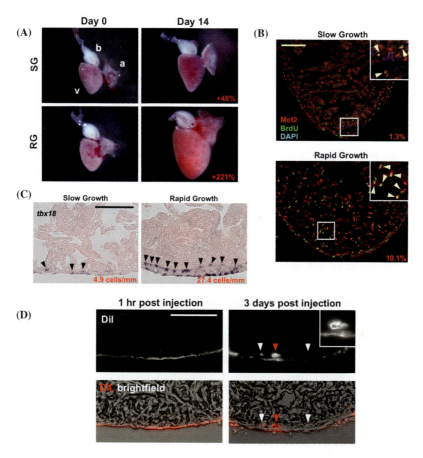

Figure 4.   Density-induced cardiac myogenesis in adult zebrafish. (A) Repre
sentative images of hearts collected from animals maintained for 14 days in low-density conditions for rapid growth (RG) and standard conditions for slow growth (SG).
Low-density conditions more than tripled ventricular size. **(B)** Ventricles from SG
and RG stained for Mef2 expression to identify cardiomyocytes (red), and for BrdU
incorporation (green). Myocyte BrdU incorporation was much greater in RG ventricles (arrowheads in insets). Nuclei are labeled with DAPI (blue). **(C)** *tbx18* is
expressed in many more epicardial cells and epicardial-derived cells (arrowheads) in
RG animals than in other groups. **(D)** DiI was carefully injected into the pericardial
sac to label the epicardium, and hearts were collected at one hour and three days post
injection. Dye is restricted to epicardial cells at one hour post injection, but at three
days, labeled cells are occasionally seen in the compact layer (arrowheads). The red
arrow indicates labeled tissue with the appearance of vascular tissue (enlarged in the
inset). Figure adapted from Ref. 49.

conditions of rapid growth, discovering that, like during injury-induced regeneration, the number of epicardial-derived cells incorporating into the ventricular wall was significantly reduced. Furthermore, long periods of Fgf inhibition in uninjured, fully mature adult zebrafish resulted in occasional spontaneous scar deposition within the ventricular wall. The authors proposed that the fibrosis was the failure of epicardial cells to efficiently contribute to the ventricular wall.[49] Taken together, these data suggest that periods of rapid growth induce a regenerative-like cardiomyocyte hyperplasia program, with concomitant supplementation of epicardial-derived cells to the ventricular wall similar to embryonic epicardial EMT.

We now understand that homeostatic growth, maintenance, and regeneration of the adult zebrafish heart employ similar cellular and molecular processes. This relationship is likely important in understanding why zebrafish can successfully supply the heart with new muscle following partial ventricular resection. For example, severe injury to the zebrafish ventricle can invoke, and perhaps magnify, an underlying maintenance program — a program that might be absent or not efficiently called upon in mammalian hearts.

5. FUTURE PROGRESS ON ZEBRAFISH HEART REGENERATION

The adult zebrafish has become an excellent model system for understanding natural regeneration and maintenance of the heart. Both of these complex events invoke similar cellular and molecular programs, with injury directing these programs to a focal injury site, while cardiogenesis is dispersed throughout the heart under growth or maintenance conditions. Future studies must focus on uncovering the earliest signals that link injury to activation of the epicardium and to muscle-building. One possible way to tackle this is by designing a forward genetic screen for heart regeneration mutants. The difficulty here lies with the ability to quickly visualize and measure levels of regeneration. Transgenic fluorescent reporter lines could potentially be utilized to expedite this within the context of a screen.

It will also be necessary to design multiple injury models in the fish to better mimic human myocardial infarction. The current model involves severe removal of tissue, and it would be interesting to determine how zebrafish respond to tissue death without mechanical disruption of the epicardial or endocardial layers. Furthermore, within the context of a mutant screen, non-invasive induction of an infarct could expedite the injury process. Recent zebrafish studies have utilized tissue-specific promoters to express bacterial nitroreductase in tissues like heart muscle.[50,51] This enzyme converts an exogenously delivered drug, metronidazole, to a cytotoxic DNA cross-linking compound. This approach could potentially be effective in adult zebrafish. An alternative to genetic screens would be a small molecule screen using compound libraries, in hopes of identifying drugs that modify cardiac regenerative capacity.

Although gene expression microarrays have been successfully performed using mRNA samples from regenerating hearts,[52] future microarray studies might focus on earlier timepoints, or on purified cardiac cell types for higher resolution information. Finally, continued use of the Cre recombinase system will not only enable marker-based lineage-tracing of interesting cardiac cells, but will also facilitate inducible, tissue-specific expression of candidate factors that may be important for regeneration.

6. SUMMARY

Acute myocardial infarction is a leading cause of death throughout the world. Although human hearts show little capacity to fix damaged cardiac tissue by regeneration, some organisms maintain this unique ability. Zebrafish in particular respond to cardiac injury with cardiomyocyte proliferation and minimal scarring. The zebrafish model system thus provides a unique model cardiac injury and regeneration to dissect with genetic and molecular tools. Recent use of these tools has identified a primary source of regenerated cardiac muscle derived from a population of cardiomyocytes.

The epicardium plays a significant role in cardiac regeneration. Epicardial cells are activated by injury to proliferate and activate embryonic gene expression in an organ-wide manner, with a portion of these cells incorporating into the injury site in an Fgf-dependent manner. Temporal and functional evidence indicate that these epicardial cells help establish new vasculature in the regenerate, aiding muscle regeneration. Future studies must examine other possible roles for the epicardium, as well as for other cardiac cell types like fibroblasts and endothelial cells. These efforts will be enabled by a continually expanding toolbox of molecular genetic approaches.

REFERENCES

1.   Dinsmore C (1991) *A History of Regeneration Research.* Cambridge University Press, Cambridge.
2.   Morgan T (1901) *Regeneration.* The Macmillan Company, New York.
3.   Spallanzani L (1768) Prodromo di un opera da imprimersi sopra la riproduzioni animali. Nella Stamperia di Giovanni Montanari, Modena. English translation by Maty M (1769) as *An Essay on Animal Reproductions.* Becket T and De Hondt PA, London.
4.   Taub R (2004) Liver regeneration: from myth to mechanism. *Nat Rev Mol Cell Biol* 5: 836–847.
5.   Stoick-Cooper CL, Moon RT, Weidinger G (2007) Advances in signaling in vertebrate regeneration as a prelude to regenerative medicine. *Genes Dev* 21: 1292–1315.
6.   Wagers AJ, Sherwood RI, Christensen JL, Weissman IL (2002) Little evidence for developmental plasticity of adult hematopoietic stem cells. *Science* 297: 2256–2259.
7.   Shizuru JA, Negrin RS, Weissman IL (2005) Hematopoietic stem and progenitor cells: clinical and preclinical regeneration of the hematolymphoid system. *Annu Rev Med* 56: 509–538.
8.   Haffter P, Granato M, Brand M, Mullins MC, *et al.* (1996) The identification of genes with unique and essential functions in the development of the zebrafish, *Danio rerio. Development* 123: 1–36.
9.   Driever W, Solnica-Krezel L, Schier AF, Neuhauss SC, *et al.* (1996) A genetic screen for mutations affecting embryogenesis in zebrafish. *Development* 123: 37–46.

10. Akimenko MA, Mari-Beffa M, Becerra J, Geraudie J (2003) Old questions, new tools, and some answers to the mystery of fin regeneration. *Dev Dyn* 226: 190–201.

11. Becker T, Wullimann MF, Becker CG, Bernhardt RR, *et al.* (1997) Axonal regrowth after spinal cord transection in adult zebrafish. *J Comp Neurol* 377: 577–595.

12. Otteson DC, Hitchcock PF (2003) Stem cells in the teleost retina: persistent neurogenesis and injury-induced regeneration. *Vision Res* 43: 927–936.

13. Poss KD, Wilson LG, Keating MT (2002) Heart regeneration in zebrafish. *Science* 298: 2188–2190.

14. Qin Z, Barthel LK, Raymond PA (2009) Genetic evidence for shared mechanisms of epimorphic regeneration in zebrafish. *Proc Natl Acad Sci USA* 106: 9310–9315.

15. Vihtelic TS, Hyde DR (2000) Light-induced rod and cone cell death and regeneration in the adult albino zebrafish (*Danio rerio*) retina. *J Neurobiol* 44: 289–307.

16. Montgomery J, Carton G, Voigt R, Baker C, *et al.* (2000) Sensory processing of water currents by fishes. *Philos Trans R Soc Lond B Biol Sci* 355: 1325–1327.

17. Harris JA, Cheng AG, Cunningham LL, MacDonald G, *et al.* (2003) Neomycin-induced hair cell death and rapid regeneration in the lateral line of zebrafish (*Danio rerio*). *J Assoc Res Otolaryngol* 4: 219–234.

18. Nechiporuk A, Raible DW (2008) FGF-dependent mechanosensory organ patterning in zebrafish. *Science* 320: 1774–1777.

19. Halloran MC, Sato-Maeda M, Warren JT, Su F, *et al.* (2000) Laser-induced gene expression in specific cells of transgenic zebrafish. *Development* 127: 1953–1960.

20. Johnson SL, Weston JA (1995) Temperature-sensitive mutations that cause stage-specific defects in Zebrafish fin regeneration. *Genetics* 141: 1583–1595.

21. Poss KD, Nechiporuk A, Hillam AM, Johnson SL, *et al.* (2002) Mps1 defines a proximal blastemal proliferative compartment essential for zebrafish fin regeneration. *Development* 129: 5141–5149.

22. Nechiporuk A, Poss KD, Johnson SL, Keating MT (2003) Positional cloning of a temperature-sensitive mutant emmental reveals a role for

sly1 during cell proliferation in zebrafish fin regeneration. *Dev Biol* 258: 291–306.

23. Makino S, Whitehead GG, Lien CL, Kim S, *et al.* (2005) Heat-shock protein 60 is required for blastema formation and maintenance during regeneration. *Proc Natl Acad Sci USA* 102: 14599–14604.

24. Whitehead GG, Makino S, Lien CL, Keating MT (2005) fgf20 is essential for initiating zebrafish fin regeneration. *Science* 310: 1957–1960.

25. Poss KD, Nechiporuk A, Stringer KF, Lee C, *et al.* (2004) Germ cell aneuploidy in zebrafish with mutations in the mitotic checkpoint gene mps1. *Genes Dev* 18: 1527–1532.

26. Wills AA, Kidd AR, 3rd, Lepilina A, Poss KD (2008) Fgfs control homeostatic regeneration in adult zebrafish fins. *Development* 135: 3063–3070.

27. Peterson RT, Shaw SY, Peterson TA, Milan DJ, *et al.* (2004) Chemical suppression of a genetic mutation in a zebrafish model of aortic coarctation. *Nat Biotechnol* 22: 595–599.

28. Mathew LK, Sengupta S, Kawakami A, Andreasen EA, *et al.* (2007) Unraveling tissue regeneration pathways using chemical genetics. *J Biol Chem* 282: 35202–35210.

29. Yoshinari N, Ishida T, Kudo A, Kawakami A (2009) Gene expression and functional analysis of zebrafish larval fin fold regeneration. *Dev Biol* 325: 71–81.

30. Hoptak-Solga AD, Nielsen S, Jain I, Thummel R, *et al.* (2008) Connexin43 (GJA1) is required in the population of dividing cells during fin regeneration. *Dev Biol* 317: 541–548.

31. Thummel R, Bai S, Sarras MP, Jr, Song P, *et al.* (2006) Inhibition of zebrafish fin regeneration using *in vivo* electroporation of morpholinos against fgfr1 and msxb. *Dev Dyn* 235: 336–346.

32. Yin VP, Thomson JM, Thummel R, Hyde DR, *et al.* (2008) Fgf-dependent depletion of microRNA-133 promotes appendage regeneration in zebrafish. *Genes Dev* 22: 728–733.

33. Lepilina A, Coon AN, Kikuchi K, Holdway JE, *et al.* (2006) A dynamic epicardial injury response supports progenitor cell activity during zebrafish heart regeneration. *Cell* 127: 607–619.

34. Ausoni S, Sartore S (2009) From fish to amphibians to mammals: in search of novel strategies to optimize cardiac regeneration. *J Cell Biol* 184: 357–364.

35. Bersell K, Arab S, Haring B, Kuhn B (2009) Neuregulin1/ErbB4 signaling induces cardiomyocyte proliferation and repair of heart injury. *Cell* 138: 257–270.

36. Engel FB, Hsieh PC, Lee RT, Keating MT (2006) FGF1/p38 MAP kinase inhibitor therapy induces cardiomyocyte mitosis, reduces scarring, and rescues function after myocardial infarction. *Proc Natl Acad Sci USA* 103: 15546–15551.

37. Yelon D, Horne SA, Stainier DY (1999) Restricted expression of cardiac myosin genes reveals regulated aspects of heart tube assembly in zebrafish. *Dev Biol* 214: 23–37.

38. Kikuchi K, Holdway JE, Werdich AA, Anderson RM, *et al.* (2010) Primary contribution to zebrafish heart regeneration by gata4(+) cardiomyocytes. *Nature* 464: 601–605.

39. Jopling C, Sleep E, Raya M, Marti M, *et al.* (2010) Zebrafish heart regeneration occurs by cardiomyocyte dedifferentiation and proliferation. *Nature* 464: 606–609.

40. Serluca FC (2008) Development of the proepicardial organ in the zebrafish. *Dev Biol* 315: 18–27.

41. Dettman RW, Denetclaw W, Jr, Ordahl CP, Bristow J (1998) Common epicardial origin of coronary vascular smooth muscle, perivascular fibroblasts, and intermyocardial fibroblasts in the avian heart. *Dev Biol* 193: 169–181.

42. Olivey HE, Compton LA, Barnett JV (2004) Coronary vessel development: the epicardium delivers. *Trends Cardiovasc Med* 14: 247–251.

43. Carlone RL, Ganagarajah M, Rathbone MP (1981) Bovine pituitary fibroblast growth factor has neurotrophic activity on newt limb regenerates and skeletal muscles *in vitro*. *Exp Cell Res* 132: 15–21.

44. Boilly B, Cavanaugh KP, Thomas D, Hondermarck H, *et al.* (1991) Acidic fibroblast growth factor is present in regenerating limb blastemas of axolotls and binds specifically to blastema tissues. *Dev Biol* 145: 302–310.

45. Mullen LM, Bryant SV, Torok MA, Blumberg B, *et al.* (1996) Nerve dependency of regeneration: the role of distal-less and FGF signaling in amphibian limb regeneration. *Development* 122: 3487–3497.

46. Lavine KJ, Yu K, White AC, Zhang X, *et al.* (2005) Endocardial and epicardial derived FGF signals regulate myocardial proliferation and differentiation *in vivo*. *Dev Cell* 8: 85–95.

47. Morabito CJ, Dettman RW, Kattan J, Collier JM, *et al.* (2001) Positive and negative regulation of epicardial-mesenchymal transformation during avian heart development. *Dev Biol* 234: 204–215.

48. Pennisi DJ, Mikawa T (2005) Normal patterning of the coronary capillary plexus is dependent on the correct transmural gradient of FGF expression in the myocardium. *Dev Biol* 279: 378–390.

49. Wills AA, Holdway JE, Major RJ, Poss KD (2008) Regulated addition of new myocardial and epicardial cells fosters homeostatic cardiac growth and maintenance in adult zebrafish. *Development* 135: 183–192.

50. Curado S, Anderson RM, Jungblut B, Mumm J, *et al.* (2007) Conditional targeted cell ablation in zebrafish: a new tool for regeneration studies. *Dev Dyn* 236: 1025–1035.

51. Pisharath H, Rhee JM, Swanson MA, Leach SD, *et al.* (2007) Targeted ablation of beta cells in the embryonic zebrafish pancreas using *E. coli* nitroreductase. *Mech Dev* 124: 218–229.

52. Lien CL, Schebesta M, Makino S, Weber GJ, *et al.* (2006) Gene expression analysis of zebrafish heart regeneration. *PLoS Biol* 4: e260.

III. Does Heart Regeneration Occur?

C. Regenerative Potential of the Mammalian Heart

May Liang-Lam and William C. Claycomb

1. INTRODUCTION

Despite lifestyle modifications and improvements in clinical and pharmacologic management, cardiovascular disease remains the leading cause of death worldwide. Statistics from the American Heart Association report 7,900,000 annual cases of myocardial infarction (MI) and 5,700,000 cases of heart failure in the United States alone.[1] Heart failure, often initiated by MI or prolonged hypertension, is a progressive disease involving loss of functional cardiac muscle cells followed by the onset of arrhythmias resulting from a disruption of the electrical conduction system. These statistics support the decades-old dogma that the adult mammalian heart has minimal capacity to regenerate,[2,3] despite recent reports that this organ harbors stem cells that could contribute to the renewal of cardiomyocytes in injured hearts.[4]

2. OVERVIEW OF THE CELL CYCLE

The cell division cycle (Figure 1) is a coordinated sequence of tightly regulated events, which consists of four distinct phases: gap 1 (G1), DNA synthesis (S), gap 2 (G2) and mitosis (M). The cell cycle is regulated by the specific associations of regulatory proteins known as cyclins, enzymes named cyclin dependent kinases (Cdks), as well as negative regulators known as cyclin dependent kinase inhibitors (CKIs).[5] Three D cyclins (D1, D2 and D3), two E cyclins (E1 and E2), two A cyclins (A1 and A2) and two B cyclins (B1 and B2) have been identified.[6] Cdks include Cdk2, Cdk4, Cdk6 and Cdc2. Two protein families of CKIs have been discovered. The INK4 family of CKIs includes p15, p16, p18 and p19, and the Cip/Kip family includes $p21^{Cip1}$, $p27^{Kip1}$ and $p57^{Kip2}$.

The gap phases represent the portions of the cell cycle during which a cell primarily grows and matures. The D cyclins are only responsive to mitogens during the G1 phase, when the mitogens stimulate the transcription, translation and stabilization of these cyclins. Following the association of the D cyclins with either Cdk4 or Cdk6, the catalytically-active holoenzyme is translocated into the nucleus, where it phosphorylates members of the tumor suppressor retinoblastoma (pRb)

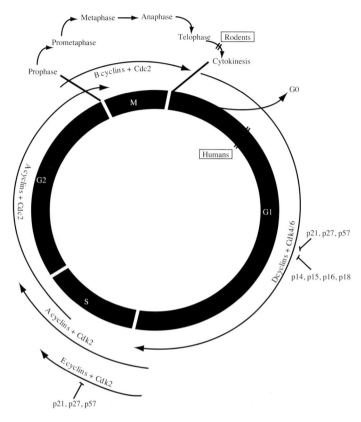

Figure 1.   **Mammalian cell cycle.** Phases of the cell cycle include G1 (gap 1), S (DNA synthesis), G2 (gap 2) and M (mitosis). M phase is further subdivided into prophase, prometaphase, metaphase, anaphase, telophase and cytokinesis. Cell cycle withdrawal in neonatal rodent cardiomyocytes occurs after the reformation of the nuclear membrane in late telophase, resulting in a majority of binucleated cells, whereas cessation of the cell cycle in human cardiomyocytes occurs in G1 or G0, resulting in predominantly mononucleated cells.

protein family.[7] This phosphorylation causes the release of pRb or its related proteins from the E2F transcription factors. E2F then activates the transcription of certain genes whose activity is required for entry of the cell into the S phase, such as the E cyclins.[8] Cyclin E-Cdk2 complexes complete the phosphorylation of pRb, allowing the cell to progress from G1 into S phase. Both cyclin E-Cdk2 and cyclin A-Cdk2 phosphorylate several components of the prereplication complex and replication machinery in S phase. Cyclin B-Cdc2 is required for the G2/M transition as well as for mitosis. Control mechanisms, or checkpoints, exist within each phase of the cell cycle to ensure that a cell has met all criteria for progression into the next phase. The primary cell cycle checkpoint in mammalian cells occurs near the G1/S transition and is called the restriction point.[9]

Mitosis is divided into prophase, prometaphase, metaphase, anaphase, telophase and cytokinesis. During prophase, the chromatin condenses into chromosomes. Prometaphase begins with the disintegration of the nuclear membrane. By metaphase, the chromosomes are aligned at the metaphase plate. The paired chromosomes then begin to separate during anaphase and move to the opposite poles of the cell. Telophase marks the stage when the sister chromatids have reached the opposite poles, the chromosomes uncoil and the nuclear membrane reforms around the separated sister chromatids, indicating the completion of karyokinesis. The cell then prepares to undergo cytokinesis, which is the cleavage into two daughter cells. Cytokinesis begins in late telophase with the contraction of a plasma membrane-anchored actomyosin ring that results in the formation of a cleavage furrow in the midline of the cell, followed by a physical separation of the cell into two daughter cells.[10]

3. STUDIES ESTABLISHING THE PHENOMENON OF CELL CYCLE WITHDRAWAL IN THE POSTNATAL MAMMALIAN HEART

The study of heart development has benefited immensely from the use of different experimental models involving both *in vivo* and *in vitro* systems. We will discuss some of these areas of research as they pertain to the question of whether terminally differentiated mammalian cardiomyocytes can be experimentally manipulated to proliferate.

3.1. Embryonic Heart Development

Early studies, relying on a combination of autoradiographic, biochemical and electron microscopic techniques, established that both

DNA synthesis and cell division occur in embryonic and very early fetal rat hearts.[11–14] It is well-established that the heart is the first functioning organ to develop and that during embryonic development and at birth rat cardiomyocytes have the ability to undergo both DNA synthesis and cell division while synthesizing, assembling and disassembling their cellular contractile proteins.[15,16] Thus, unlike skeletal muscle, differentiation and proliferation are not mutually exclusive in the differentiating cardiomyocyte.[3] A comprehensive compilation of publications involving DNA synthesis in rodent cardiomyocytes has been assembled by Soonpaa and Field.[17] Each round of DNA synthesis in the embryonic and very early fetal heart is followed by cytokinesis, or division of the cellular cytoplasm, resulting in cardiac muscle cells that are 95% to 100% mononucleated *in utero*.[17]

3.2. Postnatal Heart Development

The rate of DNA synthesis in cardiomyocytes drops dramatically at birth.[11,14] By measuring (^{3}H)thymidine incorporation in heart tissue and cardiomyocytes isolated from one day-, ten day-, 17 day-old as well as adult rats, it was determined that DNA synthesis in the cardiac muscle cell declines rapidly after birth, and essentially no longer occurs by the 17th day of development in the rat.[11,18] This cessation in DNA synthesis corresponds temporally with a simultaneous sharp decline in the activities of the DNA enzymes polymerase α and thymidine kinase, which also reach adult levels of almost zero in the 17 day-old heart.[11] BrdU incorporation analysis indicates that a second round of cardiomyocyte DNA synthesis occurs in early neonatal life, peaking at neonatal day 4 to 6 in the rat.[14] This second wave of DNA synthesis drops to undetectable levels by the 17th day of postnatal development, after which no further significant cardiomyocyte DNA synthesis is again observed for the life of the animal.[11,19] Following birth, the increased functional demand placed on the heart induces the growth pattern of the heart to change from cellular proliferation (hyperplasia) to cell enlargement of the existing cardiac muscle cells (physiological hypertrophy). Terminal cellular differentiation in this tissue thus represents an irreversible commitment to the differentiated cardiac phenotype, accompanied by permanent cell cycle withdrawal.

3.3. Uncoupling of Karyokinesis and Cytokinesis

During the embryonic and very early neonatal period of heart development in the rodent, both DNA replication and cytokinesis are ongoing in the majority of cardiomyocytes, resulting in cardiac muscle cells that are approximately 85% mononucleated.[20] At about three days after birth in the rat and mouse, cytokinesis ceases but karyokinesis continues until about 17 days after birth.[11] This phenomenon of ongoing karyokinesis in the absence of cytokinesis results in the binucleation of these cells.[21] In fact, binucleation in the adult mouse, rat, rabbit, guinea pig and cat have been observed to be approximately 80%.[20,22] Dog and cow cardiomyocytes are approximately 45% binucleated.[20] In the pig heart, myocytes have been shown to contain two, four, eight or 16 nuclei.[23]

Contrary to the high percentage of predominantly binucleated myocytes found in adult rodent hearts, 75% to 90% of adult human cardiomyocytes are mononucleated.[20,23] This implies that, unlike rodent cardiac muscle cells, which withdraw from the cell cycle and become terminally differentiated during late telophase, mononucleated human cardiomyocytes withdraw from the cell cycle following cytokinesis, in the G1 or G0 phase (Figure 1). Investigators who utilize either the rat or mouse model in their studies of cardiomyocyte cell cycle regulation should consider this very important difference when designing and interpreting their experiments if they wish to compare their results to the human cardiac muscle cell.

3.4. Diseased and Aging Hearts

Cardiomyocytes actively undergoing mitosis are rarely, if ever, reported in the healthy adult mammalian heart. The most common response of the mammalian heart to pathological stimuli such as MI, mechanical overload or ischemia is an adaptive hypertrophic response.[24] This pathological hypertrophy is associated with a low level of DNA replication, an occasional mitotic figure and re-expression of cardiac genes normally only detected during fetal development.[24,25] Using a combination of immunohistochemical techniques and confocal microscopy, Anversa's group recorded 14 mitotic figures per million myocytes in the healthy adult human heart.[26] The number of mitotic figures was observed to increase to 131 cardiomyocytes per million in idiopathic dilated cardiomyopathy and to 152 per million in the end-stage ischemic heart.[26] In an earlier study using cytophotometric methods to determine DNA content in human heart specimens, Adler and co-workers found only minimal differences between normal and infarcted hearts.[27] Rumyantsev and Kassem found 1.3% labeled nuclei in the infracted left ventricle.[28] Myocytes were only occasionally labeled in the control ventricular myocardium.[28]

The neonatal rat ventricular cardiomyocyte culture system has also been utilized to study the effects of a hypertrophic stimulus.[29] Within 30 minutes of either α- or β-adrenergic stimulation, mRNAs for the immediate early genes *c-fos* and *c-jun* are upregulated. This is followed by the reactivation of genes normally only expressed in the fetal heart,[30–33] with upregulation of atrial natriuretic factor, β-myosin heavy chain, α-skeletal muscle actin, myosin light chain-2 and α-cardiac muscle actin. This is accompanied by a downregulation of the α-myosin heavy chain gene. These changes in gene expression result in an increase in RNA and protein content as well as several-fold increase in the number of sarcomeric units, as evaluated by electron microscopy.

4. EXPERIMENTAL MANIPULATIONS OF MAMMALIAN CARDIOMYOCYTES DURING DEVELOPMENT

Mouse models, utilizing both normal and transgenic animals, have provided invaluable information toward an understanding of the molecular regulation of the mouse cardiomyocyte cell cycle by the identification of key cell cycle gene proteins. Cultured cardiomyocytes isolated from both neonatal and adult rat hearts are often used as alternatives for such studies. We highlight both *in vivo* and *in vitro* studies that aimed to increase the proliferation potential of cardiomyocyte proliferation.

4.1. Transgenic Mouse Models

4.1.1. *Oncogenes*

To assess the effect of the overexpression of simian virus (SV) 40 large T antigen (TAg) on cardiomyocytes, this oncoprotein was targeted to the atrium using the atrial natriuretic factor (ANF) promoter.[34] Immunohistochemical analyses showed high expression of TAg and hyperplasia in both atria. Overgrowth of atrial tissue resulted in cardiac conduction arrhythmias that led to premature death.[34] Cardiomyocytes were isolated from the atrial hyperplastic tissue of these hearts.[35] The cultured cells, which were named AT-1 cells,[36] exhibited characteristics typical of adult atrial cardiomyocytes and proliferated extensively in culture. These AT-1 cells could be maintained by serial propagation as ectopic subcutaneous grafts in syngeneic mice, but could not be passaged in culture.

By careful control of the culture environment, an immortalized cardiomyocyte cell line, designated HL-1, was established from the AT-1 cells.[37] These spontaneously contracting immortalized HL-1 cardiomyocytes have been extensively characterized and have been shown to have an adult cardiac muscle phenotype by electron microscopy, immunohistochemistry, RT-PCR analysis and electrophysiology.[37,38]

In a subsequent study, transgenic mice were generated in which expression of TAg, controlled by the rat α-cardiac myosin heavy chain promoter, was targeted to all cardiomyocytes of the heart.[39] Expression of TAg was observed in both atrial and ventricular cardiomyocytes in these transgenic mice and resulted in extensive hyperplasia of the cardiac muscle cells. Ventricular cardiomyocytes isolated from these hearts were designated AT-2 cells.[40] When cultured, they retained their cardiac phenotype for only a few passages before becoming totally dedifferentiated.

Jackson and co-workers analyzed the effects of the protooncogene *c-Myc* during embryonic development.[41] In these transgenic mice, increased *c-Myc* mRNA expression was found to be associated with both atrial and ventricular enlargement, which was secondary to cardiomyocyte hyperplasia. The transgenic hearts contained more than twice as many myocytes as did non-transgenic hearts. The data indicate an additional hyperplastic growth during fetal development. To test the effect of *c-Myc* overexpression in the adult mouse heart, a tamoxifen-inducible *c-Myc* was targeted to cardiomyocytes.[42,43] Activation of *c-Myc* in adult mice resulted in a marked increase in BrdU uptake, myocardial mass and wall thickness. Cyclin D, Cdk2 and Cdk4, as well as genes normally expressed in the heart during hypertrophy, such as ANF, β-MHC and α-skeletal actin, were also substantially upregulated in these hearts. This forced overexpression of *c-Myc* in the heart, leading to cell cycle re-entry in adult cardiac muscle cells, resulted in the development of severe hypertrophic cardiomyopathy and death of the animals within three weeks.[42] Proliferation of adult cardiomyocytes was not detected.

4.1.2. *Cyclins and cdks*

Transgenic mice in which the overexpression of cyclin D1, D2 or D3 was targeted to the heart were created to explore the use of these G1 cyclins in promoting myocyte cell cycle progression past the G1/S restriction point in the adult heart.[44,45] Following cardiac injury, an increase in cardiomyocyte cell number and reduction in infarct size were observed only in the hearts in which cyclin D2 was overexpressed. In these injured hearts, cyclin D2 was found to be localized within the nuclei of the cardiomyocytes, while cyclins D1 and D3 were distributed throughout the cytoplasm. It was concluded that the nuclear distribution of cyclin D2 within the cardiomyocytes thus correlated with cellular DNA synthesis. These data show that cardiomyocytes in the adult rodent heart can be induced to re-enter the cell cycle to a limited extent by overexpression of cyclin D2, and that cyclins D1, D2 and D3 are not functionally redundant in the mouse heart.

In a subsequent study, these authors subjected the cyclin D2 mice to permanent coronary artery occlusion-induced myocardial infarction and showed that a portion of the scar tissue was replaced by myocardial tissue in these hearts.[46] Also, left ventricular pressure-volume measurements documented significant improvements in cardiac function within two months of the injury, confirming that overexpression of cardiomyocyte-targeted cyclin D2 induced cellular proliferation of the cardiomyocytes in the region adjacent to the injury.

Cyclin dependent kinase 2 (cdk2) plays a critical role in the G1 to S phase checkpoint of the cell cycle (Figure 1). Overexpression of *cdk2* mRNA resulted in expression of catalytically active cdk2 protein and also significantly increased the levels of cdk4 and cyclins A, D3, and E. In addition, an increased level of DNA synthesis throughout development was detected. The ratio of heart weight to body weight in cdk2 transgenic mice was significantly increased at neonatal day 2 but not in adults compared with that of wild-type mice. Analysis of dispersed individual adult cardiomyocytes showed a 5.6-fold increase in the proportion of smaller mononuclear cardiomyocytes in the transgenic mice. Interestingly, adult transgenic ventricles expressed beta-myosin heavy chain and atrial natriuretic factor. These data suggest that overexpression of cdk2 results in a transient increase of cardiomyocyte proliferation and promotes smaller, less-differentiated mononuclear cardiomyocytes in adult hearts.[47]

Another promising target to increase cardiomyocyte proliferation is cyclin A2 as it regulates the G2/M checkpoint (Figure 1). Ectopic expression of cyclin A2 from embryonic day 8 into adulthood induces cardiac enlargement indicating an increase in cardiomyocyte proliferation. The authors demonstrated that cyclin A2 overexpression resulted in an increase in the expression of cdk1 and cdk2 and that cyclin A2 forms active complexes with these kinases. Interestingly, cyclin D2 and D3 were found to be upregulated in these transgenic animals. Furthermore, the authors detected mitotic cardiomyocytes throughout postnatal development. Several stages of mitosis were observed within cardiomyocytes and correlated with the nuclear localization of cyclin A2.[48] Subsequent studies suggest that ectopic expression of cyclin A2 in adult cardiomyocytes can also induce proliferation and improve cardiac function after an MI (see Chapter IV. A.).

4.1.3. *Others*

To test the hypothesis that telomere shortening, via telomerase downregulation, might be important in the timing of the withdrawal from the cell cycle of the ventricular cardiomyocyte during the development of the heart, transgenic mice were created in which telomerase reverse transcriptase (TERT) was overexpressed in the cardiac muscle cells.[49] In these mice, cardiac cell cycle exit was delayed during the first month after birth. In the three-month-old mouse, however, while telomerase activity remained high, all cardiac muscle cellular proliferation ceased and the cardiomyocytes underwent hypertrophy. This study demonstrates that overriding the tightly regulated cell cycle block in cardiac muscle cell proliferation during early development requires a combination of factors, only one of which may be telomerase. It also demonstrates that the inactivation of telomerase activity in the cardiac muscle cell during early development is not the cause of cell cycle withdrawal but is merely one of the many effects.

Most researchers have investigated the effect of overexpression of positive cell cycle regulators on cardiomyocyte proliferation. However, it might also be possible that the expression of negative regulators maintains the cell cycle arrest and that the inactivation of those factors enables cardiomyocyte proliferation. Recently, it has been shown that ablation of c-Abl protein, a ubiquitously expressed non-receptor tyrosine kinase involved in the development and function of many mammalian organ systems, results in premature death in animals with dramatically enlarged hearts.[50] To determine whether the enlargment of the heart is due to cardiomyocyte proliferation the authors performed several analyses. They scored the prevalence of BrdU incorporation (indicator of S phase) and phosphorylation of histone H3 (indicator of M phase). Compared with those from control littermates, ventricular cardiac tissues from *Abl* KO embryos demonstrated a 1.6-fold increase in BrdU-positive cells and a 1.9-fold increase in phospho-histone H3-positive cells. High-resolution microscopy of tissue sections co-stained with antibodies against phosphorylated histone H3 and cardiac myosin heavy chain antibodies confirmed that increased mitosis in mutant hearts was primarily confined to cardiomyocytes, rather than cardiac fibroblasts or endothelial cell. These findings reveal an essential role for c-Abl in mammalian heart growth and development and establish that loss of c-Abl results in increased cardiac proliferation during late gestation.[50]

That the cell cycle withdrawal is actively induced is further supported by the recent observation that FRNK (FAK-related non-kinase) is expressed shortly after birth when cardiomyocytes are arrested in the cell cycle and that FRNK overexpression leads to a severe ventricular non-compaction phenotype associated with reduced cardiomyocyte proliferation.[51]

4.2. Immortalized Cardiomyocyte Cell Lines

To identify possible molecular controls of cardiomyocyte proliferation several investigators have generated immortalized cardiomyocytes. In the TAg-immortalized AT-1, AT-2 and HL-1 cardiomyocyte cell lines, protein complexes from these cells were immunoprecipitated with anti-p53 antibody.[52] These studies identified protein complexes containing homologs of MRE11, NBS1 and RAD50, which play important roles in the detection and repair of DNA double-strand breaks, activation of cell cycle checkpoints and telomere maintenance. It was proposed that the interaction of SV40 TAg with the MRE11-NBS1-RAD50 pathway and with the tumor suppressors p53 and pRb ablates critical cell cycle checkpoints and that this is one of the major factors involved in the ability of this oncoprotein to immortalize cardiomyocytes. A similar study detected an apparently apoptotic protein of molecular weight 193 kDa.[53]

5. EXPERIMENTAL MANIPULATION OF MAMMALIAN CARDIOMYOCYTES DURING POSTNATAL DEVELOPMENT

Manipulation of the cell cycle during embryonic development results in overproliferation of cardiomyocytes and in some cases in an increased regenerative potential. However, these manipulations occurred before cardiomyocytes withdrew from the cell cycle. Thus, the question remains whether those manipulations in the adult mouse or human patient can still improve their regenerative capacity. For example, it has been shown that overexpression of c-myc during embryonic development results in an increase of cardiomyocyte proliferation.[41] However, overexpression of *c-Myc* in the adult heart utilizing a tamoxifen-inducible *c-Myc* failed to induce cardiomyocyte proliferation. Instead, activation of Myc resulted in polynucleation and polyploidization.[43] Following, we highlight some studies that have demonstrated that postnatal cardiomyocytes can also be induced to proliferate (for details see Chapter IV. A.).

To determine whether it is possible to induce proliferation in postnatal cardiomyocytes investigators often use *in vitro* cell culture systems. While neonatal cardiomyocyte cultures are technically easy to establish, these cultures are not suitable for long-term experiments due to the rapid proliferation and eventual overgrowth of contaminating non-cardiomyocytes.

DNA transfer techniques have been used in conjunction with cultured neonatal cardiomyocytes to determine the effects of exogenous molecules on the cardiac cell cycle. For example, it has been shown that overexpression of the receptor for FGF was associated with increases in DNA synthesis as assessed by significant increases in BrdU uptake as well as in cell number.[54] Another approach is the stimulation of the cells with growth factors and cytokines. Recently, it has been shown that the stimulation of cultured neonatal cardiomyocytes with TWEAK (TNF-related weak inducer of apoptosis) resulted in an increase in both BrdU uptake and the expression of the proliferation marker Ki67.[55] Cells undergoing mitosis and cytokinesis were also demonstrated by staining with phosphorylated histone H3 and aurora B kinase, specific markers of mitosis and the cleavage furrow for cytokinesis, respectively.[56–58] Cell counting experiments showed a significant increase in cardiomyocyte numbers following the treatment. Taken together, these data suggest that TWEAK stimulates the progression of cardiomyocytes not only into the S phase of the cell cycle but through mitosis as well.

Many studies have been performed using neonatal cardiomyocyte cultures. However, substantial physiological and morphological differences exist between neonatal cardiomyocytes and terminally differentiated adult cardiomyocytes, which can make interpretations of results problematic. Both atrial and ventricular cardiomyocytes have been successfully isolated from adult rat and monkey hearts and maintained in long-term culture.[59–61] Freshly isolated ventricular myocytes are morphologically cylindrical and striated, thereby resembling their counterparts in intact cardiac muscle. In culture, these cells flatten onto the surface of the culture flask while displaying characteristics of differentiated adult cardiomyocytes, including sarcomeric myofibrils, sarcoplasmic reticulum, intercalated discs and a highly organized transverse tubular system. When grown in primary culture, these terminally differentiated adult cardiac myocytes regain the ability to conduct a limited number of rounds of semi-conservative DNA replication, which can be further increased by growth factors.[22]

When terminally differentiated adult rat ventricular cardiac muscle cells were cultured in the presence of several growth factors such as insulin-like growth factor (IGF), epidermal growth factor (EGF), FGF, the tumor promoter phorbol ester 12-O-tetradecanoylphorbol-13-acetate (TPA) and diacylglycerol (DAG), (^{3}H)thymidine incorporation increased in a concentration-dependent manner.[62] These data demonstrate that adult ventricular cardiomyocytes that have undergone cell cycle arrest retain a potential for DNA synthesis when removed from the heart and placed in culture. They were never observed to divide, however.

In contrast, Engel and co-workers examined the occurrence of DNA synthesis, karyokinesis and mitosis in cultured ventricular cardiomyocytes isolated from 12-week-old adult rats.[63] Approximately 2% of the cardiomyocytes, which were stimulated with FGF1, were BrdU-positive. Inhibition of p38 MAPK doubled the percentage of BrdU-positive cells. In addition, karyokinesis and cytokinesis, as determined by histone H3 analysis and aurora B, respectively, were also noted. Taken together, these results indicate that inhibition of p38 MAPK and growth factors such as FGF1 work synergistically to reactivate the cell cycle in both neonatal and adult cardiomyocytes.

An increase in both BrdU incorporation and histone H3 phosphorylation were also documented in ventricular cardiomyocytes isolated from p38 null mice, suggesting that knocking out p38 MAPK resulted in both increased DNA synthesis and mitosis in ventricular myocytes *in vivo*.

In a recent study, cultured adult rat ventricular cardiomyocytes were stimulated with the extracellular growth factor neuregulin 1 (NRG1).[64] Both DNA synthesis and cytokinesis were observed in the NRG1-treated cultured adult ventricular cardiomyocytes by BrdU incorporation and aurora B kinase staining, respectively. *In vivo*, NRG1 injected into adult mice also showed both BrdU-positive and aurora B kinase-positive cardiomyocytes. NRG1 was injected into adult mice following an MI induced by ligation of the left anterior descending coronary artery. Ejection fraction was used to determine cardiac function, and infarct size was monitored microscopically.[64] Compared to control hearts, infarcts in hearts from NRG1-injected mice were approximately 50% the size of infarcts from vehicle-injected mice, and showed a significant improvement in cardiac function.

Taken together, these studies show that cardiomyocytes in culture can be manipulated by growth factors to replicate their DNA and even undergo cell division to a limited degree (for further details see Chapter IV. A.).

6. ADULT STEM CELLS FOUND IN THE MAMMALIAN HEART

Small numbers of cardiac stem or progenitor cells have recently been identified within niches thought to be distributed throughout the adult heart. Depending on the laboratory in which they were studied, the cells were found to be positive for either stem/progenitor cell markers such as the stem cell factor receptor (c-kit), stem cell antigen-1 (Sca-1) or for the transcription factor Isl-1.[65–67] The discovery of the presence of resident stem or progenitor cells within the adult heart is exciting, but it is obvious that these endogenous cells do not contribute to *de novo* cardiomyogenesis to a significant degree following an MI. It has been proposed that if these cardiac stem cells surrounding an infarct could be induced to proliferate and then to differentiate into healthy heart cells, the infarct may be reduced in size or even prevented, thereby improving and restoring cardiac function. At this point in time, unfortunately this is still science fiction. We are however getting close to this reality (for details see Chapters IV. B to IV. D. and VIII.).

7. SUMMARY AND CONCLUDING REMARKS

It has been a decades-long goal of scientists to reinitiate cell proliferation of terminally differentiated cardiomyocytes in the adult heart. One idea is that if healthy cardiac muscle cells in the border zone surrounding the infarcted myocardium could be induced to undergo endogenous regeneration, the *de novo* cells would restore cardiac function. Before such ambitious studies could proceed, however, it was first necessary and important to acquire basic information about the development of the heart. Early studies showed that DNA synthesis and cytokinesis are ongoing events in the embryonic heart. In the postnatal heart, cytokinesis and DNA replication cease as the cardiomyocytes permanently exit the cell cycle. The subsequent increase in the size of the postnatal heart then occurs predominantly as a result of cellular enlargement or hypertrophy of the existing myocytes. Since the majority of adult cardiomyocytes in the rodent are binucleated, this cell cycle block must occur during the telophase stage of mitosis. Thus, much recent research has focused on reinitiating cell proliferation in neonatal or adult cardiomyocytes through growth factor stimulation or forced overexpression of cell cycle genes in transgenic mice (see Chapter IV. A.). Transgenic and knockout mice, as well as myocytes isolated from neonatal or adult mouse hearts, are very useful experimental model systems in many of these studies. However, what the forced overexpression of cell cycle genes and subsequent reinitiation of a limited amount of DNA synthesis and cell division in the heart teaches us about the normal physiological mechanism that regulates terminal differentiation of the cardiomyocyte is not clear.

Although small samples of cardiac tissue can be obtained from biopsies, most studies conducted on the diseased human heart are obtained from the morgue following autopsies. For obvious reasons, normal cardiac tissue is not accessible for experimental controls. Thus rodents have traditionally been used as model systems to study most biological systems. However, significant differences exist between rodent and human heart development. One species-specific difference that is of particular concern is that cardiomyocytes in the adult human heart are predominantly mononucleated while those in the rodent heart are mostly binucleated. This finding raises questions as to whether cell cycle withdrawal occurs within the same cell cycle phase in these two species and whether rodents represent the most appropriate model system for studying the regulation of human cardiac muscle cell division.

It is clear that terminally differentiated adult mammalian cardiac muscle cells lack significant endogenous regeneration capacity. Through genetic manipulations, however, DNA synthesis, karyokinesis, and even cytokinesis can be accomplished to a limited degree in terminally differentiated cardiomyocytes in rodents. These genetic manipulations, although interesting, have not really led us to a better understanding of the normal molecular mechanism that restricts cardiac muscle cell division during development. We will eventually understand how Nature accomplishes this, but it will require much more creative work. Acquisition of this knowledge will allow us to regenerate the heart by selective control of cardiac muscle cell division.

ACKNOWLEDGMENTS

This work was supported by NIH Grant HL-076498.

REFERENCES

1. Lloyd-Jones DM, Wilson PW, Larson MG, Beiser A, *et al.* (2004) Framingham risk score and prediction of lifetime risk for coronary heart disease. *Am J Cardiol* 94: 20–24.

2. Rumyantsev PP (1977) Interrelations of the proliferation and differentiation processes during cardiact myogenesis and regeneration. *Int Rev Cytol* 51: 186–273.

3. Claycomb WC (1983) Cardiac muscle cell proliferation and cell differentiation *in vivo* and *in vitro*. *Adv Exp Med Biol* 161: 249–265.

4. Urbanek K, Quaini F, Tasca G, Torella D, *et al.* (2003) Intense myocyte formation from cardiac stem cells in human cardiac hypertrophy. *Proc Natl Acad Sci USA* 100: 10440–10445.

5. Koepp DM, Harper JW, Elledge SJ (1999) How the cyclin became a cyclin: regulated proteolysis in the cell cycle. *Cell* 97: 431–434.

6. Kozar K, Ciemerych MA, Rebel VI, Shigematsu H, *et al.* (2004) Mouse development and cell proliferation in the absence of D-cyclins. *Cell* 118: 477–491.

7. Baserga R (1994) Oncogenes and the strategy of growth factors. *Cell* 79: 927–930.

8. Cobrinik D (2005) Pocket proteins and cell cycle control. *Oncogene* 24: 2796–2809.

9. Pardee AB (1974) A restriction point for control of normal animal cell proliferation. *Proc Natl Acad Sci USA* 71: 1286–1290.

10. Eggert US, Mitchison TJ, Field CM (2006) Animal cytokinesis: from parts list to mechanisms. *Annu Rev Biochem* 75: 543–566.

11. Claycomb WC (1975) Biochemical aspects of cardiac muscle differentiation. Deoxyribonucleic acid synthesis and nuclear and cytoplasmic deoxyribonucleic acid polymerase activity. *J Biol Chem* 250: 3229–3235.

12. Rumyantsev PP. (1991) Differentiation of cardiomyocytes. In: Rumyantsev PP (ed.) *Growth and Hyperplasia of Cardiac Muscle Cells.* Harwood Academic Publishers, London, pp. 3–68.

13. Claycomb WC. (1991) Proliferation potential of the mammalian ventricular cardiac muscle vell. In: Oberpriller JO, Oberpriller JC, Mauro A (eds.) *The Development and Regenerative Potential of Cardiac Muscle.* Harwood Academic Publishers, London, pp. 351–364.

14. Soonpaa MH, Kim KK, Pajak L, Franklin M, *et al.* (1996) Cardiomyocyte DNA synthesis and binucleation during murine development. *Am J Physiol* 271: H2183–2189.

15. Chacko K (1972) Ultrastructural observations on mitosis in myocardial cells of the rat embryo. *Am J Anat* 135: 305–310.

16. Claycomb WC (1992) Control of cardiac muscle cell division. *Trends Cardiovasc Med* 2: 231–236.

17. Soonpaa MH, Field LJ (1998) Survey of studies examining mammalian cardiomyocyte DNA synthesis. *Circ Res* 83: 15–26.

18. Claycomb WC (1979) DNA synthesis and DNA enzymes in terminally differentiating cardiac muscle cells. *Exp Cell Res* 118: 111–114.

19. Lam ML, Bartoli M, Claycomb WC (2002) The 21-day postnatal rat ventricular cardiac muscle cell in culture as an experimental model to study adult cardiomyocyte gene expression. *Mol Cell Biochem* 229: 51–62.

20. Korecky B, Sweet S, Rakusan K (1979) Number of nuclei in mammalian cardiac myocytes. *Can J Physiol Pharmacol* 57: 1122–1129.

21. Olivetti G, Cigola E, Maestri R, Corradi D, *et al.* (1996) Aging, cardiac hypertrophy and ischemic cardiomyopathy do not affect the proportion of mononucleated and multinucleated myocytes in the human heart. *J Mol Cell Cardiol* 28: 1463–1477.

22. Claycomb WC, Bradshaw HD, Jr (1983) Acquisition of multiple nuclei and the activity of DNA polymerase alpha and reinitiation of DNA replication in terminally differentiated adult cardiac muscle cells in culture. *Dev Biol* 99: 331–337.

23. Grabner W, Pfitzer P (1974) Number of nuclei in isolated myocardial cells of pigs. *Virchows Arch B Cell Pathol* 15: 279–294.

24. Pfeffer MA, Braunwald E (1990) Ventricular remodeling after myocardial infarction. Experimental observations and clinical implications. *Circulation* 81: 1161–1172.

25. Beltrami AP, Urbanek K, Kajstura J, Yan SM, *et al.* (2001) Evidence that human cardiac myocytes divide after myocardial infarction. *N Engl J Med* 344: 1750–1757.

26. Kajstura J, Leri A, Finato N, Di Loreto C, *et al.* (1998) Myocyte proliferation in end-stage cardiac failure in humans. *Proc Natl Acad Sci USA* 95: 8801–8805.

27. Adler CP, Neuburger M, Herget GW, Muhlbach D (1997) Regeneration processes in human myocardium after acute ischaemia — quantitative determination of DNA, cell number and collagen content. *Virchows Arch* 430: 149–153.

28. Rumyantsev PP, Kassem AM (1976) Cumulative indices of DNA synthesizing myocytes in different compartments of the working myocardium

and conductive system of the rat's heart muscle following extensive left ventricle infarction. *Virchows Arch B Cell Pathol* 20: 329–342.

29. Iwaki K, Sukhatme VP, Shubeita HE, Chien KR (1990) Alpha- and beta-adrenergic stimulation induces distinct patterns of immediate early gene expression in neonatal rat myocardial cells. fos/jun expression is associated with sarcomere assembly; Egr-1 induction is primarily an alpha 1-mediated response. *J Biol Chem* 265: 13809–13817.

30. Lee HR, Henderson SA, Reynolds R, Dunnmon P, *et al.* (1988) Alpha 1-adrenergic stimulation of cardiac gene transcription in neonatal rat myocardial cells. Effects on myosin light chain-2 gene expression. *J Biol Chem* 263: 7352–7358.

31. Long CS, Ordahl CP, Simpson PC (1989) Alpha 1-adrenergic receptor stimulation of sarcomeric actin isogene transcription in hypertrophy of cultured rat heart muscle cells. *J Clin Invest* 83: 1078–1082.

32. Knowlton KU, Baracchini E, Ross RS, Harris AN, *et al.* (1991) Co-regulation of the atrial natriuretic factor and cardiac myosin light chain-2 genes during alpha-adrenergic stimulation of neonatal rat ventricular cells. Identification of cis sequences within an embryonic and a constitutive contractile protein gene which mediate inducible expression. *J Biol Chem* 266: 7759–7768.

33. Waspe LE, Ordahl CP, Simpson PC (1990) The cardiac beta-myosin heavy chain isogene is induced selectively in alpha 1-adrenergic receptor-stimulated hypertrophy of cultured rat heart myocytes. *J Clin Invest* 85: 1206–1214.

34. Field LJ (1988) Atrial natriuretic factor-SV40 T antigen transgenes produce tumors and cardiac arrhythmias in mice. *Science* 239: 1029–1033.

35. Steinhelper ME, Lanson NA, Jr, Dresdner KP, Delcarpio JB, *et al.* (1990) Proliferation *in vivo* and in culture of differentiated adult atrial cardiomyocytes from transgenic mice. *Am J Physiol* 259: H1826–H1834.

36. Delcarpio JB, Lanson NA, Jr, Field LJ, Claycomb WC (1991) Morphological characterization of cardiomyocytes isolated from a transplantable cardiac tumor derived from transgenic mouse atria (AT-1 cells). *Circ Res* 69: 1591–1600.

37. Claycomb WC, Lanson NA, Jr, Stallworth BS, Egeland DB, *et al.* (1998) HL-1 cells: a cardiac muscle cell line that contracts and retains

phenotypic characteristics of the adult cardiomyocyte. *Proc Natl Acad Sci USA* 95: 2979–2984.

38. White SM, Constantin PE, Claycomb WC (2004) Cardiac physiology at the cellular level: use of cultured HL-1 cardiomyocytes for studies of cardiac muscle cell structure and function. *Am J Physiol Heart Circ Physiol* 286: H823–829.

39. Katz EB, Steinhelper ME, Delcarpio JB, Daud AI, *et al.* (1992) Cardiomyocyte proliferation in mice expressing alpha-cardiac myosin heavy chain-SV40 T-antigen transgenes. *Am J Physiol* 262: H1867–H1876.

40. Borisov AB, Claycomb WC (1995) Proliferative potential and differentiated characteristics of cultured cardiac muscle cells expressing the SV40 T oncogene. *Ann N Y Acad Sci* 752: 80–91.

41. Jackson T, Allard MF, Sreenan CM, Doss LK, *et al.* (1990) The c-myc proto-oncogene regulates cardiac development in transgenic mice. *Mol Cell Biol* 10: 3709–3716.

42. Lee HG, Chen Q, Wolfram JA, Richardson SL, *et al.* (2009) Cell cycle re-entry and mitochondrial defects in myc-mediated hypertrophic cardiomyopathy and heart failure. *PLoS One* 4: e7172.

43. Xiao G, Mao S, Baumgarten G, Serrano J, *et al.* (2001) Inducible activation of c-Myc in adult myocardium *in vivo* provokes cardiac myocyte hypertrophy and reactivation of DNA synthesis. *Circ Res* 89: 1122–1129.

44. Pasumarthi KBS, Nakajima H, Nakajima HO, Soonpaa MH, *et al.* (2005) Targeted expression of Cyclin D2 results in cardiomyocyte DNA synthesis and infarct regression in transgenic mice. *Circ Res* 96: 110–118.

45. Soonpaa MH, Koh GY, Pajak L, Jing S, *et al.* (1997) Cyclin D1 overexpression promotes cardiomyocyte DNA synthesis and multinucleation in transgenic mice. *J Clin Invest* 99: 2644–2654.

46. Hassink RJ, Pasumarthi KB, Nakajima H, Rubart M, *et al.* (2008) Cardiomyocyte cell cycle activation improves cardiac function after myocardial infarction. *Cardiovasc Res* 78(1): 18–25.

47. Liao HS, Kang PM, Nagashima H, Yamasaki N, *et al.* (2001) Cardiac-specific overexpression of cyclin-dependent kinase 2 increases smaller mononuclear cardiomyocytes. *Circ Res* 88: 443–450.

48. Chaudhry HW, Dashoush NH, Tang H, Zhang L, *et al.* (2004) Cyclin A2 mediates cardiomyocyte mitosis in the postmitotic myocardium. *J Biol Chem* 279: 35858–35866.

49. Oh H, Taffet GE, Youker KA, Entman ML, *et al.* (2001) Telomerase reverse transcriptase promotes cardiac muscle cell proliferation, hypertrophy, and survival. *Proc Natl Acad Sci USA* 98: 10308–10313.

50. Qiu Z, Cang Y, Goff SP (2010) c-Abl tyrosine kinase regulates cardiac growth and development. *Proc Natl Acad Sci USA* 107: 1136–1141.

51. DiMichele LA, Hakim ZS, Sayers RL, Rojas M, *et al.* (2009) Transient expression of FRNK reveals stage-specific requirement for focal adhesion kinase activity in cardiac growth. *Circ Res* 104: 1201–1208.

52. Lanson NA, Jr, Egeland DB, Royals BA, Claycomb WC (2000) The MRE11-NBS1-RAD50 pathway is perturbed in SV40 large T antigen-immortalized AT-1, AT-2 and HL-1 cardiomyocytes. *Nucleic Acids Res* 28: 2882–2892.

53. Daud AI, Lanson NA, Jr, Claycomb WC, Field LJ (1993) Identification of SV40 large T-antigen-associated proteins in cardiomyocytes from transgenic mice. *Am J Physiol* 264: H1693–1700.

54. Sheikh F, Fandrich RR, Kardami E, Cattini PA (1999) Overexpression of long or short FGFR-1 results in FGF-2-mediated proliferation in neonatal cardiac myocyte cultures. *Cardiovasc Res* 42: 696–705.

55. Novoyatleva T, Diehl F, van Amerongen MJ, Patra C, *et al.* (2010) TWEAK is a positive regulator of cardiomyocyte proliferation. *Cardiovasc Res* 85: 681–690.

56. Hans F, Dimitrov S (2001) Histone H3 phosphorylation and cell division. *Oncogene* 20: 3021–3027.

57. Terada Y, Tatsuka M, Suzuki F, Yasuda Y, *et al.* (1998) AIM-1: a mammalian midbody-associated protein required for cytokinesis. *EMBO J* 17: 667–676.

58. Yokoyama T, Goto H, Izawa I, Mizutani H, *et al.* (2005) Aurora-B and Rho-kinase/ROCK, the two cleavage furrow kinases, independently regulate the progression of cytokinesis: possible existence of a novel cleavage furrow kinase phosphorylates ezrin/radixin/moesin (ERM). *Genes Cells* 10: 127–137.

59. Jacobson SL (1977) Culture of spontaneously contracting myocardial cells from adult rats. *Cell Struct Funct* 2: 1–9.

60. Claycomb WC (1985) Long-term culture and characterization of the adult ventricular and atrial cardiac muscle cell. *Basic Res Cardiol* 80(Suppl 2): 171–174.

61. Claycomb WC, Moses RL (1985) Culture of atrial and ventricular cardiac muscle cells from the adult squirrel monkey *Saimiri sciureus. Exp Cell Res* 161: 95–100.

62. Moses RL, Claycomb WC (1989) Differentiated membrane specializations and myofibrillar breakdown and recovery in cultured adult cardiac myocytes treated with TPA and diacylglycerol. *J Cell Sci* 93(Pt 1): 95–105.

63. Engel FB, Schebesta M, Duong MT, Lu G, *et al.* (2005) p38 MAP kinase inhibition enables proliferation of adult mammalian cardiomyocytes. *Genes Dev* 15: 1175–1187.

64. Bersell K, Arab S, Haring B, Kuhn B (2009) Neuregulin1/ErbB4 signaling induces cardiomyocyte proliferation and repair of heart injury. *Cell* 138: 257–270.

65. Beltrami AP, Barlucchi L, Torella D, Baker M, *et al.* (2003) Adult cardiac stem cells are multipotent and support myocardial regeneration. *Cell* 114: 763–776.

66. Oh H, Bradfute SB, Gallardo TD, Nakamura T, *et al.* (2003) Cardiac progenitor cells from adult myocardium: homing, differentiation, and fusion after infarction. *Proc Natl Acad Sci USA* 100: 12313–12318.

67. Bu L, Jiang X, Martin-Puig S, Caron L, *et al.* (2009) Human ISL1 heart progenitors generate diverse multipotent cardiovascular cell lineages. *Nature* 460: 113–117.

IV. Repopulation of the Heart with New Cardiomyocytes

A. Induction of Cardiomyocyte Proliferation

Machteld J. van Amerongen and Felix B. Engel

1. INTRODUCTION

Ischemic heart disease is among the leading causes of death worldwide. The primary cause of death is reduced heart function due to the loss of heart muscle cells, the cardiomyocytes. Natural compensatory processes of the injured heart in mammals are limited to hypertrophy (increase in cell size) of the remaining cardiomyocytes as cardiomyocytes *in vivo* stop to proliferate during the first weeks after birth (see Chapters II. and III. C.). The loss of cardiomyocytes triggers pathological remodeling, which leads to further loss of cardiomyocytes, to an increase in interstitial fibrosis and replacement of necrotic regions with fibrotic scar tissue and results eventually in heart failure.[1] Thus, a major goal is to develop strategies to generate new cardiomyocytes to promote myocardial regeneration. One approach is to induce the remaining cardiomyocytes to proliferate, as we will discuss in this chapter.

2. CELL CYCLE ACTIVITY IN HEALTHY AND DISEASED HEARTS

It is commonly believed that mammalian cardiomyocytes lose their ability to proliferate shortly after birth because the mammalian heart does not regenerate and cardiomyocyte proliferation has neither been detected in the adult healthy heart nor after injury. However, since the mid-19th century scientists have reported intermediate stages of cardiomyocyte proliferation. These studies suggested that adult mammalian cardiomyocytes might have retained the ability to proliferate and raised the hope that it might be possible to generate new cardiomyocytes after injury by inducing their proliferation.

2.1. Healthy Hearts

The heart increases in size and weight from birth to adulthood. For example, the wet weight of rat hearts increases from around 23 to 1200 mg in just a month.[2] The human male heart increases from around 75 g at an age between one to five years to 415 g at an age of 80 years.[3] This increase in heart mass is correlated with an increase in cardiomyocyte cell size over time.[4,5] Therefore, the conclusion from these studies was that the increase in heart weight after birth is due to cardiomyocyte hypertrophy and not proliferation.

The heart pumps constantly blood with a frequency of around 80 (human) to 600 (mice) beats per minute at rest. Thus, the mammalian heart contracts 40 to 315 million times per year. It is difficult to imagine that cardiomyocytes can have a lifespan of years in mice and rats or even decades in humans despite being subjected to such a constant high workload. Therefore, additional studies have been performed to determine whether there is any turnover of cardiomyocytes during adulthood. Several investigators performed labeling assays using deoxyribonucleotide analogues (tritiated thymidine or bromodeoxyuridine (BrdU)) in mice and rats and some found that a number of adult cardiomyocytes of a healthy heart synthesize DNA, which is a required step for cell proliferation. The main discrepancy between these studies is the frequency at which DNA synthesis occurs (0% to 2.46% in adult rats). These data are summarized and discussed by Soonpaa and Field.[6] An important conclusion from those studies is, however, that DNA synthesis occurs in fully differentiated adult cardiomyocytes even though it is a rather rare event.

Incorporation of deoxyribonucleotide analogues occurs also during DNA repair. In addition, it is possible that BrdU-labeled cardiomyocytes are derived from stem cells that proliferated before they differentiated into cardiomyocytes. Therefore, other studies investigated whether adult cardiomyocytes can progress into mitosis to demonstrate that adult mammalian cardiomyocytes can proliferate. Anversa and Kajstura reviewed these studies and provided additional evidence that cardiomyocytes also undergo mitosis in the healthy adult mouse heart.[7,8] Nevertheless, polyploidization and polynucleation (due to endomitosis or failed cytokinesis) also require mitosis and to our knowledge there is so far no compelling evidence that the fate of adult mitotic cardiomyocytes is cell division. Meckert and co-workers have provided evidence that in injured hearts the fate of mitotic cardiomyocytes is at least in part polynucleation or polyploidization.[9]

The lifespan of mice and rats, which are usually investigated, is very short compared to humans. Thus, it has been speculated that human adult cardiomyocytes might have a higher proliferative capacity. Recently, Bergmann and co-workers took advantage of the "Limited Nuclear Test Ban Treaty" in 1963. They utilized the fact that carbon-14 concentrations in the atmosphere sharply increased in 1955 after aboveground nuclear bomb tests and dropped exponentially after

1963. Humans take up carbon-14 with their food and incorporate it in their DNA. Thus, a person born before 1955 should have cardiomyocytes with low carbon-14 content if cardiomyocytes stop to proliferate after birth. Instead, it has been found that cardiomyocytes had much higher carbon-14 content and thus must have synthesized DNA in the years before and even after the increase of atmospheric carbon-14 content increased.[10] The authors could exclude in their study the possibility that the observed increase in carbon-14 in the DNA is due to polyploidization by measuring the DNA content per nuclei. Furthermore, they argue that DNA damage and repair as well as fusion with other proliferating cells occurs too rarely to explain their carbon-14 data. Finally, they conclude that there is a very slow but steady turnover of cardiomyocytes during adult life (around 1% per year), which demonstrates that human cardiomyocytes have a lifespan of decades. However, this study cannot answer the question of whether the turnover is stem cell-based or due to cardiomyocyte proliferation. Using a genetic fate mapping approach Hsieh and co-authors showed that at least in mice there is no stem cell-based turnover in the adult heart.[11] In addition, Field and co-workers performed in mice a four-hour chase experiment with tritiated thymidine excluding the possibility that stem cells contribute to positive cardiomyocyte counts. Interestingly, they also observed an annual turnover rate of 1% (labeling index of 0.0005% × (24-hour per day/ 4-hour pulse) × 365 days).[12] Taken together, these data suggest that there might be a proliferation-based turnover of cardiomyocytes in healthy adult hearts. However, a direct evidence for adult cardiomyocyte cell division in adult healthy hearts is still missing.

2.2. Myocardial Infarction and Pathophysiological Hypertrophy

Even though the mammalian heart cannot regenerate scientists were interested to determine whether at least some cardiomyocytes or a subpopulation are induced to proliferate under pathophysiological conditions. If that were the case it could be possible to promote this process to achieve cardiac regeneration.

The ability of mammalian cardiomyocytes to divide has been suggested by many investigators since the end of the 19th century. Rumyantsev, stated in a 1977 review that "a limited percentage" of adult mammalian ventricular myocytes can overcome the most rigid restraints regarding the resumption of DNA synthesis and "pass through all phases of the mitotic cycle."[13] Several decades ago, the German pathologist Linzbach reported increased numbers of myocyte nuclei and cardiac-muscle fibers in hypertrophied human hearts larger than 500 g.[14] However, these early studies were limited by the available technology and could not prove that cell division occurs but subsequent studies have further supported their hypothesis using molecular biology techniques. Studies regarding the cell cycle machinery revealed that after myocardial infarction or during hypertrophy the expression of the cell cycle inhibitors p21 and p27 is downregulated.[15,16] In addition, the expression and activity of positive cell cycle regulators like cdk/cyclin complexes[17,18] and the processivity factor for DNA polymerase delta, PCNA,[19,20] are upregulated. Meanwhile there have been several reports describing also the detection of mitotic figures in the mammalian failing and infarcted heart.[7,21,22] However, it is unclear whether these mitotic cardiomyocytes complete cell division or whether they are undergoing endomitosis.[9] Nevertheless, these data indicate that after injury cardiomyocytes attempt to enter the cell cycle. Thus, it appears worthwhile to elucidate the molecular mechanisms regulating cardiomyocyte proliferation and to work towards the development of strategies to enhance cardiomyocyte proliferation to improve cardiac function after injury.

3. INDUCTION OF CELL CYCLE PROGRESSION *IN VITRO*

Studying cardiomyocyte proliferation during postnatal development and disease suggested that it might be possible to induce cell division of an adult mammalian cardiomyocyte. Consequently, it has been attempted to enhance the proliferative capacity of postnatal cardiomyocytes using growth factors, cell cycle activators, viral oncoproteins, extracellular matrix proteins, and small molecules.

In order to undergo cell division a cell has first to duplicate its DNA. A cell-free system has been developed to determine if the DNA of adult rat cardiomyocytes could be replicated.[23] In this system it has been shown that the DNA of over 75% of isolated adult rat cardiomyocyte nuclei can be replicated when incubated with extracts of cells arrested in S phase, the cell cycle phase in which DNA is synthesized.

DNA replication is followed by the so-called G2 phase in which cells prepare for mitosis. Studies of proliferating fetal cardiomyocytes have revealed that during mitosis the contractile apparatus that enables muscle contraction is disassembled in an ordered process.[24] At the same time they have to assemble contractile structures for chromosome segregation (spindle apparatus) and cytokinesis (contractile ring). It has been argued that these processes of assembly and disassembly of contractile structures interfere with each other. Thus, it has been concluded that the increased maturity and complexity of the contractile apparatus in postnatal cardiomyocytes prevents cytokinesis.[25,26]

3.1. Neonatal Cardiomyocytes

3.1.1. *Cell cycle control*

Cell cycle-related genes have been manipulated during early cardiac development in many studies to elucidate which mechanisms establish or can prevent the cell cycle arrest in cardiomyocytes after birth (see Chapter III. C.). For example, overexpression of the proto-oncogene c-myc *in vivo* resulted in both atrial and ventricular enlargement as a result of cardiomyocyte proliferation. The transgenic hearts contained more than twice as many cardiomyocytes as nontransgenic hearts.[27] This and other studies raised the hope that overexpression of genes that regulate cardiomyocyte proliferation during development could be used to induce proliferation of adult cardiomyocytes. Unfortunately, *de novo* expression of c-myc in postnatal cardiomyocytes *in vitro* or *in vivo* failed to promote cardiomyocyte proliferation.[28] The reason is unclear. It is known that many positive cell cycle regulators are downregulated during development, but the molecular mechanisms regulating cell cycle arrest are still poorly understood. In conclusion, it is regarding a regenerative therapy important to identify factors that induce proliferation in postnatal cardiomyocyte in which the cell cycle arrest already took place.

Two decades ago the first attempts to induce neonatal cardiomyocyte proliferation *in vitro* employed viral proteins. Neonatal rat cardiomyocytes overexpressing the SV 40 large T antigen displayed increased DNA synthesis, mitotic figures and even proliferation.[29] Subsequent studies revealed that viral onco-proteins interfere with the members of the retinoblastoma gene family that regulate cell cycle exit and cell differentiation via E2F transcription factors.[30] Genetic manipulations of cell cycle genes resulted also in cell cycle re-entry of neonatal cardiomyocytes, however, often failed to induce proliferation (Table 1). The problem might be that cell cycle genes are usually periodically and not constantly expressed during the cell cycle. In addition, the progression during the cell cycle is tightly regulated. This might be the reason why overexpression of a single gene like E2F1 allows cell cycle re-entry and progression through S phase but its continuous high expression prevents progression into mitosis and leads to apoptosis.[31] However, it has been reported that, for example, overexpression of Cyclin D2 results in induction of neonatal cardiomyocyte cell division.[32]

3.1.2. *Extracellular signals*

It has been shown that a variety of growth factors and small molecule inhibitors can reactivate DNA synthesis in postnatal cardiomyocytes[8,33–37] (Table 2). Stimulation with a combination of FGF1 and p38 MAP kinase inhibitor SB203580,[33] the Wnt inhibitor BIO,[34] Periostin,[37] TWEAK[38] or Neuregulin[33] resulted in marked induction of DNA synthesis (>60%), mitosis (>4%) and cell division. In addition, it has been demonstrated that the PI3 kinase[33,37] and GSK-3beta[34] pathways are important for induction of cardiomyocyte proliferation.

Taken together, the above-mentioned studies have proven that neonatal cardiomyocytes (three days old) are not terminally differentiated and can be induced to proliferate. This provides hope for pediatric heart patients.

Table 1.   Genetic Manipulations that Enhance Cardiomyocyte Cell Cycle Re-Entry.

Gene	Age/Promoter	Cell Cycle	Species	Reference
	in vitro			
β-catenin	neo	S = 11%	Rat	Hahn JY et al., J Biol Chem (2006)
Cdc5, E1A/E1B	neo	M = 0.42%, cell count	Rat	Williams SD et al., Gene Ther (2006)
Cyc B1/cdc2	neo	cell count	Rat	Bicknell K et al., Biochem J (2004)
Cyc D2	neo	S > 40%, M, cell count	Rat	Busk PK et al., Exp Cell Res (2005)
	21 days	S > 30%, cell count	Rat	
E1A	neo	S > 60%	Rat	Liu Y et al., J Cell Biol (1996)
E2F1/E1B	neo	S > 30%	Rat	Kirshenbaum LA et al., Dev Biol (1996)
E2F1/IGF1	neo	S = 23%	Rat	Von Harsdorf R et al., Circ Res (1999)
E2F2	neo	S > 40%, BrdU-Hoechst assay	Rat	Ebelt H et al., Circ Res (2005)
E2F4	neo	S > 20%, BrdU-Hoechst assay	Rat	Ebelt H et al., Circ Res (2005)
FGF2	neo	S, M > 8%, cell count	Rat	Pasumarthi KB et al., Circ Res (1996)
FGFR1	neo	S, cell count	Rat	Sheikh F et al., Cardiovasc Res (1999)
c-myc	adult	S, M	Rat	Xiao G et al., Circ Res (2001)
TAG	neo	M, cell count	Rat	Sen A et al., J Biol Chem (1988)
tsTAG A58	adult	S, cell count	Rat	Miller C et al., Mol Cell Biochem (1994)

(*Continued*)

Table 1. (*Continued*)

Gene	Age/Promoter	Cell Cycle	Species	Reference
Cyc D1/cdk4/Skp2	*in vivo* adenovirus	S > 3%, Ki67 > 3%,	Rat	Tamamori-Adachi M et al., *Cardiovasc Res* (2008)
p73DN	adenovirus	S = 1.94%, C = 0.04%	Mouse	Ebelt H et al., *J Mol Cell Cardiol* (2008)
E2F1/p53−/−	adenovirus	S	Rat	Agah R et al., *J Clin Invest* (1997)
E2F2	adenovirus	S (0.028%), M (0.005%)	Mouse	Ebelt H et al., *Cardiovasc Res* (2008)
AT1 receptor	αMHC	hyperplasia (embryonic)	Mouse	Hein L et al., *PNAS* (1997)
Bcl2	αMHC	S, M, cell count	Mouse	Limana et al., *PNAS* (2002)
calmodulin	ANF	hyperplasia	Mouse	Gruver CL et al., *Endocrinology* (1993)
Cdk2	αMHC	S, PCNA, mononuclear cells	Mouse	Liao H-S et al., *Circ Res* (2001)
Cyc A2	αMHC	PCNA, M (neonatal)	Mouse	Chaudry et al., *J Biol Chem* (2004)
Cyc D1	αMHC	S, multinucleation	Mouse	Soonpaa MH et al., *J Clin Invest* (1997)
Cyc D2	αMHC	S	Mouse	Pasumarthi KB et al., *Circ Res* (2005)
Cyc D3	αMHC	S	Mouse	Pasumarthi KB et al., *Circ Res* (2005)
IGF1B	αMHC	S, cell count	Mouse	Reiss K et al., *PNAS* (1996)
c-myc	αMHC	cell count	Mouse	Jackson T et al., *Mol Cell Biol* (1990)
mycER	αMHC	S, M	Mouse	Xiao G et al., *Circ Res* (2001)
p193DN/p53DN	αMHC	S = 0.5% (BMI)	Mouse	Nakajima H et al., *Circ Res* (2004)
SV40 T antigen	ANF	hyperplasia (atrium)	Mouse	Field LJ, *Science* (1988)
SV40 T antigen	Protamine	hyperplasia (atrium)	Mouse	Behringer RR et al., *PNAS* (1988)
SV40 T antigen	αMHC	S, M, hyperplasia	Mouse	Katz EB et al., *Am J Physiol* (1992)
Telomerase RT	αMHC	S, M, hyperplasia	Mouse	Oh H et al., *PNAS* (2001)
TSC2 mutant	αMHC	S	Mouse	Pasumarthi KB et al., *Circ Res* (2000)

neo = Neonatal; S = DNA synthesis; M = mitosis; C = cytokinesis; BMI = border zone after infarct.

Table 2.   Effects of Exogenously Added Molecules on Cell Cycle Activity in Primary Cardiomyocyte Cultures.

Gene	Age	Cell Cycle	Species	Reference
BIO	neo	S (>60%),	Rat	Tseng AS *et al., Chem Biol* (2006)
EPO	neo	S	Rat	Wald MR *et al., J Cell Physiol* (1996)
FBS	neo	S (>35%),	Rat	Engel FB *et al., Genes Dev* (2005)
FGF1	neo	S (>60%), M (>1%)	Rat	Engel FB *et al., Genes Dev* (2005)
	adult	S (>2%), M (>0.04%)	Rat	Engel FB *et al., Genes Dev* (2005)
	adult	S	Newt	Soonpaa MH *et al., Cell Tissue Res* (1994)
FGF1+ p38i	neo	S (>70%), M (>5%), C (yes), cell count	Rat	Engel FB *et al., Genes Dev* (2005)
	adult	S (>4%), M (>0.14%)	Rat	Engel FB *et al., Genes Dev* (2005)
FGF2	neo	S (>20%),	Rat	Engel FB *et al., Genes Dev* (2005)
	neo	cell count	Rat	Kardami E, *Mol Cell Biochem* (1990)
	adult	S	Newt	Soonpaa MH *et al., Cell Tissue Res* (1994)
FGF4	neo	S (>40%),	Rat	Engel FB *et al., Genes Dev* (2005)
FGF6	neo	S (>40%),	Rat	Engel FB *et al., Genes Dev* (2005)
FGF8b	neo	S (>25%),	Rat	Engel FB *et al., Genes Dev* (2005)
FGF9	neo	S (>30%),	Rat	Engel FB *et al., Genes Dev* (2005)
FGF17	neo	S (>30%),	Rat	Engel FB *et al., Genes Dev* (2005)

(*Continued*)

Table 2. (*Continued*)

Gene	Age	Cell Cycle	Species	Reference
IL1β	neo	S (>40%), M (>1.5%)	Rat	Engel FB *et al.*, *Genes Dev* (2005)
NRG1-beta1	neo	S (>45%),	Rat	Engel FB *et al.*, *Genes Dev* (2005)
	adult	S (0.4%), M (3%), C (0.6%)	Rat	Bersell K *et al.*, *Cell* (2009)
	adult	M (32.6%), C (0.6%) (mononucleated)	Rat	Bersell K *et al.*, *Cell* (2009)
PE	neo	S (>20%),	Rat	Engel FB *et al.*, *Genes Dev* (2005)
Periostin	neo	S (>25%), C (yes), cell count	Rat	Kuhn B *et al.*, *Nat Med* (2007)
	adult	S (>1%), C (>0.4 %)	Rat	Kuhn B *et al.*, *Nat Med* (2007)
TGFβ1	neo	S (>15%),	Rat	Engel FB *et al.*, *Genes Dev* (2005)
TPA	adult	S	Rat	Claycomb WC *et al.*, *Dev Biol* (1988)
	adult	S	Newt	Soonpaa MH *et al.*, *Cell Tissue Res* (1994)
TWEAK	neo	S (>35%), M (>1.5%), C (yes), cell count	Rat	Novoyatleva T *et al.*, *Cardiovasc Res* (2010)
	adult	S (>35%, after overexpression of its receptor)	Rat	Novoyatleva T *et al.*, *Cardiovasc Res* (2010)

neo = Neonatal; S = DNA synthesis; M = mitosis; C = cytokinesis.

3.2. Adult Cardiomyocytes

In the past several theories have been formulated to support the assumption that adult cardiomyocytes cannot be induced to proliferate. First, adult cardiomyocytes have a more mature and compact contractile apparatus than neonatal cardiomyocytes. Thus, it has been hypothesized that from an energetic point of view it is unlikely that an adult cardiomyocyte would disassemble its contractile apparatus to enable cell division. Second, it has been argued that proliferation cannot be tolerated as transient sarcomere disassembly during mitosis prevents cellular contraction and would thus cause arrhythmias. Third, another theory assumes that hypertrophy and proliferation are mutually exclusive. As the heart has to adapt to different workloads it grows and shrinks in size. This can be easily accomplished by hypertrophy maintaining the contractile apparatus, cell-to-cell contact and electric coupling. However, regulating organ size by cardiomyocyte proliferation and consequently apoptosis appears more challenging. Despite these arguments several groups devoted their research to explore possibilities to induce adult cardiomyocyte proliferation.

3.2.1. *Cell cycle control*

Several investigators have studied the expression patterns of cell cycle genes during development. Neonatal cardiomyocytes still express all necessary factors to proliferate but cell cycle progression is blocked. In contrast, in adult cardiomyocytes most factors that drive the cell cycle are downregulated.[39,40] Thus, it appears even less likely that overexpression of single cell cycle factors could promote adult cardiomyocyte proliferation.

Several studies have demonstrated that at least a subpopulation of adult cardiomyocytes of mice overexpressing positive cell cycle activators under the control of the alpha myosin heavy chain (αMHC) promoter proliferate (Table 1). However, the αMHC promoter enables expression in cardiomyocytes before the cell cycle arrest is established. Interestingly, Busk and co-workers have shown that adenoviral overexpression of Cyclin D2 in isolated wild type cardiomyocytes from 21-day-old rats induced DNA synthesis *in vitro* in more than 30%

of the cells.[32] In addition, they reported an increased amount of cells (< two-fold). Unfortunately, the authors did not provide any evidence for the induction of mitosis or cell division and their experimental design did not exclude that the increased cell number is due to increased survival and/or better attachment. Induction of cell division of fully matured adult mammalian cardiomyocytes utilizing cell cycle genes has to our knowledge not been described (see Table 1).

3.2.2. *Extracellular signals*

It appears that growth factors or extracellular matrix proteins have much less effect on adult cardiomyocytes than on neonatal cells. A possible explanation for this phenomenon is the fact that many receptors like FGFR1 and FN14 are downregulated in adult cardiomyocytes.[38,41] Nevertheless, the use of growth factors and small molecule inhibitors proved that in adult cardiomyocytes it is possible to induce cell cycle activation and progression through all cell cycle phases including cytokinesis.[33,36,37] Interestingly, it has been observed that binucleated cardiomyocytes can divide *in vitro* forming one binucleated and two mononucleated cells.[33] However, the induction efficiency of DNA synthesis, mitosis and cytokinesis is with less than 1% rather low (Table 1). Thus, it has been speculated that there might be a subpopulation of cardiomyocytes with a higher potential for proliferation. For example, the heart consists of cardiomyocytes with different amounts of nuclei and polyploidy. Although a specific subpopulation has so far not been identified it appears that mononucleated cells exhibit a higher ability to re-enter the cell cycle. Kuhn and co-workers have found that Neuregulin induced more than 30% of mononucleated cardiomyocytes but only 1% of binucleated cardiomyocytes to undergo karyokinesis.[36]

Despite all the above-mentioned studies it is still not widely accepted that mature adult cardiomyocytes can be induced to proliferate. One of the reasons are the techniques used to demonstrate proliferation. As described above assays for DNA synthesis and mitosis were used. However, these assays cannot distinguish between proliferation and polyploidization or polynucleation (due to failed cytokinesis or endomitosis). The next level was to visualize cytokinesis using antibodies against

Aurora B. Aurora B is a kinase that is involved in the attachment of the mitotic spindle to the centromere, which regulates chromosome segregation during mitosis. During anaphase Aurora B localizes to the central spindle between the chromosomes. After chromosome segregation the cytoplasm is separated (cytokinesis) to form two daughter cells. For this purpose a contractile ring forms at the cell cortex, which is adjacent to the cell membrane at the position of the cleavage plane. Constriction of this contractile ring results in membrane ingression until the formation of the midbody to which Aurora B localizes.[42] Finally abscission cleaves the midbody resulting in two daughter cells (Figure 1, black pathway). However, shortly after introduction of Aurora B staining in the cardiomyocyte field it has been shown that serum stimulation, which results in binucleation, shows the same expression patterns.[43] In addition, it has been shown in megakaryocytes that Aurora B also exhibits its function during endomitosis.[44] As shown in Figure 1 there are several indicators that indicate whether cell division occurs. (1) Mitotic chromosomes in an intact nuclear membrane are typical for endomitosis.[9] An additional indication for endomitosis is the lack of or the only partial disassembly of the sarcomeres during mitosis.[24] (2) If cardiomyocytes enter mitosis and progress to anaphase they exhibit specific Aurora B and Anillin patterns. Anillin is a marker for the contractile ring necessary for cytoplasmic cleavage. In contrast to Aurora B, it behaves differently during polynucleation and cell division. It has been shown that after serum stimulation in cardiomyocytes Anillin localizes during anaphase to the cell equator (like Aurora B) instead of to the cell cortex. (3) Cardiomyocytes with an anaphase like central spindle (Aurora B, green) have telophase nuclei. (4) Another sign of failing cell division is that cleavage furrow ingression is not symmetrical.[43] Instead of abscission and formation of daughter cells the cardiomyocyte undergoes polynucleation due to degression of the cleavage furrow. The phenomenon of furrow regression had been previously shown in HeLa cells caused by chromosome nondisjunction.[45] However, Kuhn and co-workers have recently provided the unambiguous prove that adult cardiomyocytes can divide using live cell imaging after overexpression of Histone H2B under control of the chicken troponin T promoter.[36]

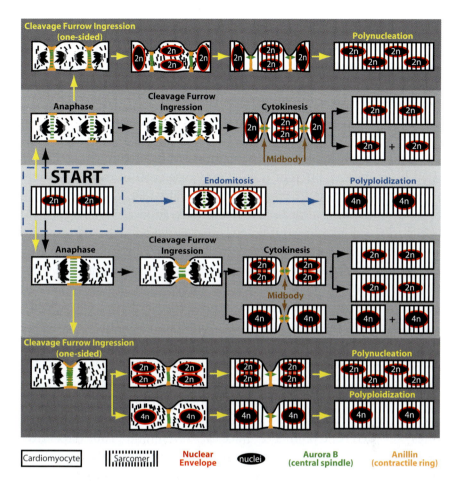

Figure 1. **Cellular fates of mitotic cardiomyocytes.** This schematic drawing illustrates several theoretical cellular fates of binucleated adult cardiomyocytes (START). After replicating their DNA, these cells can enter two forms of mitosis: normal mitosis and endomitosis. **Blue pathway:** Chromosome segregation during endomitosis occurs in an intact nuclear envelope (red) and results in polyploidization. There is no need for the disassembly of the sarcomeres. **Black and yellow pathway:** During normal mitosis the nuclear envelope is degraded and the sarcomeres (striation) disassemble. Not only the central spindle is formed (Aurora B, green) as during endomitosis but also the contractile ring is assembled (Anillin, orange). **Black pathway:** During mitosis Anillin (orange) localizes to the cell cortex adjacent to the cell membrane and a cleavage furrow is formed that symmetrically ingresses until abcission physically separates the cell. Depending on the orientation of the nuclei and thus on the formation of one or two cleavage furrows binucleated cells can divide in two

Taken together, it has been shown that adult mononucleated as well as binucleated cardiomyocytes can undergo DNA synthesis, mitosis (paralleled by transient disassembly of the contractile apparatus[33] as it has been described for fetal cardiomyocytes[24]) and cytokinesis.

4. THERAPEUTIC INVENTIONS

Induction of cardiomyocyte proliferation can be used to repair a damaged heart in two different ways. First, expansion of cardiomyocytes *in vitro* can be used to engineer tissue grafts (see Chapter VI. B.). Second, induction of cardiomyocyte proliferation *in vivo* might provide a minimal invasive strategy for heart patients.

Since 2005 several studies have been published indicating that enhancement of cardiomyocyte proliferation after myocardial infarction can improve heart function. Field and co-workers have generated mice that overexpress Cyclin D2 under the cardiomyocyte-specific alpha myosin heavy chain (αMHC) promoter. Their studies provided important insights in the possibility of heart regeneration based on cardiomyocyte proliferation despite the fact that αMHC drives expression of Cyclin D2 before the cell cycle arrest in cardiomyocytes is established.[46,47] Adult cardiomyocytes of αMHC-Cyclin D2 mice showed sustained cell cycle activity. Importantly, adult cardiomyocytes of αMHC-Cyclin D2 mice at seven and 150 days after myocardial infarction showed a significant higher DNA synthesis activity (around 1%) in the peri-infarct area than cardiomyocytes in control animals (>0.1%). In addition, a significant increase in mitotic

Figure 1. (*Continued*)   binucleated or one binucleated and two mononucleated cells. **Yellow pathway:** This pathway summarizes observations describing mitotic cardiomyocytes that did not complete cell division. (1) Cleavage furrow ingression is asymmetrical. (2) Anillin (orange) is distributed along the cell equator and overlaps with the central spindle (Aurora B, green). (3) One often observes the formation of nuclei although the central spindle (Aurora B, green) is typical for anaphase. (4) When the central spindle has condensed (Aurora B, two green triangles) the cytoplasm has not yet been separated even though the sarcomeres (striation) have been reassembled. It is likely that the cleavage furrow degresses and that mitosis results in polynucleation and polyploidisation instead of cell division.

cardiomyocytes and an increase in number of cardiomyocytes from day 7 to day 150 after myocardial infarction have been reported. The presence of regenerative growth was further supported by the observation that there was no significant difference in scar size and cardiac function between control and αMHC-Cyclin D2 mice at seven days after myocardial infarction. In contrast, 180 days after injury infarct size was markedly reduced and cardiac function almost restored. To further facilitate therapeutical applications it would be useful to use a promoter that enables induction of Cyclin D2 expression after myocardial infarction.

Woo and co-workers investigated the ability of cyclin A2 to improve cardiac function after injury.[48] They expressed Cyclin A2 under the cytomegalovirus promoter (it enables ubiquitous expression) by adenoviral delivery immediately after coronary ligation. They found that this therapy induced massive cardiomyocyte proliferation in the borderzone (40.1% PCNA, 46.3% Ki67, 44.2% BrdU, >12% H3P-positive compared to control: 9.3% PCNA, 20.4% Ki67, 5.2% BrdU, <0.5% H3P-positive). In comparison to the Cyclin D2 study these data are surprising as the authors of the Cyclin A2 study observed in their controls five times more BrdU-positive cardiomyocytes than in the αMHC-Cyclin D2 mice. The differences between the Cyclin A2 and the Cyclin D2 study might be explained by the following. The Cyclin A2 study relies on indirect immunofluorescence staining to allocate nuclei to cardiomyocytes, a suboptimal technique because of autofluorescence in infarcted tissue. In addition, it has been shown that cytoplasmic markers (here: α-sarcomeric actin) are insufficient to allocate nuclear stain to a specific cell type. In order to avoid these problems Field and co-workers crossed their transgenic mice with αMHC-nLAC mice that allow the precise identification of cardiomyocyte nuclei. Despite these discrepancies, Chaudhry and co-workers demonstrated that their therapy resulted in increased myofilament density and improved function. To further support their theory they investigated in a subsequent study the ability of αMHC-Cyclin A2 mice to recover after myocardial infarction.[49] In this study they observed induction rates similar to the Cyclin D2 study. In addition, they also performed a temporal analysis of functional recovery.

In contrast to previous studies, they utilized magnetic resonance imaging. This technique allows monitoring heart function in the same animal over time. One week after injury αMHC-Cyclin A2 and control animals showed the same decrease in heart function. Yet, in contrast to control mice, the αMHC-Cyclin A2 mice improved their function over time. Taken together, these studies demonstrated that it is possible to improve cardiac function by manipulating the cell cycle control in cardiomyocytes. However, transgenic animals are weak indicators for a possible therapy as expression of the genes start before establishment of the cell cycle arrest. In contrast, the adenoviral approach can be translated to a therapy for human patients. However, the recent years have demonstrated that gene therapy approaches are difficult to put into practice. Thus, recent studies try to induce cardiomyocyte proliferation *in vivo* by applying extracellular factors to the heart.

In 2006 it had been tested whether FGF1/p38 inhibitor stimulation after myocardial infarction can promote cardiomyocyte proliferation *in vivo* and can improve cardiac function. The p38 inhibitor was injected intraperitoneally in rats immediately after coronary artery ligation every three days for a month. FGF1 was embedded by Lee and co-workers in a nanofiber solution and injected in the heart adjacent to the ischemic myocardium. Lee and co-workers had shown before that this strategy allows continuous release of growth factors for at least two weeks.[50] Seven days after myocardial infarction a significant increase in mitotic cardiomyocytes (>0.3%) compared to animals without treatment (0.1%, sham group: 0.02%) was observed. Cardiac function was determined at one day, two weeks and three months after therapy by echocardiography. The data demonstrated that FGF1/p38 inhibitor treatment reduces scarring and rescues function after injury. However, the fact that treatment was started immediately after injury and that cardiac function at day 1 after infarction was not significantly different between sham-operated animals and FGF1/p38 inhibitor treated animals suggested that our therapy induces also mechanisms that prevent damage to the heart. This was to be expected as FGF1 is known to exhibit pleiotropic effects.[51] In conclusion, this study suggested that treatment improves heart function after myocardial injury by the combined effects on cardiomyocyte proliferation, angiogenesis and apoptosis. It will be interesting to see in future experiments whether FGF1/p38 inhibitor treatment can rescue heart function also when it is applied several days after injury.

Kuhn and co-workers investigated the therapeutical potential of Neuregulin 1 (NRG1).[36] Instead of injecting NRG1 immediately after coronary ligation they started treatment one week later and injected NRG1 intraperitoneally once a day. This design is relevant as it simulates the case in which a patient refers to a hospital several hours after an infarct. The authors describe cardiomyocytes after treatment in all cell cycle phases suggesting regenerative growth. They report that the population of mononucleated cardiomyocytes has a higher potential of proliferation than binucleated cardiomyocytes. Moreover, they genetically marked single differentiated cardiomyocytes before treatment and observed cluster formation after NRG1 stimulation. Based on their proliferation assays they concluded that NRG1 stimulation resulted in the generation of around 690,000 cardiomyo-cytes over a time period of 12 weeks (around 20% of total cardiomyocytes per heart according to their data). In addition, the hearts had a smaller scar size. However, the scarring was still significant, there was no sign of new tissue growth inside the scar as reported for Cyclin D2[47] and ejection fraction was not improved over time. One week after myocardial infarction before treatment ejection fraction was not significantly different compared to the control group (33.99% vs. 32.23%). Twelve weeks after treatment ejection fraction decreased in the control group to 22.34% whereas ejection fraction was maintained in the NRG1-treated group (32.69%). These data suggest that NRG1 can protect the heart from remodelling and further functional worsening.

Several studies have been performed claiming that improved cardiac function after injury is at least partially due to induction of cardiomyocyte proliferation (Table 3). Some of these studies utilized Aurora B stainings to detect cardiomyocytes undergoing cytokinesis (see Figure 1 regarding critical analysis of Aurora B as cytokinesis marker). However, in none of these studies the experiments provided proof that newly formed cardiomyocytes are at least to some extent

Table 3.   Effects of Exogenously Added Molecules on Cell Cycle Activity and Heart Function *In Vivo*.

Gene	Delivery	Cell Cycle	Heart Function	Species	Reference
Cyc A2	adenovirus	S, PCNA, Ki67	dP/dtmax 2500 to 3000	Rat	Woo YJ et al., Circulation (2006)
	transgenic	S, M	EF: 23 to 40%	Mouse	Cheng RK et al., Circ Res (2007)
Cyc D1/cdk4/ Skp2	adenovirus	M, C	FS: 11 to 20%	Rat	Tamamori-Adachi M et al., Cardiovasc Res (2008)
Cyc D2	transgenic	S, M, new tissue	dP/dtmax/EDV 520 to 1172	Mouse	Hassink RJ et al., Cardiovasc Res (2008) Pasumarthi KB et al., Circ Res (2005)
FGF1/p38	nanofiber	S, M	FS: 34 to 50%	Rat	Engel FB et al., PNAS (2006)
FGF5	adenovirus	M, Ki67	ND	Pig	Suzuki G et al., Circ Res (2005)
LIF	plasmid	Ki67	EF: 32 to 51% FS: 54 to 70%	Mouse	Zou Y et al., Circulation (2003)
NRG1	intraperitoneal injection	S, M, C	EF: 34 to 32%	Mouse	Bersel K et al., Dev Cell (2009)
Periostin	gelfoam	S, M, C	FS: 25 to 33% EF: 53 to 66%	Rat	Kuhn B et al., Nat Med (2007)
VEGF165	AAV	PCNA	ND	Dog	Ferrarini M et al., Circ Res (2006)
	plasmid	M, Ki67	ND	Sheep	Vera Janavel G et al., Gene Ther (2006)

AAV = Adeno-associated virus; S = DNA synthesis; M = mitosis; C = cytokinesis; EF = ejection fraction; FS = fractional shortening.

responsible for improved heart function. The studies relied on estimations based on histological techniques and mathematical algorithms, which are often not conclusive, as we have recently reviewed elsewhere.[8] As indicated above, improved cardiac function after treatment can be due to several reasons besides induction of cardiomyocyte proliferation. The field would greatly benefit from the existence of a mouse model that enabled us to genetically prevent induction of cardiomyocyte proliferation. For example, one could use mice that express postnatally a cell cycle inhibitor, like p27. In that case the therapy could be performed in parallel in these mice and the impact of cardiomyocyte proliferation on improved heart function could be determined. Whereas it is unlikely that the effects of cardiac-specific overexpression of Cyclin A2 or Cyclin D2 are not due to an increased proliferation potential of cardiomyocytes it is very likely that extracellular factors have additional effects that can improve cardiac function independently of cell cycle induction.[8] Some studies listed in Table 3 describe additional effects of the investigated factor. For example, the capillary density was increased after treatment with Periostin (four-fold),[37] FGF1/p38 inhibitor treatment (>2.5-fold)[52] or a plasmid encoding the 165 isoform of human vascular endothelial growth factor (pVEGF165, 2.6-fold).[53] In addition, some of these factors have previously been studied in the context of cardiac injury independently of proliferation illustrating their pleiotrophic effects. For example, FGF1 is known to act on many different cell types and affects apoptosis and angiogenesis. Li and co-workers compared the effect of FGF1 and a non-mitogenic form of FGF1 in an ischemia/reperfusion model.[54] They demonstrated that this form has no pro-proliferative effect on several cell types but it improved heart function significantly. Another good example for a pleitrophic factor is NRG1. Pentassuglia and co-workers have recently reviewed the effects of NRG1 signaling in the heart. They summarized the effects of NRG1 described *in vitro* concluding that "a number of processes appear to be regulated by Nrg-1/ErbB signaling, including cell growth, myofilament structure and organization, survival, myocyte-matrix coupling, glucose uptake, and angiogenesis."[55] They also emphasized the role of ErbB signaling in the sympathetic tone. Due to the variety of functions and the fact that ErbB receptors are downregulated after cardiac injury it had already been speculated that NRG1 stimulation could improve heart function. In accordance to this speculation, Liu and co-workers have shown in 2006 that induction of the NRG1/ErbB signaling by injection of a recombinant form of NRG1 improves cardiac performance, attenuated pathological changes, and prolonged survival in several different cardiac injury models.[56]

Taken together, several treatments have been reported to induce cardiomyocyte proliferation and functional improvement after myocardial injury. However, none of these studies has unambiguously demonstrated that induction of cardiomyocyte proliferation caused improved cardiac function. Thus, it will be important in the future to repeat these studies with animals in which cardiomyocytes have been genetically disabled to undergo proliferation.

5. SUMMARY AND CONCLUDING REMARKS

For many years, it was assumed that the adult mammalian cardiomyocyte is a terminally differentiated and thus post mitotic cell. However, in recent years evidence has been accumulated showing that adult cardiomyocytes can be stimulated to re-enter the cell cycle progressing through mitosis and cytokinesis (Figure 2). The ability to divide has even been demonstrated by live cell imaging.[36] Consequently, the main question is no longer whether adult cardiomyocytes can proliferate but whether we can exploit this knowledge to improve heart function in humans after myocardial injury.

The studies described in this chapter provide hope especially for patients with congenital heart defect (CHD) as neonatal mammalian cardiomyocytes can be efficiently induced to proliferate. Unfortunately, although CHD is the leading cause of birth defect related death,[57,58] cardiac regeneration studies appear to focus on model systems simulating cardiac injury only in the adult population.

Induction of adult cardiomyocyte proliferation is not yet very efficient. Despite this fact several investigators moved into *in vivo* models. There are several good reasons for that. (1) Even a low efficiency in inducing proliferation might be sufficient over time to improve

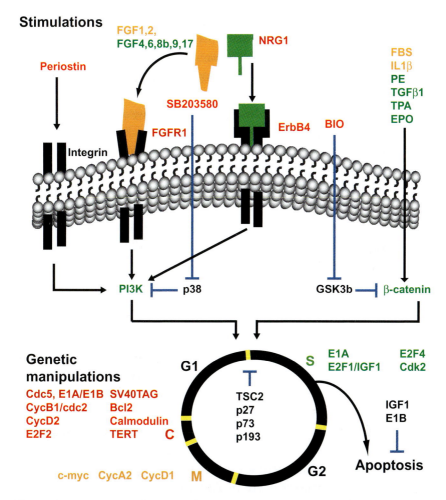

Figure 2.   Factors regulating cardiomyocyte proliferation.   Schematic summary of the most important factors and signaling pathways known to influence cell cycle control in cardiomyocytes. The color of a factor corresponds to the color of the cell cycle phase it can induce. Black arrows indicate positive stimulation. Blue lines symbolize blocking activity. G1: Gap 1 phase, S: DNA synthesis, M: mitosis, C: cytokinesis.

significantly heart function. (2) Cardiomyocytes might behave differently *in vivo* and *in vitro* due to their environment, i.e. neighboring cells and extracellular matrix. Several studies have demonstrated that factors that induce adult cardiomyocyte proliferation can improve heart function (Table 3). However, these studies have flaws regarding clinical application. (1) Manipulation started before establishment of the cell cycle arrest (Cyclin D2, Cyclin A2). (2) Treatment started immediately after injuring the heart (Cyclin A2, FGF1/p38 inhibitor, Periostin). (3) Treatment performed several days after injury did not result in functional improvement, but at least prevented further decline of cardiac function (NRG1).

In light of the above-mentioned studies, it is intriguing that after cardiac injury adult cardiomyocytes undergo hypertrophy and increase their ploidy rather than progressing towards karyokinesis and cytokinesis. It has been observed that inhibition of the p38 signaling pathway enables cardiomyocyte proliferation.[33] This kinase is needed for hypertrophy[59] and thus it might be possible that hypertrophy and proliferation are mutually exclusive in cardiomyocytes. An intriguing hypothesis is that hypertrophy is preferred because it allows the heart to easily adapt to different workloads, a constant requirement through-out men's evolutionary history, whereas myocardial infarction is in comparison a rather recent disease mostly due to a wrong diet. Other hypothesized reasons such as inability of adult cardiomyocytes to dis-assemble their contractile apparatus and proliferation-induced arrhythmia could be excluded as adult cardiomyocyte division has been proven[33,36] and no arrhythmia has been described in the *in vivo* experiments after induction of cardiomyocyte proliferation (Table 3).

In our opinion it is mandatory to study the following questions to further improve the effectiveness of therapies for cardiac injury based on induction of cardiomyocyte proliferation. (1) How is cardiomyo-cyte proliferation during development regulated? (2) Which mechanisms control the decision to undergo polyploidization instead of karyokinesis? (3) Which mechanisms orchestrate sarcomere disas-sembly and cytokinesis? (4) How can we increase the efficiency of adult cardiomyocyte proliferation? (5) Which genetic model can link increased cardiomyocyte proliferation and improved cardiac function? (6) Do the described therapies also increase heart function when treatment starts after injury has been established?

In conclusion, investigators have provided evidence suggesting that induction of cardiomyocyte proliferation is a possibility to restore cardiac function after injury. However, the reports are still few and have not been confirmed independently. In addition, it has so far not been demonstrated that the cause for improved cardiac function is the induction of cardiomyocyte proliferation. Also stem cell therapy was believed to regenerate the heart by providing new cardiomyocytes to the injured heart. Nowadays, it is generally accepted that the injected stem cells provide cytokines and growth factors and only infrequently differentiate into cardiomyocytes.[60] Thus, we should be careful with the interpretation of the published data but in our opinion we can nonetheless be optimistic. We should continue to devote our time to improve the efficiency of cardiomyocyte proliferation induction and provide further evidence that cardiomyocyte proliferation can be exploited to regenerate the heart.

ACKNOWLEDGMENTS

This work was supported by a grant from the Alexander von Humboldt Foundation (Sofja Kovalevskaja Award to F. B. E.), the DFG (EN 453/9-1), the Excellence Cluster Cardio-Pulmonary System (DFG) and the International Research Training Group: Protecting the Heart from Ischemia (PROMISE, DFG).

REFERENCES

1. Dorn GW, 2nd (2009) Novel pharmacotherapies to abrogate postinfarction ventricular remodeling. *Nat Rev Cardiol* 6: 283–291.
2. van den Hoff MJ, Deprez RH, Monteiro M, de Boer PA, *et al.* (1997) Developmental changes in rat cardiac DNA, RNA and protein tissue base: implications for the interpretation of changes in gene expression. *J Mol Cell Cardiol* 29: 629–639.
3. Linzbach AJ, Akuamoa-Boateng E (1973) Changes in the aging human heart. I. Heart weight in the aged. *Klin Wochenschr* 51: 156–163.
4. Fraticelli A, Josephson R, Danziger R, Lakatta E, *et al.* (1989) Morphological and contractile characteristics of rat cardiac myocytes from maturation to senescence. *Am J Physiol* 257: H259–265.

5. Bishop SP, Hine P (1975) Carciac muscle cytoplasmic and nuclear development during canine neonatal growth. *Recent Adv Stud Cardiac Struct Metab* 8: 77–98.

6. Soonpaa MH, Field LJ (1998) Survey of studies examining mammalian cardiomyocyte DNA synthesis. *Circ Res* 83: 15–26.

7. Anversa P, Kajstura J (1998) Ventricular myocytes are not terminally differentiated in the adult mammalian heart. *Circ Res* 83: 1–14.

8. van Amerongen MJ, Engel FB (2008) Features of cardiomyocyte proliferation and its potential for cardiac regeneration. *J Cell Mol Med* 12: 2233–2244.

9. Meckert PC, Rivello HG, Vigliano C, Gonzalez P, *et al.* (2005) Endomitosis and polyploidization of myocardial cells in the periphery of human acute myocardial infarction. *Cardiovasc Res* 67: 116–123.

10. Bergmann O, Bhardwaj RD, Bernard S, Zdunek S, *et al.* (2009) Evidence for cardiomyocyte renewal in humans. *Science* 324: 98–102.

11. Hsieh PC, Segers VF, Davis ME, MacGillivray C, *et al.* (2007) Evidence from a genetic fate-mapping study that stem cells refresh adult mammalian cardiomyocytes after injury. *Nat Med* 13: 970–974.

12. Soonpaa MH, Field LJ (1997) Assessment of cardiomyocyte DNA synthesis in normal and injured adult mouse hearts. *Am J Physiol* 272: H220–226.

13. Rumyantsev PP (1977) Interrelations of the proliferation and differentiation processes during cardiact myogenesis and regeneration. *Int Rev Cytol* 51: 186–273.

14. Linzbach AJ (1960) Heart failure from the point of view of quantitative anatomy. *Am J Cardiol* 5: 370–382.

15. Burton PB, Yacoub MH, Barton PJ (1999) Cyclin-dependent kinase inhibitor expression in human heart failure. A comparison with fetal development. *Eur Heart J* 20: 604–611.

16. Li JM, Brooks G (1997) Downregulation of cyclin-dependent kinase inhibitors p21 and p27 in pressure-overload hypertrophy. *Am J Physiol* 273: H1358–1367.

17. Reiss K, Cheng W, Giordano A, De Luca A, *et al.* (1996) Myocardial infarction is coupled with the activation of cyclins and cyclin-dependent kinases in myocytes. *Exp Cell Res* 225: 44–54.

18. Li JM, Poolman RA, Brooks G (1998) Role of G1 phase cyclins and cyclin-dependent kinases during cardiomyocyte hypertrophic growth in rats. *Am J Physiol* 275: H814–822.

19. Quaini F, Cigola E, Lagrasta C, Saccani G, *et al.* (1994) End-stage cardiac failure in humans is coupled with the induction of proliferating cell nuclear antigen and nuclear mitotic division in ventricular myocytes. *Circ Res* 75: 1050–1063.

20. Beltrami CA, Di Loreto C, Finato N, Rocco M, *et al.* (1997) Proliferating cell nuclear antigen (PCNA), DNA synthesis and mitosis in myocytes following cardiac transplantation in man. *J Mol Cell Cardiol* 29: 2789–2802.

21. Kajstura J, Leri A, Finato N, Di Loreto C, *et al.* (1998) Myocyte proliferation in end-stage cardiac failure in humans. *Proc Natl Acad Sci USA* 95: 8801–8805.

22. Beltrami AP, Urbanek K, Kajstura J, Yan SM, *et al.* (2001) Evidence that human cardiac myocytes divide after myocardial infarction. *N Engl J Med* 344: 1750–1757.

23. Engel FB, Hauck L, Cardoso MC, Leonhardt H, *et al.* (1999) A mammalian myocardial cell-free system to study cell cycle reentry in terminally differentiated cardiomyocytes. *Circ Res* 85: 294–301.

24. Ahuja P, Perriard E, Perriard JC, Ehler E (2004) Sequential myofibrillar breakdown accompanies mitotic division of mammalian cardiomyocytes. *J Cell Sci* 117: 3295–3306.

25. Li F, Wang X, Bunger PC, Gerdes AM (1997) Formation of binucleated cardiac myocytes in rat heart: I. Role of actin-myosin contractile ring. *J Mol Cell Cardiol* 29: 1541–1551.

26. Li F, Wang X, Gerdes AM (1997) Formation of binucleated cardiac myocytes in rat heart: II. Cytoskeletal organisation. *J Mol Cell Cardiol* 29: 1553–1565.

27. Jackson T, Allard MF, Sreenan CM, Doss LK, *et al.* (1990) The c-myc proto-oncogene regulates cardiac development in transgenic mice. *Mol Cell Biol* 10: 3709–3716.

28. Xiao G, Mao S, Baumgarten G, Serrano J, *et al.* (2001) Inducible activation of c-Myc in adult myocardium *in vivo* provokes cardiac myocyte hypertrophy and reactivation of DNA synthesis. *Circ Res* 89: 1122–1129.

29. Sen A, Dunnmon P, Henderson SA, Gerard RD, *et al.* (1988) Terminally differentiated neonatal rat myocardial cells proliferate and maintain specific differentiated functions following expression of SV40 large T antigen. *J Biol Chem* 263: 19132–19136.

30. Qin XQ, Chittenden T, Livingston DM, Kaelin WG, Jr. (1992) Identification of a growth suppression domain within the retinoblastoma gene product. *Genes Dev* 6: 953–964.

31. Kirshenbaum LA, Abdellatif M, Chakraborty S, Schneider MD (1996) Human E2F-1 reactivates cell cycle progression in ventricular myocytes and represses cardiac gene transcription. *Dev Biol* 179: 402–411.

32. Busk PK, Hinrichsen R, Bartkova J, Hansen AH, *et al.* (2005) Cyclin D2 induces proliferation of cardiac myocytes and represses hypertrophy. *Exp Cell Res* 304: 149–161.

33. Engel FB, Schebesta M, Duong MT, Lu G, *et al.* (2005) p38 MAP kinase inhibition enables proliferation of adult mammalian cardiomyocytes. *Genes Dev* 15: 1175–1187.

34. Tseng AS, Engel FB, Keating MT (2006) The GSK-3 inhibitor BIO promotes proliferation in mammalian cardiomyocytes. *Chem Biol* 13: 957–963.

35. Laguens RP, Crottogini AJ (2009) Cardiac regeneration: the gene therapy approach. *Expert Opin Biol Ther* 9: 411–425.

36. Bersell K, Arab S, Haring B, Kuhn B (2009) Neuregulin1/ErbB4 signaling induces cardiomyocyte proliferation and repair of heart injury. *Cell* 138: 257–270.

37. Kuhn B, Del Monte F, Hajjar RJ, Chang YS, *et al.* (2007) Periostin induces proliferation of differentiated cardiomyocytes and promotes cardiac repair. *Nat Med* 13: 962–969.

38. Novoyatleva T, Diehl F, van Amerongen MJ, Patra C, *et al.* (2010) TWEAK is a positive regulator of cardiomyocyte proliferation. *Cardiovasc Res* 85: 681–690.

39. Kang MJ, Koh GY (1997) Differential and dramatic changes of cyclin-dependent kinase activities in cardiomyocytes during the neonatal period. *J Mol Cell Cardiol* 29: 1767–1777.

40. Pasumarthi KB, Field LJ (2002) Cardiomyocyte cell cycle regulation. *Circ Res* 90: 1044–1054.

41. Jin Y, Pasumarthi KB, Bock ME, Lytras A, *et al.* (1994) Cloning and expression of fibroblast growth factor receptor-1 isoforms in the mouse heart: evidence for isoform switching during heart development. *J Mol Cell Cardiol* 26: 1449–1459.

42. Wheatley SP, Carvalho A, Vagnarelli P, Earnshaw WC (2001) INCENP is required for proper targeting of Survivin to the centromeres and the anaphase spindle during mitosis. *Curr Biol* 11: 886–890.

43. Engel FB, Schebesta M, Keating MT (2006) Anillin localization defect in cardiomyocyte binucleation. *J Mol Cell Cardiol* 41: 601–612.

44. Geddis AE, Kaushansky K (2004) Megakaryocytes express functional Aurora-B kinase in endomitosis. *Blood* 104: 1017–1024.

45. Shi Q, King RW (2005) Chromosome nondisjunction yields tetraploid rather than aneuploid cells in human cell lines. *Nature* 437: 1038–1042.

46. Hassink RJ, Pasumarthi KB, Nakajima H, Rubart M, *et al.* (2008) Cardiomyocyte cell cycle activation improves cardiac function after myocardial infarction. *Cardiovasc Res* 78(1): 18–25.

47. Pasumarthi KBS, Nakajima H, Nakajima HO, Soonpaa MH, *et al.* (2005) Targeted expression of cyclin D2 results in cardiomyocyte DNA synthesis and infarct regression in transgenic mice. *Circ Res* 96: 110–118.

48. Woo YJ, Panlilio CM, Cheng RK, Liao GP, *et al.* (2006) Therapeutic delivery of cyclin A2 induces myocardial regeneration and enhances cardiac function in ischemic heart failure. *Circulation* 114: I206–213.

49. Cheng RK, Asai T, Tang H, Dashoush NH, *et al.* (2007) Cyclin A2 induces cardiac regeneration after myocardial infarction and prevents heart failure. *Circ Res* 100: 1741–1748.

50. Hsieh PC, Davis ME, Gannon J, Macgillivray C, *et al.* (2006) Controlled delivery of PDGF-BB for myocardial protection using injectable self-assembling peptide nanofibers. *J Clin Invest* 116: 237–248.

51. Duda DG, Jain RK (2005) Pleiotropy of tissue-specific growth factors: from neurons to vessels via the bone marrow. *J Clin Invest* 115: 596–598.

52. Engel FB, Hsieh PC, Lee RT, Keating MT (2006) FGF1/p38 MAP kinase inhibitor therapy induces cardiomyocyte mitosis, reduces scarring, and rescues function after myocardial infarction. *Proc Natl Acad Sci USA* 103: 15546–15551.

53. Vera Janavel G, Crottogini A, Cabeza Meckert P, Cuniberti L, *et al.* (2006) Plasmid-mediated VEGF gene transfer induces cardiomyogenesis and reduces myocardial infarct size in sheep. *Gene Ther* 13: 1133–1142.

54. Li XK, Lin ZF, Li Y, Hu S, *et al.* (2007) Cardiovascular protection of nonmitogenic human acidic fibroblast growth factor from oxidative damage *in vitro* and *in vivo*. *Cardiovasc Pathol* 16: 85–91.

55. Pentassuglia L, Sawyer DB (2009) The role of Neuregulin-1beta/ErbB signaling in the heart. *Exp Cell Res* 315: 627–637.

56. Liu X, Gu X, Li Z, Li X, *et al.* (2006) Neuregulin-1/erbB-activation improves cardiac function and survival in models of ischemic, dilated, and viral cardiomyopathy. *J Am Coll Cardiol* 48: 1438–1447.

57. Hoffman JI, Kaplan S (2002) The incidence of congenital heart disease. *J Am Coll Cardiol* 39: 1890–1900.

58. Reamon-Buettner SM, Spanel-Borowski K, Borlak J (2006) Bridging the gap between anatomy and molecular genetics for an improved understanding of congenital heart disease. *Ann Anat* 188: 213–220.

59. Liang Q, Molkentin JD (2003) Redefining the roles of p38 and JNK signaling in cardiac hypertrophy: dichotomy between cultured myocytes and animal models. *J Mol Cell Cardiol* 35: 1385–1394.

60. Guan K, Hasenfuss G (2007) Do stem cells in the heart truly differentiate into cardiomyocytes? *J Mol Cell Cardiol* 43: 377–387.

IV. Repopulation of the Heart with New Cardiomyocytes

B. Multipotent Progenitor Cells in Regenerative Cardiovascular Medicine

Jason T. Lam, Alessandra Moretti and Karl-Ludwig Laugwitz

1. INTRODUCTION

Cardiovascular disease remains the leading cause of mortality in the Western world.[1] The loss of viable cardiac muscle cells is a major underlying cause of human heart failure. The lack of sufficient cardiac muscle mass ultimately leads to pathological cardiac remodeling, chamber dilation, associated increases in wall stress, and the onset of cardiac dysfunction. In recent years the concept of using stem or progenitor cells as the basis for cardiovascular therapy has gained considerable interest.

One potential approach to mitigate heart failure progression and/or influence partial reversal of important clinical and physiological endpoints is to augment the number of functional, electrically coupled cardiomyocytes in the failing myocardium. A number of pre-clinical studies have attempted to repair damaged myocardium by transplantation of exogenous cells. The source of these cells has been fetal cardiomyocytes, skeletal muscle myoblasts and bone marrow-derived cells.[2–5] Positive studies in rodents have resulted in a number of trials being performed in patients with acute myocardial infarction or end-stage heart failure.[6–10] Although some encouraging results have emerged from these clinical trials, the field is divided on whether the improvement of cardiac function correlates to long-term benefit and whether this improvement is restricted to certain subcategories of patients.[11] The main problem with determining how to optimize these transplantation approaches is that currently little is known about the fate of the transplanted cells and the molecular mechanisms behind the beneficial effects post engraftment.

In animal studies with bone marrow-derived cells, improvements in cardiac function seem to be related to increased angiogenesis, or to alterations in remodeling of scar tissue. Whether bone marrow-derived cells can differentiate towards the myocytic lineage is still controversial, and even if this happens, the occurrence has very low frequency.[12–14] Only in the case of transplantation with cardiomyocytes, but not with skeletal muscle-derived myoblasts, functional coupling of donor cells to endogenous myocardium has been demonstrated.[15,16] To achieve functional integration and provide no arrhythmogenic substrate with the cell transplant, it is likely that cardiomyocytes or cardiomyogenic precursors will represent the most promising cell source.

The discovery of multipotent cardiovascular precursor cells not only in mammalian embryos and postnatal rodent and human hearts but also as intermediate stages during differentiation of embryonic stem (ES) cells offers exciting opportunities in the field of cardiac translational medicine. This brief review will focus on cardiac progenitors from embryonic and adult heart and from *in vitro* differentiated pluripotent stem cells. In the context of regenerative cardiovascular applications, the expansion of cardiovascular precursors in culture is potentially the most efficient way of generating large numbers of cardiac cells for future cell therapy. Moreover, induced pluripotent stem (iPS) cell technology could provide an ideal source to obtain patient-specific cardiac precursors and mature cardiovascular cells with disease-causing mutations for molecular and genetic screens and for autologous cell replacement.

Only with a comprehensive understanding of the biological properties of cardiac progenitor cells and the mechanisms of their lineage-specific differentiation we will be able to realize the true promise of cardiovascular cell-based therapy.

2. DIVERSE CARDIOVASCULAR CELL LINEAGES ARISE FROM MULTIPOTENT EMBRYONIC PROGENITOR POPULATIONS

Mammalian cardiogenesis requires the generation of diverse muscle and non-muscle cell lineages: atrial/ventricular cardiomyocytes, conduction system cells, smooth muscle/endothelial cells of the coronary arteries and veins, endocardial cells, valvular components and connective tissue. In this regard, three major sources of heart cell precursors in the embryo have been identified: 1) the cardiogenic mesoderm, 2) the cardiac neural crest, and 3) the proepicardial organ. Each of these represents spatially and temporally distinct pools of embryonic heart progenitors, which are known to give rise to distinct cardiac structures and cell components. The cardiogenic mesoderm, which consists of two populations or fields of cardiac precursors — known as first (FHF) and second (SHF) heart fields — with distinct contributions to the mature heart, forms the linear heart tube and ultimately the majority of the working myocardium in the ventricular and atrial chambers. The cardiac neural crest migrates into the heart at later stages of development, and gives rise to the vascular smooth muscle of the aortic arch, ductus arteriosus, and the great vessels. Additionally, neural crest contributes to essential components of the cardiac autonomic nervous system.[17,18] The mesenchyme portion of the developing heart and the majority of epicardial cells are derived from the proepicardium.[19–21] For a comprehensive overview of the origin and lineage relationship of cardiac cell types, the readers are directed to Chapter II. or more detailed reviews.[22–24]

One of the earliest steps in cardiogenesis is the formation of the cardiac crescent, which is derived from cells of the mesoderm that become instructed to adopt a cardiac fate in response to signals from adjacent tissues.[25,26] While entering the cardiac mesodermal stage of development, heart-forming progenitors migrate away from the primitive streak to the lateral-cranial parts of the embryo and form the so-called cardiogenic regions. During this transition, they start to downregulate the T-box transcription factor Brachyury T (Bry), the earliest mesodermal marker, and activate the mesoderm posterior 1 (Mesp1) gene (Figure 1).[27] Mesp1$^+$ cells encompass all cardiac precursors and expression of Mesp1 is downregulated as the cells organize to form the cardiac crescent. Mesp1$^+$ progenitors have not completely committed yet to the cardiac fate as some also give rise to the paraxial mesoderm and skeletal muscle of the head.[28] At the crescent stage heart progenitors irreversibly commit to the cardiac lineage by expressing the key developmental transcription factors of the homeodomain family Islet-1 (Isl1) and Nkx2-5.[24] While Isl1$^+$ cardiac progenitors retain the capacity for vascular differentiation and give rise to endothelial and smooth muscle cells of the aorta, pulmonary artery and the proximal coronary tree, the Nkx2-5$^+$ precursors become more lineage-restricted to cardiomyocytes and smooth muscle cells during the looping stage of cardiac development (Figure 1).[29] Recent work has also identified a set of multipotent proepicardial progenitor cells expressing the T-box transcription factor Tbx18 and the Wilms tumor suppressor protein (Wt1) contributing to coronary smooth muscle, cardiac fibroblasts and some atrial and ventricular myocardium (Figure 1).[30,31]

Transgenic expression of fluorescent proteins under the control of the lineage markers described above (Isl1, Nkx2-5, Mesp1, Tbx18 and Wt1) in combination with fluorescence-activated cell sorting (FACS) has allowed to elucidate, at least partially, the biological phenotypes of these different populations of embryonic cardiac progenitors.[32–35] Depending on the markers used and the developmental stage of the lineage examined, embryonic precursor cells may

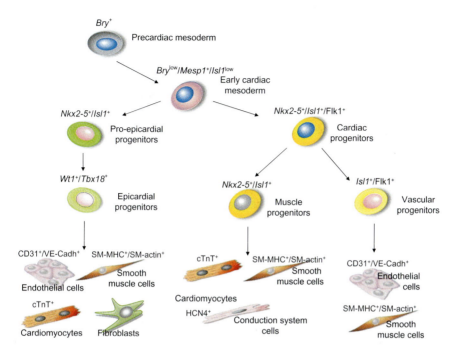

Figure 1.   Hierarchy model of cardiac progenitor cells and their lineage specification. Brachyury T (Bry) expression marks the precardiac mesodermal cells of the primitive streak, which will later give rise to the myocardial cells of the heart. The ingression of these early mesodermal cells through the primitive streak during gastrulation coincides with the expression of Mesp1. As these early cardiac mesodermal cells contribute to the developing heart, their transcriptional program determines the lineage specification it will ensue. Cardiac progenitors expressing Nkx2-5/Isl1/Flk1 give rise to all three lineages of the heart: endothelial, smooth muscle cells, and cardiomyocytes. It appears, as progenitor cells become lineage restricted, so do their transcriptional signature. Muscle progenitors, marked by Nkx2-5/Isl1, and vascular progenitors, marked by Isl1/Flk1, are bipotent precursors, giving rise to two cell lineages of the heart. Recent studies have reported on the identification of epicardial progenitor cells that are marked by Wt1/Tbx 18. These cells have been shown to give rise to endothelial, smooth muscle cells, cardiomyocytes, and fibroblasts of the heart. It remains to be determined whether the epicardial progenitors are derived from the same Nkx2-5/Isl1 muscle progenitor cells. Bry: brachyury T, cTNT: cardiac troponin T, HCN4: hyperpolarization-activated cation channel 4, SM-MHC: smooth muscle myosin heavy chain, VE-Cadh: VE-Cadherin, Mesp1: mesoderm posterior 1, Wt1: Wilms tumor suppressor protein 1, Tbx18: T-box transcription factor 18, Flk1: vascular endothelial growth factor receptor 2.

be bi-potent or tri-potent and may spontaneously differentiate into one or more of the four main cardiac cell lineages: atrial/ventricular cardiomyocytes, vascular smooth muscle cells, endothelial cells, and cardiac fibroblasts. To date it has not been possible to isolate and characterize the developmental potential of specifically purified progenitors of the FHF, SHF, and the proepicardium, because of absence of molecular markers unique to each precursor field. So far, the earliest molecular pan-cardiac markers for both SHF and FHF myocardial cell lineages are the transcriptions factors Mesp1/2 and Fgf8, which are expressed in the early cardiac mesoderm at the primitive-streak stage.[36,37] Nkx2-5 is also expressed in both lineages at later stages. It is still controversial whether Isl1 is specific for the SHF or also transiently expressed in the FHF.[24,33] In this regard, retrospective clonal analysis, a powerful approach to tag single cells during development, has demonstrated a common embryonic origin for some cells in the FHF and SHF.[22] In addition, Wt1$^+$/Tbx18$^+$ epicardial cells are derived from pro-epicardial progenitors that express Isl1 and Nkx2-5, suggesting that they may share a common developmental origin with the cardiogenic mesoderm progenitors.[31] The identification of specific sets of markers for the FHF and SHF lineages would be extremely valuable, since it is unclear whether cardiac regeneration of the left ventricle, a FHF-derived structure, will require purified populations of first heart field progenitors.

The mesodermal marker Brachyury T has been used, in combination with the cell surface marker KDR1 (Flk1), known also as the VEGF receptor subtype-2, to enrich for a population of multipotent cardiac progenitors arising during murine ES cell differentiation.[38] Recent work from the same group has demonstrated that a comparable progenitor is present during human cardiogenesis.[39]

Collectively, these studies redefine our understanding of mammalian heart development: multipotent cardiovascular progenitor cells contribute to the formation of a functional organ by making lineage fate decisions at a single cell level. This clonal model of heart lineage diversification would be similar to other stem cell-mediated systems such as blood, skin and intestine where a few multipotent stem cells generate large numbers of differentiated progeny by a process of amplification and progressive lineage restriction.[40–42] Uncovering the molecular pathways that control the formation and differentiation of cardiovascular progenitors into specific mature progeny will be critical for unlocking the potential of stem cell therapy for cardiovascular degenerative diseases, such as heart failure, conduction system disorders, and congenital heart disease (CHD).

3. RESIDENT CARDIAC STEM/PROGENITOR CELL POPULATIONS OF THE FETAL AND ADULT HEART

The search for "novel heart muscle forming cells" resident in the postnatal/adult myocardium was based on several instigating rationales: 1) an emerging model that tissue resident progenitor cells, seen as developmental remnants, share some common signatures of "stemness" although already restricted to differentiate into cell lineages of the organ in which they reside;[43] 2) the inability of skeletal myoblasts to transdifferentiate into the myocytic lineage and to generate an electromechanically coupled syncytium with the host myocardium;[44,45] and 3) the disappointment about the claimed plasticity of bone marrow-derived cells.[13,14,46] Since the beginning of 2000, increasing evidence indicates that the myocardium harbours several different types of progenitors that can re-enter the cell cycle and differentiate to mature cardiomyocytes.

Taking advantage of the fact that Isl1 expression is downregulated in most cardiac precursor cells as they differentiate, recent studies have utilized inducible *Isl1*-Cre and knock-in *Isl1*-nlacZ mice to analyze the timing of Isl1[+] progenitor migration into the looping heart and the distinct subdomains of the heart that they colonize during embryonic development.[47,48] Interestingly, a subset of Isl1[+] undifferentiated progenitors remains embedded in the embryonic heart after its formation and a few cells are still detectable after birth in compartments that arise from Isl1[+] lineage precursors during cardiac development. Tamoxifen-inducible Cre/*lox* technology has enabled this novel postnatal Isl1[+] cell population and its progeny to be selectively marked at a defined time and purified to relative homogeneity.[47] The ability of these cells to self-renew *in vitro* on a cardiac mesenchymal feeder layer and to be stimulated to differentiate into fully mature functional cardiomyocytes suggests that they represent native cardiac progenitors, remnants of the embryonic $Isl1^+$ precursors.[47] Postnatal Isl1-expressing cells can be detected in mouse, rat and human myocardium. While the pivotal role of $Isl1^+$ cardiovascular progenitors in the formation of the major heart lineages is clear, the importance of these cells in endogenous programs of cardiovascular repair is still unknown. Cardiomyocytes rarely seem to enter the cell cycle after birth and consequently the heart has a very limited regenerative capacity after injury. Given that the number of $Isl1^+$ cells within the heart is vanishingly small after the postnatal window, it is unlikely that they play a role in the regeneration of adult working myocardium. The location of $Isl1^+$ cells in the embryonic/fetal cardiac structures that are associated with CHDs (outflow tract, inter-atrial septum, etc.) suggests a possible role of these cells in regenerative pathways that are activated in these specific heart compartments in response to hemodynamic changes after birth.[32,48] Uncovering the stimuli that might lead to the *in vivo* mobilization of $Isl1^+$ cardiovascular progenitor cells will be of importance for understanding their contribution to endogenous regeneration.

To date, no validated cardiac-specific surface markers are available for the unambiguous isolation of cardiac stem/progenitor cells. In this regard, other cell populations have been identified based on the expression of cell surface proteins or properties characteristic of bone marrow hematopoietic stem cells, including Abcg2, a member of the family of ATP-binding cassette transporters responsible for the dye efflux in SP (side population) cells, stem cell antigen-1 (Sca-1), CD34, and the tyrosine kinase receptor c-kit.[49–51] Though established in the hematopoietic lineage as markers for stem cells, these cell surface antigens are expressed widely in germ cells, neurocrest derivatives, and melanocytes.[52] The $c-kit^+$ cardiac stem cells were reported to be clonogenic and self-renewing and capable of *in vitro* differentiation into cardiomyocytes, smooth muscle cells, and endothelium. Both $c-kit^+$ and $sca-1^+$ stem/progenitor cells have been shown to engraft in the heart and differentiate *in situ* when transplanted into the peri-infarct region in rodents. Although the exact stimuli within the host myocardium supporting the differentiation of these cells are largely unknown, it appears that at least some of them can adopt a differentiated cardiomyocyte phenotype by fusing with endogenous myocytes.[50] Recently the isolation of adult human cardiac stem cells based on the expression of c-KIT has been reported.[53,54] These cells appear to be phenotypically similar to their murine counterparts and, following transplantation into the infarcted myocardium of nude rats, they were shown to differentiate into cardiomyocytes as well as smooth muscle and endothelial cells.[53] To date, our understanding of the molecular profile and biological properties of these cells remain limited. Furthermore, the exact lineage relationship between these adult cardiac progenitor populations and the embryonic cardiac precursors is currently unknown. To this end, genetic strategies of cell fate mapping or cell ablation using different conditional Cre reporter strains need to be employed during cardiac injury to indelibly mark the progenitor populations of interest.

How many cardiomyocytes are generated in the healthy heart after birth by these newly identified adult cardiogenic cell types? What are their contributions in replacing dying or dead cells in a diseased myocardium? Do they play any role in physiological repair mechanisms after heart injury? Those questions still need to be addressed. Since adult cardiac stem cells are expected to be rare, any possible use for transplantation and achievement of detectable level of engraftments will require their extensive *in vitro* expansion. This necessity for long-term culture can epigenetically modify their gene expression and their biological properties.[55] Hence, investigators from different groups are actively working to identify signaling molecules, e.g. Notch, Wnts, and BMPs, which regulate self-renewal of these adult cardiac stem cells with the hope that "growth factor infusions" may stimulate the expansion of the endogenous cells *in situ* and provide an alternative approach for stem cell therapy. A recent study has shown that adult epicardial progenitor cells can be reactivated by injection of thymosin β4 resulting in neovascularization.[56]

4. NUCLEAR REPROGRAMMING AS A TECHNIQUE TO OBTAIN HUMAN MULTIPOTENT CARDIOVASCULAR PROGENITOR CELLS

The identification of developmental and resident cardiac progenitors holds great potential not only for cell replacement therapy but also for the generation of model systems to study molecular pathways controlling self-renewal, specification, and differentiation in cardiogenesis (Figure 2). However, a major limitation related to these cells is their low number *in vivo* and the difficulties for human fetal tissue to become the cell origin for clinical cardiac regeneration. To overcome this problem, an alternative option is to isolate cardiac progenitors from murine and human ES cells, a tissue source that can recapitulate the developmental characteristics of an embryo. The similarity between ES cells and embryos in their stage-specific gene expression and developmental potency enables one to apply principles of embryonic development to enhance ES cell differentiation into specific cell types.[57] Recent studies have used a developmental approach to identify multipotent cardiovascular progenitors arising during mouse ES-cell differentiation. Expression of marker genes under the control of Isl1, Nkx2-5, or Bry has allowed the isolation and clonal amplification *in vitro* of these different progenitor populations, which display multipotentiality to differentiate in all three cardiovascular lineages *in vitro*.[32,34,38] It has recently been shown that during directed specification of human ES cells into the cardiac lineage, a population of KDR^{low}/cKIT^{neg} cardiovascular progenitors develops, which can generate cardiomyocytes, endothelial and smooth muscle cells *in vitro* and after transplantation *in vivo*. The injection of these cells into infarcted hearts of immuno-compromised mice resulted in a significant improvement in cardiac function two weeks after transplantation.[39] Taken together, these studies suggest that ES cells represent a renewable source of functional cardiac progenitors.

For translational studies, the availability of autologous immune-matched cells is essential; hence the accessibility to patient-specific ES-like cells will circumvent the need for post transplant

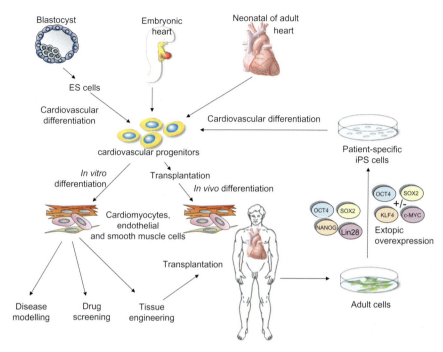

Figure 2. Sources and strategies for cardiovascular progenitors. Cardiovascular progenitors have been identified and isolated from ES cells, embryonic, and postnatal hearts. Recent nuclear reprogramming technology allows the generation of patient-specific pluripotent cells as an additional source to obtain cardiovascular progenitor cells. Their multipotency presents them as ideal model system to study specification and differentiation in respect to disease modeling, drug screening, or tissue engineering.

immunosuppression. In this regard, one of the most significant scientific breakthroughs is the identification of defined factors sufficient to reprogram differentiated somatic cells into a pluripotent ES-like state.[58] After this landmark discovery, many studies have demonstrated that iPS cells not only share morphological and epigenetic similarities with ES cells, but also functional potential *in vitro* and *in vivo* (teratoma/chimera formation and germline transmission).[59–64] iPS technology is not restricted to only healthy, young individuals, but is applicable also to patients with different age and diseases. Recent reports have illustrated examples of establishing iPS cells from skin fibroblasts of an 82-year-old woman with amyotrophic lateral sclerosis, a child with spinal muscular atrophy, and patients with a variety of both simple and complex genetic disorders.[65–67]

iPS cells offer the possibility to produce a large number of cell types with patients' own genetic background, which raises exciting new prospects for biomedical research and regenerative medicine. In respect to the cardiovascular field, iPS cells may represent a valuable source of multipotent cardiovascular progenitors and mature cardiovascular cell types. These iPS-derived cells can form the basis to study pathogenesis of cardiovascular diseases, to perform molecular and genetic screens to enable patient-specific drug design, and to engineer cardiovascular tissue *in vitro*. Additionally, they could serve as autologous cell type candidates for regenerative therapy to treat heart diseases (Figure 2). Several studies have already confirmed that transplantation of ES-derived cardiomyocytes into a normal or injured myocardium results in improved cardiac function. However, the size of engraftment seems to diminish over time as immune rejection of donor cells eventually takes its toll. Even in immuno-compromised mice, the functional improvement observed was only transient, suggesting that stable integration of grafts into the host tissue requires the formation of blood vessels to supply oxygen and nutrients.[68] To this end, iPS-derived multipotent cardiovascular progenitors may offer an advantage over merely cardiomyocytes as they should be able to proliferate and differentiate into diverse mature cardiovascular lineages, thus contributing to both remuscularization and revascularization.

Nonetheless, there are obstacles that need to be overcome before iPS-derived cardiac progenitors can be used in the clinical setting. The initial works have used lenti- or retroviral vectors to deliver the reprogramming factors. However, random integration into the host genome poses risk for cancer formation and disruption of endogenous gene expression. A recent improvement in the human reprogramming protocol includes the consolidation of multiple viral vectors into one polycistronic vector, hence decreasing the number of integration sites.[69,70] Ongoing work using phenotypic screens continues to identify small molecules or engineer cell-permeable factors that can recapitulate the reprogramming effects of these transcription factors.[71] In addition, the possibility of creating human leukocyte antigen (HLA)-haplotyped iPS cell banks may circumvent the time line requirement for generating and validating new cells for transplantation based therapies.[72] For further details see also Chapter IV. D.

The prospect to obtain cardiovascular progenitors from iPS cells offers optimism in the field of cell replacement therapy to treat cardiovascular diseases. However, we must learn from previous cell replacement experiments and raise the stringency of criteria we use in differentiation protocols for multipotent cells, in purification techniques for the differentiated progeny, and in our analysis of the engraftment and cardiac performance post transplantation.

5. SUMMARY

Modern translational medicine is based on the progressive study of pathways and principles from animal model organisms to clinical research in humans. To date, most of our knowledge of human disease is based on studies in patients themselves. However, many of the most critical and puzzling human cardiovascular disorders cannot be adequately studied because specific human cardiovascular cell types, such as cardiomyocytes, coronary endothelial and smooth muscle cells, cannot be obtained. Two crucial steps towards reaching the goal of studying specific cardiovascular cell types from patients with various forms of congenital and/or acquired heart diseases have recently been made: 1) the identification of multipotent cardiovascular progenitor cells not only in mammalian embryos and postnatal (adult) hearts but also as an intermediate stage during differentiation of ES cells; and 2) the breakthrough discovery that adult somatic cells can be reprogrammed to pluripotency by ectopic expression of few defined transcription factors — iPS cell technology. Therefore, iPS cells could be an ideal source to obtain patient-specific cardiac progenitors and in turn large number of cardiovascular cells with the disease-causing mutation. This will represent a powerful *in vitro* system to study pathogenesis of cardiovascular diseases at the cellular level and perform molecular and genetic screens to enable patient-specific drug design. Additionally, in the context of regenerative medicine, autologous lineage specific multipotent progenitors could ideally have an advantage over more developed cardiovascular cells, as they should be able to proliferate and at the same time, differentiate into working myocardial cells and vascular endothelial and smooth muscle cells *in situ*.

ACKNOWLEDGMENTS

The authors would like to thank members of the laboratory for their helpful discussions and comments. We apologize to those colleagues whose work is not mentioned here due to space limitations. The authors of this work are supported by a Marie Curie Excellence Team Grant from the European Research Council (MEXT-23208), the German Research Foundation (La 1238 3-1/4-1), and the Ministry for Research and Education (01 GN 0826).

REFERENCES

1. Lloyd-Jones D, Adams R, Carnethon M, De Simone G, *et al.* (2008) Heart disease and stroke statistics — 2009 update. A report from the American Heart Association Statistics Committee and Stroke Statistics Subcommittee. *Circulation* 121(7): 948–954.
2. Laflamme MA, Chen KY, Naumova AV, Muskheli V, *et al.* (2007) Cardiomyocytes derived from human embryonic stem cells in pro-survival factors enhance function of infarcted rat hearts. *Nat Biotechnol* 25: 1015–1024.
3. Murry CE, Keller G (2008) Differentiation of embryonic stem cells to clinically relevant populations: lessons from embryonic development. *Cell* 132: 661–680.
4. Rosenzweig A (2006) Cardiac cell therapy — mixed results from mixed cells. *N Engl J Med* 355: 1274–1277.
5. Rubart M, Field LJ (2006) Cardiac regeneration: repopulating the heart. *Annu Rev Physiol* 68: 29–49.
6. Assmus B, Schachinger V, Teupe C, Britten M, *et al.* (2002) Transplantation of progenitor cells and regeneration enhancement

in acute myocardial infarction (TOPCARE-AMI). *Circulation* 106: 3009–3017.

7. Menasche P (2003) Myoblast-based cell transplantation. *Heart Fail Rev* 8: 221–227.

8. Schachinger V, Erbs S, Elsasser A, Haberbosch W, *et al.* (2006) Intracoronary bone marrow-derived progenitor cells in acute myocardial infarction. *N Engl J Med* 355: 1210–1221.

9. Strauer BE, Brehm M, Zeus T, Kostering M, *et al.* (2002) Repair of infarcted myocardium by autologous intracoronary mononuclear bone marrow cell transplantation in humans. *Circulation* 106: 1913–1918.

10. Wollert KC, Meyer GP, Lotz J, Ringes-Lichtenberg S, *et al.* (2004) Intracoronary autologous bone-marrow cell transfer after myocardial infarction: the BOOST randomised controlled clinical trial. *Lancet* 364: 141–148.

11. Murry CE, Field LJ, Menasche P (2005) Cell-based cardiac repair: reflections at the 10-year point. *Circulation* 112: 3174–3183.

12. Balsam LB, Wagers AJ, Christensen JL, Kofidis T, *et al.* (2004) Haematopoietic stem cells adopt mature haematopoietic fates in ischaemic myocardium. *Nature* 428: 668–673.

13. Murry CE, Soonpaa MH, Reinecke H, Nakajima H, *et al.* (2004) Haematopoietic stem cells do not transdifferentiate into cardiac myocytes in myocardial infarcts. *Nature* 428: 664–668.

14. Nygren JM, Jovinge S, Breitbach M, Sawen P, *et al.* (2004) Bone marrow-derived hematopoietic cells generate cardiomyocytes at a low frequency through cell fusion, but not transdifferentiation. *Nat Med* 10: 494–501.

15. Reffelmann T, Dow JS, Dai W, Hale SL, *et al.* (2003) Transplantation of neonatal cardiomyocytes after permanent coronary artery occlusion increases regional blood flow of infarcted myocardium. *J Mol Cell Cardiol* 35: 607–613.

16. Rubart M, Pasumarthi KB, Nakajima H, Soonpaa MH, *et al.* (2003) Physiological coupling of donor and host cardiomyocytes after cellular transplantation. *Circ Res* 92: 1217–1224.

17. Epstein JA, Buck CA (2000) Transcriptional regulation of cardiac development: implications for congenital heart disease and DiGeorge syndrome. *Pediatr Res* 48: 717–724.

18. Kirby ML, Gale TF, Stewart DE (1983) Neural crest cells contribute to normal aorticopulmonary septation. *Science* 220: 1059–1061.

19. Dettman RW, Denetclaw W, Jr, Ordahl CP, Bristow J (1998) Common epicardial origin of coronary vascular smooth muscle, perivascular fibroblasts, and intermyocardial fibroblasts in the avian heart. *Dev Biol* 193: 169–181.

20. Manner J, Perez-Pomares JM, Macias D, Munoz-Chapuli R (2001) The origin, formation and developmental significance of the epicardium: a review. *Cells Tissues Organs* 169: 89–103.

21. Moore AW, McInnes L, Kreidberg J, Hastie ND, *et al.* (1999) YAC complementation shows a requirement for Wt1 in the development of epicardium, adrenal gland and throughout nephrogenesis. *Development* 126: 1845–1857.

22. Buckingham M, Meilhac S, Zaffran S (2005) Building the mammalian heart from two sources of myocardial cells. *Nat Rev Genet* 6: 826–835.

23. Kelly RG, Buckingham ME (2002) The anterior heart-forming field: voyage to the arterial pole of the heart. *Trends Genet* 18: 210–216.

24. Laugwitz KL, Moretti A, Caron L, Nakano A, *et al.* (2008) Islet1 cardiovascular progenitors: a single source for heart lineages? Development 135: 193–205.

25. Harvey RP (2002) Patterning the vertebrate heart. *Nat Rev Genet* 3: 544–556.

26. Srivastava D, Olson EN (2000) A genetic blueprint for cardiac development. *Nature* 407: 221–226.

27. Solloway MJ, Harvey RP (2003) Molecular pathways in myocardial development: a stem cell perspective. *Cardiovasc Res* 58: 264–277.

28. Saga Y, Miyagawa-Tomita S, Takagi A, Kitajima S, *et al.* (1999) MesP1 is expressed in the heart precursor cells and required for the formation of a single heart tube. *Development* 126: 3437–3447.

29. Wu SM, Chien KR, Mummery C (2008) Origins and fates of cardiovascular progenitor cells. *Cell* 132: 537–543.

30. Cai CL, Liang X, Shi Y, Chu PH, *et al.* (2003) Isl1 identifies a cardiac progenitor population that proliferates prior to differentiation and contributes a majority of cells to the heart. *Dev Cell* 5: 877–889.

31. Zhou B, Ma Q, Rajagopal S, Wu SM, *et al.* (2008) Epicardial progenitors contribute to the cardiomyocyte lineage in the developing heart. *Nature* 454: 109–113.

32. Moretti A, Caron L, Nakano A, Lam JT, *et al.* (2006) Multipotent embryonic isl1+ progenitor cells lead to cardiac, smooth muscle, and endothelial cell diversification. *Cell* 127: 1151–1165.

33. Prall OW, Menon MK, Solloway MJ, Watanabe Y, *et al.* (2007) An Nkx2-5/Bmp2/Smad1 negative feedback loop controls heart progenitor specification and proliferation. *Cell* 128: 947–959.

34. Wu SM, Fujiwara Y, Cibulsky SM, Clapham DE, *et al.* (2006) Developmental origin of a bipotential myocardial and smooth muscle cell precursor in the mammalian heart. *Cell* 127: 1137–1150.

35. Bondue A, Lapouge G, Paulissen C, Semeraro C, *et al.* (2008) Mesp1 acts as a master regulator of multipotent cardiovascular progenitor specification. *Cell Stem Cell* 3: 69–84.

36. Kitajima S, Takagi A, Inoue T, Saga Y (2000) MesP1 and MesP2 are essential for the development of cardiac mesoderm. *Development* 127: 3215–3226.

37. Saga Y, Kitajima S, Miyagawa-Tomita S (2000) Mesp1 expression is the earliest sign of cardiovascular development. *Trends Cardiovasc Med* 10: 345–352.

38. Kattman SJ, Huber TL, Keller GM (2006) Multipotent flk-1+ cardiovascular progenitor cells give rise to the cardiomyocyte, endothelial, and vascular smooth muscle lineages. *Dev Cell* 11: 723–732.

39. Yang L, Soonpaa MH, Adler ED, Roepke TK, *et al.* (2008) Human cardiovascular progenitor cells develop from a KDR+ embryonic-stem-cell-derived population. *Nature* 453: 524–528.

40. Weissman IL (2000) Stem cells: units of development, units of regeneration, and units in evolution. *Cell* 100: 157–168.

41. Fuchs E (2008) Skin stem cells: rising to the surface. *J Cell Biol* 180: 273–284.

42. Barker N, van de Wetering M, Clevers H (2008) The intestinal stem cell. *Genes Dev* 22: 1856–1864.

43. Ramalho-Santos M, Yoon S, Matsuzaki Y, Mulligan RC, *et al.* (2002) "Stemness": transcriptional profiling of embryonic and adult stem cells. *Science* 298: 597–600.

44. Leobon B, Garcin I, Menasche P, Vilquin JT, *et al.* (2003) Myoblasts transplanted into rat infarcted myocardium are functionally isolated from their host. *Proc Natl Acad Sci USA* 100: 7808–7811.

45. Rubart M, Soonpaa MH, Nakajima H, Field LJ (2004) Spontaneous and evoked intracellular calcium transients in donor-derived myocytes following intracardiac myoblast transplantation. *J Clin Invest* 114: 775–783.

46. Wagers AJ, Weissman IL (2004) Plasticity of adult stem cells. *Cell* 116: 639–648.

47. Laugwitz KL, Moretti A, Lam J, Gruber P, *et al.* (2005) Postnatal isl1+ cardioblasts enter fully differentiated cardiomyocyte lineages. *Nature* 433: 647–653.

48. Sun Y, Liang X, Najafi N, Cass M, *et al.* (2007) Islet 1 is expressed in distinct cardiovascular lineages, including pacemaker and coronary vascular cells. *Dev Biol* 304: 286–296.

49. Beltrami AP, Barlucchi L, Torella D, Baker M, *et al.* (2003) Adult cardiac stem cells are multipotent and support myocardial regeneration. *Cell* 114: 763–776.

50. Oh H, Bradfute SB, Gallardo TD, Nakamura T, *et al.* (2003) Cardiac progenitor cells from adult myocardium: homing, differentiation, and fusion after infarction. *Proc Natl Acad Sci USA* 100: 12313–12318.

51. Martin CM, Meeson AP, Robertson SM, Hawke TJ, *et al.* (2004) Persistent expression of the ATP-binding cassette transporter, Abcg2, identifies cardiac SP cells in the developing and adult heart. *Dev Biol* 265: 262–275.

52. Besmer P, Manova K, Duttlinger R, Huang EJ, *et al.* (1993) The kit-ligand (steel factor) and its receptor c-kit/W: pleiotropic roles in gametogenesis and melanogenesis. *Dev Suppl*: 125–137.

53. Bearzi C, Rota M, Hosoda T, Tillmanns J, *et al.* (2007) Human cardiac stem cells. *Proc Natl Acad Sci USA* 104: 14068–14073.

54. Smith RR, Barile L, Cho HC, Leppo MK, *et al.* (2007) Regenerative potential of cardiosphere-derived cells expanded from percutaneous endomyocardial biopsy specimens. *Circulation* 115: 896–908.

55. Yamanaka S (2007) Strategies and new developments in the generation of patient-specific pluripotent stem cells. *Cell Stem Cell* 1: 39–49.

56. Smart N, Risebro CA, Melville AA, Moses K, *et al.* (2007) Thymosin beta4 induces adult epicardial progenitor mobilization and neovascularization. *Nature* 445: 177–182.

57. Wobus AM, Boheler KR (2005) Embryonic stem cells: prospects for developmental biology and cell therapy. *Physiol Rev* 85: 635–678.

58. Takahashi K, Yamanaka S (2006) Induction of pluripotent stem cells from mouse embryonic and adult fibroblast cultures by defined factors. *Cell* 126: 663–676.

59. Aasen T, Raya A, Barrero MJ, Garreta E, *et al.* (2008) Efficient and rapid generation of induced pluripotent stem cells from human keratinocytes. *Nat Biotechnol* 26: 1276–1284.

60. Aoi T, Yae K, Nakagawa M, Ichisaka T, *et al.* (2008) Generation of pluripotent stem cells from adult mouse liver and stomach cells. *Science* 321: 699–702.

61. Hanna J, Markoulaki S, Schorderet P, Carey BW, *et al.* (2008) Direct reprogramming of terminally differentiated mature B lymphocytes to pluripotency. *Cell* 133: 250–264.

62. Kim JB, Zaehres H, Wu G, Gentile L, *et al.* (2008) Pluripotent stem cells induced from adult neural stem cells by reprogramming with two factors. *Nature* 454: 646–650.

63. Silva J, Barrandon O, Nichols J, Kawaguchi J, *et al.* (2008) Promotion of reprogramming to ground state pluripotency by signal inhibition. *PLoS Biol* 6: e253.

64. Stadtfeld M, Brennand K, Hochedlinger K (2008) Reprogramming of pancreatic beta cells into induced pluripotent stem cells. *Curr Biol* 18: 890–894.

65. Park IH, Arora N, Huo H, Maherali N, *et al.* (2008) Disease-specific induced pluripotent stem cells. *Cell* 134: 877–886.

66. Dimos JT, Rodolfa KT, Niakan KK, Weisenthal LM, *et al.* (2008) Induced pluripotent stem cells generated from patients with ALS can be differentiated into motor neurons. *Science* 321: 1218–1221.

67. Ebert AD, Yu J, Rose FF, Jr, Mattis VB, *et al.* (2009) Induced pluripotent stem cells from a spinal muscular atrophy patient. *Nature* 457: 277–280.

68. van Laake LW, Passier R, Doevendans PA, Mummery CL (2008) Human embryonic stem cell-derived cardiomyocytes and cardiac repair in rodents. *Circ Res* 102: 1008–1010.

69. Carey BW, Markoulaki S, Hanna J, Saha K, *et al.* (2009) Reprogramming of murine and human somatic cells using a single polycistronic vector. *Proc Natl Acad Sci USA* 106: 157–162.
70. Sommer CA, Stadtfeld M, Murphy GJ, Hochedlinger K, *et al.* (2008) iPS cell generation using a single lentiviral stem cell cassette. *Stem Cells* 27: 543–9.
71. Huangfu D, Maehr R, Guo W, Eijkelenboom A, *et al.* (2008) Induction of pluripotent stem cells by defined factors is greatly improved by small-molecule compounds. *Nat Biotechnol* 26: 795–797.
72. Nakatsuji N, Nakajima F, Tokunaga K (2008) HLA-haplotype banking and iPS cells. *Nat Biotechnol* 26: 739–740.

IV. Repopulation of the Heart with New Cardiomyocytes

C. Circulating Progenitor Cells

Masamichi Koyanagi and Stefanie Dimmeler

1. INTRODUCTION

In the human body, adult stem/progenitor cells reside in stem cell niches (i.e. in bone marrow) and are mobilized into the circulation in response to appropriate signals. Whereas the mobilization of tissue-derived or vessel-associated progenitor cells is unclear it has been shown that various cytokines as well as neurohumoral activation induce mobilization of bone marrow-derived progenitor cells. Thereby, hematopoietic progenitor cells (HPCs) and endothelial progenitor cells (EPCs) but also mesenchymal stem/stromal cells (MSCs) or other stem/progenitor cells have been reported to be mobilized into the circulation depending on the induction signal. In this chapter, we summarize and discuss the characteristics, mobilization, and therapeutic potential of these mobilized circulating progenitor cells.

2. HEMATOPOIETIC AND ENDOTHELIAL PROGENITOR CELLS

Hematopoietic and endothelial progenitor cells are mobilized by a variety of stimuli including cytokines such as granulocyte-colony stimulating factor (G-CSF) or stromal-derived factor-1 (SDF-1). However, individual cell populations may also be selectively affected by specific cytokines. Whereas HPCs are clearly defined, the identification of EPCs is still under debate. Therefore, the next chapters first discuss the characterization of EPCs by FACS or culture assays before summarizing the mobilizing factors.

2.1. Characterization of Endothelial Progenitor Cells by FACS

In 1997, Asahara and co-workers reported the isolation of putative EPCs from human peripheral blood.[1] These cells were positive for the hematopoietic stem cell marker CD34 and endothelial markers such as VEGF-R2 (KDR). Isolated CD34+ cells differentiated to endothelial cells *in vitro* and incorporated into newly formed vessels during angiogenesis *in vivo*.[1] These pioneering studies suggested that bone marrow-derived cells can participate in vascular repair. Indeed, transplantation of hematopoietic stem cells was shown to give rise to both blood cells and vascular endothelium;[2–4] even when a single stem cell was used for transplantation.[5] Moreover, fractionation of cord blood- or bone marrow-derived CD34+ cells into cells that express or lack KDR revealed that specifically CD34+KDR+ cells give rise to hematopoietic and endothelial progeny. Interestingly, the endothelial differentiation conditions required a minimal cell number of 250 cells/well indicating the requirement of cell-cell-communication for endothelial colony formation activity.[3] The contribution of bone marrow-derived circulating cells to the endothelium was further supported by gender-mismatch allogeneic bone marrow and stem-cell transplantation as well as heart transplantations in humans.[6,7]

In search for valid markers to identify human EPCs, other investigators used the hematopoietic marker CD133.[8,9] Based on these studies, EPCs have been characterized as CD133+KDR+ or CD133+CD34+KDR+ cells. However, recent publications challenged the use of these markers by demonstrating that CD133+CD34+KDR+ cells isolated from umbilical cord blood or G-CSF mobilized blood did not form endothelial colonies *in vitro*.[10] The endothelial colony forming activity was attributed to a rare population of CD34+CD45– cells.[10] Likewise, CD34+CD45– but not CD34+CD45+ cells were shown to give rise to endothelial outgrowing colonies.[11] However, other studies revealed different results. Injected CD34-CD133+KDR+ cells were shown to have a higher endothelial repair capacity in an arterial injury mouse model.[12]

At first glance these studies appear discrepant, however, one should keep in mind that the culture conditions *in vitro* may not reflect the environment *in vivo*. Particularly after injury the cytokine milieu may modify the cell fate and injected cells may have additional effects, e.g. by paracrine mechanisms.

2.2. Characterization and Isolation of Endothelial Progenitor Cells by Culture Assays

In addition to the measurement of surface markers, different *in vitro* colony assays have been used to identify EPCs in the past (Figure 1). The first assay was designed with the aim to enrich EPCs for cell therapy of ischemic disease and is based on the cultivation of peripheral blood mononuclear cells on fibronectin or fibronectin/gelatin matrix.[13] This EPC culture assay, which was later termed "early EPC culture assay," yields cells that express endothelial markers such as VEGF-R2, vWF and CD31, but co-express myeloid markers including CD45 and CD14[13–15] suggesting that these cells may share features with myeloid cells. Subsequent studies slightly modified the

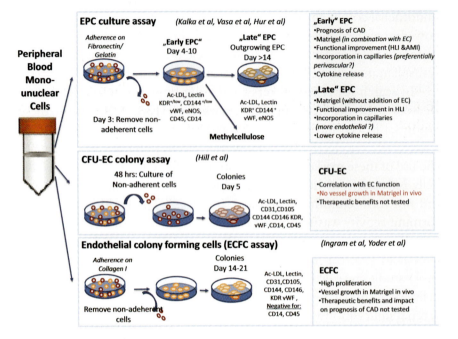

Figure 1. *In vitro* **colony assays to identify EPC.** Comparison of assay designed to enrich EPC for cell therapy from peripheral blood mononuclear cells.

assay in order to exclude contamination with macrophages and mature endothelial cells, which might have been shed of the vessel wall.[16] Thereby, the fibronectin-adherent cells were discharged and only the suspension was used for further cultivation.[16] This CFU-EC assay was subsequently commercialized in a slightly modified version (Endocult Assay, StemCell Technologies).

Recent studies assessed the characteristic of cultivated colonies in more detail and demonstrated that CFU-ECs are not derived from CD14+ monocytes. However, depletion of CD14+ cells prevented the formation of CFU-ECs indicating that this assay depends on contaminating monocytes (or their soluble factors).[17] The origin of the CFU-ECs appears to be hematopoietic, but final proof of the origin, e.g. by gender mismatched bone marrow transplantation is missing.

Several studies document a close relationship of different cultivated "EPCs" with monocytes. It is important to note that monocytes have a high plasticity. A subset was shown to co-express endothelial markers[18] and isolated CD14+ cells, which were cultivated under endothelial differentiation promoting conditions, increased neovascularization in a hind limb ischemia model.[14] Isolated CD14+KDR+ cells showed a preferential endothelial regeneration activity compared to CD14+KDR– cells[19] supporting that myeloid cells contribute to endothelial regeneration or repair *in vivo*.

Vascular endothelial cells can indeed arise from transplanted common myeloid progenitor (CMP) cells or granulocyte/macrophage progenitors (GMPs) independent of cell fusion.[20] Interestingly, the transplantation of hematopoietic stem cells results in a long-term reconstitution of vascular endothelial cells.[20] Other studies have demonstrated that myeloid cells are required by VEGF to support neovascularization mediated by bone marrow cell-derived paracrine factors.[21] These bone marrow-derived circulating cells are recruited and retained by VEGF and SDF-1, respectively, and are positive for CXCR4, VEGF-R1, CD45, CX3CR1, and the myeloid marker CD11b. Two other studies also described the recruitment of VEGF-R1$^+$ hematopoietic progenitors that augmented revascularization and metastasis.[22,23] Moreover, immune cells such as monocytes, macrophages, dendritic cells, and B lymphocytes have been found to be recruited to areas of neoangiogenesis, suggesting that the co-mobilization of different subtypes of bone marrow mononuclear cells may contribute to neo-vessel formation either via the release of paracrine factors (VEGF-R1$^+$ cells) or via physical incorporation (VEGF-R2$^+$ cells). *Ex vivo* cultured EPCs which co-express CD14+ and CD45+ cells also were shown to act as pro-angiogenic cells after transplantation in ischemia models *in vivo*,[21,24] in addition (or alternatively) to their capacity to differentiate into endothelial cells.[20,25] The question is whether these myeloid EPCs represent a unique subset of monocytes and indeed exhibit a progenitor cell characteristic or act in a paracrine manner similar to the well-known activity of monocytes in arteriogenesis.

Some studies have suggested that subsets of monocytes express some stem cell markers and may exhibit some plasticity.[26,27] Therefore, it might be that the cultured myeloid cell populations are contaminated by "real" progenitor cells. Alternatively, the culture conditions may also modulate or partially reprogram the cells *ex vivo*.

In contrast to the EPC culture assays described above, late outgrowing endothelial cells showed a different marker expression and were long term proliferating.[28] Outgrowing endothelial cells were derived from peripheral blood mononuclear cells, which were plated on collagen-coated dishes in endothelial growth medium and were grown for up to 60 days. These cells lacked the expression of myeloid markers and were not derived from mature circulating endothelial cells (CECs), which might be shed of the vessel wall, but were bone marrow-derived as demonstrated in cultures obtained of bone marrow-mismatched transplanted patients. These data indicated that the outgrowing endothelial cells derive from circulating bone marrow-derived angioblasts. Ingram and Yoder used a similar isolation procedure yielding endothelial colony forming cells (ECFCs), expressing various endothelial markers, formed vessel-like structures in matrigel and lack expression of CD14 and CD45.[29] However, in their hands, the outgrowing cells were not clonally related to CFU-ECs and were likely not of hematopoietic origin. The authors speculate that these cells may derive from vessels.

When thinking about the use of the endothelial progenitor cells identified by colony assays or by FACS as biomarker for the risk of atherosclerotic diseases, the specific problem is to distinguish between circulating mature endothelial cells (CEC), vessel- or tissue-derived endothelial cells and bone marrow-derived angioblasts. While the myeloid subtypes express CD14 and/or CD45, all other cells may share the expression of similar endothelial markers including CD34, which is also expressed on endothelial cells.

2.3. Mobilization

One of the strongest and best studied mobilizing factor of hematopietic progenitor cells (HPCs) and EPCs is G-CSF.[30] Circulating EPCs are increased by G-CSF treatment in mice and humans.[31] Ripa and co-workers also demonstrated that G-CSF mobilized CD34+ cells during acute myocardial infarction.[32] The mechanism by which G-CSF mobilizes cells into circulation is partially due to reducing CXCR4 expression on HPC and SDF-1 in bone marrow.[33] Indeed, SDF-1 and its receptor CXCR4 play a major role in mobilizing EPCs and HPCs.[34] Recently, Pitchford and co-workers investigated whether specific cytokines individually mobilize distinct sets of progenitor cells such as HPCs, EPCs, and stromal progenitor cells (SPCs) from the bone marrow.[35] G-CSF pretreatment of mice in response to CXCR4 antagonist was shown to efficiently mobilize HPC, whereas pretreatment with VEGF reduced HPCs migratory capacity *in vitro* and suppressed their mobilization *in vivo*. In contrast, VEGF pretreatment enhanced EPC mobilization via VEGFR2 in response to CXCR4 antagonist. Furthermore, SPC mobilization was detected when the CXCR4 antagonist was administered to mice pretreated with VEGF, but not G-CSF.

Erythropoietin (Epo) is another mobilizating cytokine. Epo increased EPC mobilization and contributes to the improved recovery after hind limb ischemia[36] and post infarction heart failure.[37] Furthermore, serum levels of Epo and VEGF were significantly associated with the number and function of circulating EPCs.[36] Further experiments demonstrated that chronic treatment of Epo increased EPC proliferation and adhesion capacity in chronic heart failure patients.[38] Another cytokine, angiopoietin-1[34] also increased EPC mobilization. Moreover, a variety of other factors such as statins, the Wnt antagonist dickkopf or exercise mobilize EPCs or HPCs (Figure 2).

Endogenous mobilization of progenitor cells occurs under stress situation such as limb ischemia[31] and acute myocardial infarction,[39] which induce both induced increased circulating HPC and EPC levels. Moreover, cardiac surgery increased CXCR4 and mobilizes CD34+ cells,[40] and EPCs[41] and CD133+KDR+ cells rise in response to vascular trauma[42] in association with increased VEGF. The mechanism of stress-induced mobilization may be in part mediated by hypoxia, which induces systemic increases in VEGF or SDF-1. In the bone marrow, it was reported that Nox2 is required for hypoxia-induced mobilization of EPCs.[43] However, beside cytokines,

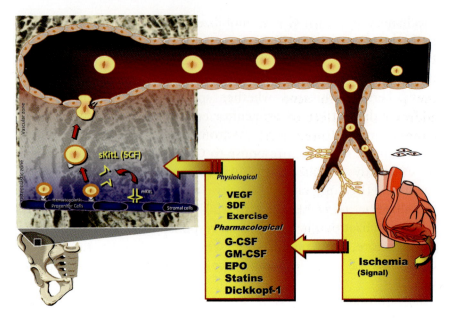

Figure 2.   Mobilization of progenitor cells. Schematic representation of the signals and factors that induce the mobilization of progenitor cells and their release in the circulation.

neurohumoral activation also may contribute to the mobilization occurring after ischemia.[44,45]

2.4. Therapeutic Potential

A lot of basic and clinical studies demonstrated favorable contribution of neovascularization of EPCs in physiological and pathological situation, and after cell transplantation. Thereby, EPCs or HPCs or bone marrow mononuclear cells (BMCs) that contain HPCs and EPCs improve the recovery after ischemia after hind limb ischemia models as well as in experimental acute myocardial infarction models (for review see Ref. 25). In addition, clinical studies in patients with acute myocardial infarction reported an improved recovery of left ventricular ejection fraction in some but not all studies (for review see Ref. 46). Although it is discussed whether the increase in ejection fraction (about 3%–5%, see recent meta-analysis) has a clinically meaningful impact, and the duration of the effect is not clear, clinical two-year follow-up suggests that there might be a benefit of bone marrow-derived cell therapy for the patients with acute myocardial infarction.[47] Some studies addressed the effect of selected populations of EPC-like cells by either using cultured early EPCs[48] or CD34+CXCR4+ cells.[49] These two studies reported similar effects of the selected EPCs versus the total BMCs. However, in chronic heart failure, cultured early EPCs showed a reduced effect compared to BMCs[50] indicating that the acute "healing" effect in the acute infarct by EPC setting may be distinct from the effects required to improve ejection fraction in patients with chronic ischemia. Isolated CD34+ or CD133+ cells were used in combination with bypass surgery,[51] in small phase I studies in patients with chronic ischemia or in patients with refractory angina.[52] Particularly, in patients with refractory angina, CD34+ showed promising effects that have recently been confirmed in two phase II studies.[53,54] Finally, in patients with limb ischemia, such as arteriosclerosis obliterans (ASO) or Buerger's disease cell therapy with HPCs or EPCs might be a promising option. Several clinical trials have been done and some are now on-going in limb ischemia.[55–57]

Besides direct injection of progenitor cells to ischemic region, controlling mobilization and homing to specific region by low-energy shock wave would be one interesting option to treat patients. Aicher and co-workers demonstrated that low-energy shock wave, which did not damage tissue, increased local SDF-1 gene expression and increased local EPC number and improved functional recovery after hind limb ischemia.[58] Clinical trials of G-CSF treatment in AMI patients in order to increase bone marrow progenitor cells into circulation failed to improve cardiac function when compared to the placebo group (meta analysis[59]). Probably, merely increasing progenitor cells in the circulation is not sufficient to stimulate the local response. Therefore, increased homing by low-energy shock wave or combining G-CSF with other strategies such as inhibition of CD26[60] might be future therapeutic options for treatment.

3. MESENCHYMAL STEM/PROGENITOR CELLS

3.1. Characterization

Mesenchymal stem cells or mesenchymal stromal cells (MSCs) comprise the second major progenitor pool in the bone marrow. However, although MSCs were initially discovered in the bone marrow,[61] various recent studies demonstrated that MSCs are also present in other tissues like adipose tissue and umbilical cord blood. Since MSCs comprise a heterogeneous cell population, the *International Society for Cellular Therapy* proposed minimal criteria to define MSCs.[62] According to this criterion, MSCs 1) can adhere to plastic, 2) express mesenchymal markers such as CD105, CD73 and CD90, and lack expression of CD45, CD34, CD14 or CD11b, CD79a or CD19 and HLA-DR, and 3) can differentiate to at least osteoblasts, adipocytes and chondroblasts under specific *in vitro* differentiating conditions. However, this criterion may be shared by different types of stem cells. For example, multipotent adult stem cells (MAPCs[63]) obtained from bone marrow, unrestricted somatic stem cells (USSCs[64]) from umbilical cord blood and mesoangioblasts[65] are negative for hematopoietic markers but share some criteria of MSCs (i.e. CD73$^+$). Therefore, it is controversial whether these MAPC, USSCs, or mesoangioblasts are subsets of MSC or are unique and different cell types. Recently, instead of mesenchymal stem/stromal cells, the term "multipotent stromal cells" has been proposed to be used.[62]

3.2. Mobilization

MSCs reside in the bone marrow niche or mesenchymal tissue and can be mobilized by specific stimuli. Zvaifler and co-workers demonstrated that MSCs were present in buffy coats of normal human blood,[66] however, the incidence of circulating MSCs under physiological conditions is quite low.[67] Colony stimulating factors such as G-CSF or granulocyte-monocyte (GM)-CSF treatment may mobilize MSCs. Fernandez and co-workers collected GM-CSF-mobilized blood and identified after culturing non-hematopoietic MSC.[68] On the other hand, other groups failed to culture MSCs after GM-CSF treatment.[69,70] Ripa and co-workers demonstrated that G-CSF increased CD34 cells in the circulation after acute myocardial infarction, but the fraction of CD45−/CD34− mesenchymal cells decreased.[32] Interestingly, the authors found an inverse correlation between number of circulating mesenchymal stem cells and changes in ejection fraction. They speculated that MSCs might home to the infarcted heart and contributed to the improvement of cardiac function.

Severe stress conditions might also potentially mobilize MSCs. In acute myocardial infarction, bone marrow-derived MSCs were shown to be mobilized and were reported to differentiate to cardiomyocytes.[71,72] One paper reported that hypoxia also mobilizes MSCs,[73] whereas the mechanism of mobilization by hypoxia is not clear. Operation or trauma may also mobilize stem cells into circulation, although, most studies only report an increase in hematopoietic CD34+ cells or EPCs.[40–42] However, skeletal muscle injury was reported to induce mobilization of MSCs.[74] Overall, some studies suggest that cytokine or stress can induce the mobilization of MSC, however, it appears less efficient and more variable compared to the induction of HPC or EPC mobilization and the underlying mechanisms have not been fully elucidated.

3.3. Therapeutic Potential

MSCs can differentiate not only to osteoblasts, adipocytes and chondroblasts, but also into other mesodermal cardiovascular cells such as cardiomycytes[75] and endothelial cells.[76,77] In addition, MSCs secrete pro-angiogenic cytokines (similar to EPCs). When compared to hematopoietic progenitor cells or EPCs, MSCs appear to have a higher potency to differentiate into cardiomyocytes. Indeed, Fukuda and co-workers demonstrated that MSCs more favorably differentiate into cardiomyocytes in a model in bone marrow chimera in which MSCs or HPCs were labeled with the fluorescent marker EGFP. After G-CSF treatment they found that only the EGFP-MSC group showed GFP-positive cardiomyocytes in the heart after acute myocardial infarction, whereas no EGFP-positive cardiac myocytes was detected in the EGFP-HPC mice.[71] Because of this potential cardiac differentiation capacity and safety concerns that the larger size of MSCs may induce thrombotic events,[78] MSCs were mostly intramuscularly injected in chronic heart failure. One clinical trial, however, treated patients with acute infarction with intravenous infusion of allogeneic MSCs.[79] The study concluded that the therapy was safe and that one specific subset of patients with large infarct showed an increased ejection fraction. One may consider using MSCs in combination with hematopoietic progenitor cells in order to simultaneously improve neovascularization and cardiac repair. Indeed, Zhang and co-workers demonstrated better outcome in combination of MSCs with EPCs after isoproterenol-induced myocardial injury.[80]

4. SUMMARY AND CONCLUDING REMARKS

In summary, circulating HPCs and EPCs can be mobilized by various factors and injection of isolated cells have promising effects in models of ischemia and in phase II clinical trials. Whereas the identification and characterization of EPCs is still a matter of debate, several different preparations of EPCs (either using FACS sorted subpopulations or different culture assays) showed beneficial effects in experimental ischemia models. Whether these effects on the recovery of ischemia are due to paracrine effects and/or the physical formation of capillaries may vary depending on the cell type but also the specific model used. In general, early EPCs or myeloid populations of EPC appear to be very efficient sources of cytokines, whereas late outgrowing EPCs may be more efficient in forming vessels. Although MSCs were detected in the circulating blood in some studies, most experiments so far used bone marrow or tissue to isolate MSCs for therapy of ischemia. The functional capacities of isolated circulating MSCs are unclear.

REFERENCES

1. Asahara T, Murohara T, Sullivan A, Silver M, *et al.* (1997) Isolation of putative progenitor endothelial cells for angiogenesis. *Science* 275: 964–967.

2. Bailey AS, Jiang S, Afentoulis M, Baumann CI, *et al.* (2004) Transplanted adult hematopoietic stems cells differentiate into functional endothelial cells. *Blood* 103: 13–19.

3. Pelosi E, Valtieri M, Coppola S, Botta R, *et al.* (2002) Identification of the hemangioblast in postnatal life. *Blood* 100: 3203–3208.

4. Larrivee B, Niessen K, Pollet I, Corbel SY, *et al.* (2005) Minimal contribution of marrow-derived endothelial precursors to tumor vasculature. *J Immunol* 175: 2890–2899.

5. Grant MB, May WS, Caballero S, Brown GA, *et al.* (2002) Adult hematopoietic stem cells provide functional hemangioblast activity during retinal neovascularization. *Nat Med* 8: 607–612.

6. Thiele J, Varus E, Wickenhauser C, Kvasnicka HM, *et al.* (2004) Mixed chimerism of cardiomyocytes and vessels after allogeneic bone marrow and stem-cell transplantation in comparison with cardiac allografts. *Transplantation* 77: 1902–1905.

7. Minami E, Laflamme MA, Saffitz JE, Murry CE (2005) Extracardiac progenitor cells repopulate most major cell types in the transplanted human heart. *Circulation* 112: 2951–2958.

8. Shi Q, Rafii S, Wu MH, Wijelath ES, *et al.* (1998) Evidence for circulating bone marrow-derived endothelial cells. *Blood* 92: 362–367.

9. Gehling UM, Ergun S, Schumacher U, Wagener C, *et al.* (2000) *In vitro* differentiation of endothelial cells from AC133-positive progenitor cells. *Blood* 95: 3106–3112.

10. Case J, Mead LE, Bessler WK, Prater D, *et al.* (2007) Human CD34+AC133+VEGFR-2+ cells are not endothelial progenitor cells but distinct, primitive hematopoietic progenitors. *Exp Hematol* 35: 1109–1118.

11. Timmermans F, Van Hauwermeiren F, De Smedt M, Raedt R, *et al.* (2007) Endothelial outgrowth cells are not derived from CD133+ cells or CD45+ hematopoietic precursors. *Arterioscler Thromb Vasc Biol* 27: 1572–1579.

12. Friedrich EB, Walenta K, Scharlau J, Nickenig G, *et al.* (2006) CD34−/CD133+/VEGFR-2+ endothelial progenitor cell subpopulation with potent vasoregenerative capacities. *Circ Res* 98: e20–25.

13. Kalka C, Masuda H, Takahashi T, Kalka-Moll WM, *et al.* (2000) Transplantation of *ex vivo* expanded endothelial progenitor cells for

therapeutic neovascularization. *Proc Natl Acad Sci USA* 97: 3422–3427.

14. Urbich C, Heeschen C, Aicher A, Dernbach E, *et al.* (2003) Relevance of monocytic features for neovascularization capacity of circulating endothelial progenitor cells. *Circulation* 108: 2511–2516.

15. Kawamoto A, Gwon HC, Iwaguro H, Yamaguchi JI, *et al.* (2001) Therapeutic potential of *ex vivo* expanded endothelial progenitor cells for myocardial ischemia. *Circulation* 103: 634–637.

16. Hill JM, Zalos G, Halcox JP, Schenke WH, *et al.* (2003) Circulating endothelial progenitor cells, vascular function, and cardiovascular risk. *N Engl J Med* 348: 593–600.

17. Rohde E, Malischnik C, Thaler D, Maierhofer T, *et al.* (2006) Blood monocytes mimic endothelial progenitor cells. *Stem Cells* 24: 357–367.

18. Schmeisser A, Garlichs CD, Zhang H, Eskafi S, *et al.* (2001) Monocytes coexpress endothelial and macrophagocytic lineage markers and form cord-like structures in Matrigel under angiogenic conditions. *Cardiovasc Res* 49: 671–680.

19. Elsheikh E, Uzunel M, He Z, Holgersson J, *et al.* (2005) Only a specific subset of human peripheral-blood monocytes has endothelial-like functional capacity. *Blood* 106: 2347–2355.

20. Bailey AS, Willenbring H, Jiang S, Anderson DA, *et al.* (2006) Myeloid lineage progenitors give rise to vascular endothelium. *Proc Natl Acad Sci USA* 103: 13156–13161.

21. Grunewald M, Avraham I, Dor Y, Bachar-Lustig E, *et al.* (2006) VEGF-induced adult neovascularization: recruitment, retention, and role of accessory cells. *Cell* 124: 175–189.

22. Jin DK, Shido K, Kopp HG, Petit I, *et al.* (2006) Cytokine-mediated deployment of SDF-1 induces revascularization through recruitment of CXCR4+ hemangiocytes. *Nat Med* 12: 557–567.

23. Kaplan RN, Riba RD, Zacharoulis S, Bramley AH, *et al.* (2005) VEGFR1-positive haematopoietic bone marrow progenitors initiate the pre-metastatic niche. *Nature* 438: 820–827.

24. Rehman J, Li J, Orschell CM, March KL (2003) Peripheral blood "endothelial progenitor cells" are derived from monocyte/ macrophages and secrete angiogenic growth factors. *Circulation* 107: 1164–1169.

25. Urbich C, Dimmeler S (2004) Endothelial progenitor cells: characterization and role in vascular biology. *Circ Res* 95: 343–353.
26. Romagnani P, Annunziato F, Liotta F, Lazzeri E, *et al.* (2005) CD14+CD34 low cells with stem cell phenotypic and functional features are the major source of circulating endothelial progenitors. *Circ Res* 97: 314–322.
27. Zhao Y, Glesne D, Huberman E (2003) A human peripheral blood monocyte-derived subset acts as pluripotent stem cells. *Proc Natl Acad Sci USA* 100: 2426–2431.
28. Lin Y, Weisdorf DJ, Solovey A, Hebbel RP (2000) Origins of circulating endothelial cells and endothelial outgrowth from blood. *J Clin Invest* 105: 71–77.
29. Yoder MC, Mead LE, Prater D, Krier TR, *et al.* (2007) Redefining endothelial progenitor cells via clonal analysis and hematopoietic stem/progenitor cell principals. *Blood* 109: 1801–1809.
30. Aicher A, Zeiher AM, Dimmeler S (2005) Mobilizing endothelial progenitor cells. *Hypertension* 45: 321–325.
31. Takahashi T, Kalka C, Masuda H, Chen D, *et al.* (1999) Ischemia- and cytokine-induced mobilization of bone marrow-derived endothelial progenitor cells for neovascularization. *Nat Med* 5: 434–438.
32. Ripa RS, Haack-Sorensen M, Wang Y, Jorgensen E, *et al.* (2007) Bone marrow derived mesenchymal cell mobilization by granulocyte-colony stimulating factor after acute myocardial infarction: results from the Stem Cells in Myocardial Infarction (STEMMI) trial. *Circulation* 116: I24–30.
33. Levesque JP, Hendy J, Takamatsu Y, Simmons PJ, *et al.* (2003) Disruption of the CXCR4/CXCL12 chemotactic interaction during hematopoietic stem cell mobilization induced by GCSF or cyclophosphamide. *J Clin Invest* 111: 187–196.
34. Moore MA, Hattori K, Heissig B, Shieh JH, *et al.* (2001) Mobilization of endothelial and hematopoietic stem and progenitor cells by adenovector-mediated elevation of serum levels of SDF-1, VEGF, and angiopoietin-1. *Ann N Y Acad Sci* 938: 36–45; discussion 45–37.
35. Pitchford SC, Furze RC, Jones CP, Wengner AM, *et al.* (2009) Differential mobilization of subsets of progenitor cells from the bone marrow. *Cell Stem Cell* 4: 62–72.

36. Heeschen C, Aicher A, Lehmann R, Fichtlscherer S, *et al.* (2003) Erythropoietin is a potent physiologic stimulus for endothelial progenitor cell mobilization. *Blood* 102: 1340–1346.
37. Westenbrink BD, Lipsic E, van der Meer P, van der Harst P, *et al.* (2007) Erythropoietin improves cardiac function through endothelial progenitor cell and vascular endothelial growth factor mediated neovascularization. *Eur Heart J* 28: 2018–2027.
38. George J, Goldstein E, Abashidze A, Wexler D, *et al.* (2005) Erythropoietin promotes endothelial progenitor cell proliferative and adhesive properties in a PI 3-kinase-dependent manner. *Cardiovasc Res* 68: 299–306.
39. Massa M, Rosti V, Ferrario M, Campanelli R, *et al.* (2005) Increased circulating hematopoietic and endothelial progenitor cells in the early phase of acute myocardial infarction. *Blood* 105: 199–206.
40. Mieno S, Ramlawi B, Boodhwani M, Clements RT, *et al.* (2006) Role of stromal-derived factor-1alpha in the induction of circulating CD34+CXCR4+ progenitor cells after cardiac surgery. *Circulation* 114: I186–192.
41. Roberts N, Xiao Q, Weir G, Xu Q, *et al.* (2007) Endothelial progenitor cells are mobilized after cardiac surgery. *Ann Thorac Surg* 83: 598–605.
42. Gill M, Dias S, Hattori K, Rivera ML, *et al.* (2001) Vascular trauma induces rapid but transient mobilization of VEGFR2(+)AC133(+) endothelial precursor cells. *Circ Res* 88: 167–174.
43. Schroder K, Kohnen A, Aicher A, Liehn EA, *et al.* (2009) NADPH oxidase Nox2 is required for hypoxia-induced mobilization of endothelial progenitor cells. *Circ Res* 105: 537–544.
44. Spiegel A, Shivtiel S, Kalinkovich A, Ludin A, *et al.* (2007) Catecholaminergic neurotransmitters regulate migration and repopulation of immature human CD34+ cells through Wnt signaling. *Nat Immunol* 8: 1123–1131.
45. Katayama Y, Battista M, Kao WM, Hidalgo A, *et al.* (2006) Signals from the sympathetic nervous system regulate hematopoietic stem cell egress from bone marrow. *Cell* 124: 407–421.
46. Dimmeler S, Burchfield J, Zeiher AM (2008) Cell-based therapy of myocardial infarction. *Arterioscler Thromb Vasc Biol* 28: 208–216.

47. Assmus B, Rolf A, Erbs S, Elsasser A, *et al.* (2009) Clinical outcome 2 years after intracoronary administration of bone marrow-derived progenitor cells in acute myocardial infarction. *Circ Heart Fail* 3: 89–96.

48. Assmus B, Schachinger V, Teupe C, Britten M, *et al.* (2002) Transplantation of progenitor cells and regeneration enhancement in acute myocardial infarction (TOPCARE-AMI). *Circulation* 106: 3009–3017.

49. Tendera M, Wojakowski W, Ruzyllo W, Chojnowska L, *et al.* (2009) Intracoronary infusion of bone marrow-derived selected CD34+ CXCR4+ cells and non-selected mononuclear cells in patients with acute STEMI and reduced left ventricular ejection fraction: results of randomized, multicentre Myocardial Regeneration by Intracoronary Infusion of Selected Population of Stem Cells in Acute Myocardial Infarction (REGENT) Trial. *Eur Heart J* 30: 1313–1321.

50. Assmus B, Honold J, Schachinger V, Britten MB, *et al.* (2006) Transcoronary transplantation of progenitor cells after myocardial infarction. *N Engl J Med* 355: 1222–1232.

51. Stamm C, Westphal B, Kleine HD, Petzsch M, *et al.* (2003) Autologous bone-marrow stem-cell transplantation for myocardial regeneration. *Lancet* 361: 45–46.

52. Losordo DW, Dimmeler S (2004) Therapeutic angiogenesis and vasculogenesis for ischemic disease: part II: cell-based therapies. *Circulation* 109: 2692–2697.

53. Losordo DW, Schatz RA, White CJ, Udelson JE, *et al.* (2007) Intramyocardial transplantation of autologous CD34+ stem cells for intractable angina: a phase I/IIa double-blind, randomized controlled trial. *Circulation* 115: 3165–3172.

54. van Ramshorst J, Bax JJ, Beeres SL, Dibbets-Schneider P, *et al.* (2009) Intramyocardial bone marrow cell injection for chronic myocardial ischemia: a randomized controlled trial. *JAMA* 301: 1997–2004.

55. Tateishi-Yuyama E, Matsubara H, Murohara T, Ikeda U, *et al.* (2002) Therapeutic angiogenesis for patients with limb ischaemia by autologous transplantation of bone-marrow cells: a pilot study and a randomised controlled trial. *Lancet* 360: 427–435.

56. Matoba S, Tatsumi T, Murohara T, Imaizumi T, *et al.* (2008) Long-term clinical outcome after intramuscular implantation of bone marrow

mononuclear cells (Therapeutic Angiogenesis by Cell Transplantation [TACT] trial) in patients with chronic limb ischemia. *Am Heart J* 156: 1010–1018.

57. Kawamoto A, Katayama M, Handa N, Kinoshita M, *et al.* (2009) Intramuscular transplantation of G-CSF-mobilized CD34(+) cells in patients with critical limb ischemia: a phase I/IIa, multicenter, single-blinded, dose-escalation clinical trial. *Stem Cells* 27: 2857–2864.

58. Aicher A, Heeschen C, Sasaki K, Urbich C, *et al.* (2006) Low-energy shock wave for enhancing recruitment of endothelial progenitor cells: a new modality to increase efficacy of cell therapy in chronic hind limb ischemia. *Circulation* 114: 2823–2830.

59. Zohlnhofer D, Dibra A, Koppara T, de Waha A, *et al.* (2008) Stem cell mobilization by granulocyte colony-stimulating factor for myocardial recovery after acute myocardial infarction: a meta-analysis. *J Am Coll Cardiol* 51: 1429–1437.

60. Zaruba MM, Theiss HD, Vallaster M, Mehl U, *et al.* (2009) Synergy between CD26/DPP-IV inhibition and G-CSF improves cardiac function after acute myocardial infarction. *Cell Stem Cell* 4: 313–323.

61. Friedenstein AJ, Piatetzky S, II, Petrakova KV (1966) Osteogenesis in transplants of bone marrow cells. *J Embryol Exp Morphol* 16: 381–390.

62. Horwitz EM, Le Blanc K, Dominici M, Mueller I, *et al.* (2005) Clarification of the nomenclature for MSC: The International Society for Cellular Therapy position statement. *Cytotherapy* 7: 393–395.

63. Aranguren XL, Luttun A, Clavel C, Moreno C, *et al.* (2007) *In vitro* and *in vivo* arterial differentiation of human multipotent adult progenitor cells. *Blood* 109: 2634–2642.

64. Kogler G, Sensken S, Airey JA, Trapp T, *et al.* (2004) A new human somatic stem cell from placental cord blood with intrinsic pluripotent differentiation potential. *J Exp Med* 200: 123–135.

65. Cossu G, Bianco P (2003) Mesoangioblasts — vascular progenitors for extravascular mesodermal tissues. *Curr Opin Genet Dev* 13: 537–542.

66. Zvaifler NJ, Marinova-Mutafchieva L, Adams G, Edwards CJ, *et al.* (2000) Mesenchymal precursor cells in the blood of normal individuals. *Arthritis Res* 2: 477–488.

67. Kuznetsov SA, Mankani MH, Gronthos S, Satomura K, *et al.* (2001) Circulating skeletal stem cells. *J Cell Biol* 153: 1133–1140.

68. Fernandez M, Simon V, Herrera G, Cao C, *et al.* (1997) Detection of stromal cells in peripheral blood progenitor cell collections from breast cancer patients. *Bone Marrow Transplant* 20: 265–271.

69. Lazarus HM, Haynesworth SE, Gerson SL, Caplan AI (1997) Human bone marrow-derived mesenchymal (stromal) progenitor cells (MPCs) cannot be recovered from peripheral blood progenitor cell collections. *J Hematother* 6: 447–455.

70. Wexler SA, Donaldson C, Denning-Kendall P, Rice C, *et al.* (2003) Adult bone marrow is a rich source of human mesenchymal 'stem' cells but umbilical cord and mobilized adult blood are not. *Br J Haematol* 121: 368–374.

71. Fukuda K, Fujita J (2005) Mesenchymal, but not hematopoietic, stem cells can be mobilized and differentiate into cardiomyocytes after myocardial infarction in mice. *Kidney Int* 68: 1940–1943.

72. Kawada H, Fujita J, Kinjo K, Matsuzaki Y, *et al.* (2004) Nonhematopoietic mesenchymal stem cells can be mobilized and differentiate into cardiomyocytes after myocardial infarction. *Blood* 104: 3581–3587.

73. Rochefort GY, Delorme B, Lopez A, Herault O, *et al.* (2006) Multipotential mesenchymal stem cells are mobilized into peripheral blood by hypoxia. *Stem Cells* 24: 2202–2208.

74. Ramirez M, Lucia A, Gomez-Gallego F, Esteve-Lanao J, *et al.* (2006) Mobilisation of mesenchymal cells into blood in response to skeletal muscle injury. *Br J Sports Med* 40: 719–722.

75. Makino S, Fukuda K, Miyoshi S, Konishi F, *et al.* (1999) Cardiomyocytes can be generated from marrow stromal cells *in vitro. J Clin Invest* 103: 697–705.

76. Prockop DJ (1997) Marrow stromal cells as stem cells for non-hematopoietic tissues. *Science* 276: 71–74.

77. Nagaya N, Fujii T, Iwase T, Ohgushi H, *et al.* (2004) Intravenous administration of mesenchymal stem cells improves cardiac function in rats with acute myocardial infarction through angiogenesis and myogenesis. *Am J Physiol Heart Circ Physiol* 287: H2670–2676.

78. Vulliet PR, Greeley M, Halloran SM, MacDonald KA, *et al.* (2004) Intra-coronary arterial injection of mesenchymal stromal cells and microinfarction in dogs. *Lancet* 363: 783–784.

79. Hare JM, Traverse JH, Henry TD, Dib N, *et al.* (2009) A randomized, double-blind, placebo-controlled, dose-escalation study of intravenous adult human mesenchymal stem cells (prochymal) after acute myocardial infarction. *J Am Coll Cardiol* 54: 2277–2286.

80. Zhang X, Wei M, Zhu W, Han B (2008) Combined transplantation of endothelial progenitor cells and mesenchymal stem cells into a rat model of isoproterenol-induced myocardial injury. *Arch Cardiovasc Dis* 101: 333–342.

IV. Repopulation of the Heart with New Cardiomyocytes

D. Pluripotent Stem Cells for Cardiac Repair

Kara White Moyes, Scott D. Lundy and Michael A. Laflamme

1. INTRODUCTION

The adult mammalian heart is one of the least regenerative organs, and the muscle lost to ischemic injury is typically replaced with non-contractile scar tissue. In recent years, cell-based therapies have attracted considerable excitement, as they represent a potential means by which the lost myocardium could be restored. Early proof-of-concept for this approach came from preclinical studies in which committed cardiomyocytes from fetal or neonatal sources were transplanted in rodent infarct models.[1–3] The transplanted cardiomyocytes formed stable intra-cardiac implants of new muscle and favorably affected indices of ventricular function such as ejection fraction. While the precise mechanisms underlying the beneficial effects of cardiomyocyte transplantation remain unclear, intravital imaging studies later confirmed that the grafts were activated during systole in synchrony with adjacent host myocardium.[4,5] Such observations raise hope that cardiomyogenic cell therapies can contribute new force-generating units to the infarcted heart, while perhaps also mediating some or all of the indirect, "paracrine" beneficial effects that have been attributed to the transplantation of non-myogenic cell types (see Refs. 6 and 7 and elsewhere in this publication).

Given their limited availability and lack of capacity for expansion, the fetal and neonatal cardiomyocytes employed in early preclinical studies appear impractical for clinical applications (see Chapter IV.A.). By contrast, human pluripotent stem cells represent an essentially inexhaustible source of committed but immature human cardiomyocytes. Pluripotent human embryonic stem cells (hESCs) have been available for just over a decade,[8] and during that time, the field has made significant advances toward the generation of homogeneous preparations of hESC-derived cardiomyocytes (hESC-CMs) for both basic science and transplantation studies. The more recently described induced pluripotent stem cells (iPSCs) have a phenotype that closely resembles that of ESCs, and guided differentiation approaches developed with ESCs are being rapidly applied to this new stem cell type. Because iPSCs can be derived by reprogramming somatic cells from the adult human, they avoid ethical concerns related to the destruction of human embryos and allow for the possibility of autologous cell transplantation, thereby circumventing the two principal objections that have been raised to hESC-based cell therapies. That said, the preclinical development of cardiac therapies utilizing either of these cell types remains at an early stage. Here, we discuss the derivation and culture of ESCs and iPSCs, approaches to the generation of cardiomyocytes and other useful differentiated cell types from both, as well as the preclinical experience and remaining challenges to their use in cardiac repair. Finally, although murine ESCs have been available for a much longer period, we now have substantial data from studies with *human* ESCs and iPSCs, so we will limit our discussion to the latter species whenever possible.

2. UNDIFFERENTIATED PLURIPOTENT STEM CELLS: DERIVATION, PHENOTYPE, AND CULTURE

2.1. Embryonic Stem Cells

ESCs are derived from the inner cell mass of preimplantation stage blastocysts, and they demonstrate the defining properties of pluripotency (i.e. potential to differentiate into cell types from all three embryonic germ layers) and unlimited capacity for self-renewal. Murine ESCs were first isolated in the early 1980s,[9,10] and their *in vitro* differentiation potential was later described in detail by Doetschman and co-workers.[11] Murine ESCs can be maintained in the undifferentiated state by two-dimensional culture on a feeder layer of mouse embryonic fibroblasts (MEFs). When this feeder layer is removed and the cells are switched to suspension culture in the presence of fetal calf serum, they spontaneously form embryoid bodies (EBs), three-dimensional aggregates that include tissue elements from all three embryonic germ layers (including mesoderm-derived cardiomyocytes). Of course, the pluripotency of murine ESCs is best illustrated by their capacity to contribute to multiple organs and tissues in chimeric mice, including germ cells.[12]

In 1998, Thomson and co-workers reported the successful derivation of hESCs,[8] work that built on their prior experience with non-human primate ESCs.[13] As with their murine counterparts, hESCs are derived from the inner cell mass of blastocysts; in this case obtained from donated human embryos that were generated for reproductive purposes but were no longer needed. The pluripotency of hESCs was rigorously confirmed by their capacity to form trilineage EBs *in vitro* and teratomas (benign tumors including tissue elements from all three embryonic germ layers) *in vivo* following injection into immunodeficient mice.[8] While the latter behavior represents a convincing proof of pluripotency, it also underscores a major problem associated with the transplantation of undifferentiated pluripotent stem cells — the risk of uncontrolled tumor formation.

Although there are important differences between ESCs from different species,[14] all ESCs have a number of phenotypic properties in common. Undifferentiated ESC cultures have a distinct morphological appearance, as they form compact colonies of cells with a large nuclear-to-cytoplasmic ratio and prominent nucleoli. ESCs express high levels of alkaline phosphatase, surface markers such as the stage-specific embryonic antigens, and transcription factors associated with self-renewal including Oct4, Sox2, Nanog, and Rex1.[8,15,16] ESCs also express high levels of telomerase,[8,17] a ribonucleic acid-protein enzymatic complex that adds hexameric repeats to the ends of chromosomal DNA (telomeres), thereby compensating for the loss that occurs during DNA replication. Telomerase is expressed by many neoplastic and immortalized cell types, and it is generally indicative of a high replicative capacity. Consistent with this, undifferentiated ESCs can be expanded indefinitely without undergoing senescence, while retaining the capacity to differentiate into somatic cell types that exhibit normal telomere shortening and senescence in culture.[18]

Undifferentiated hESCs have usually been maintained using techniques adapted from murine ESC culture, i.e. direct growth on a layer of irradiated MEFs, which release incompletely defined factors that support the proliferation and self-renewal of the hESCs. This approach is undesirable for cell transplantation for two reasons: 1) the feeder cells must be completely eliminated from the cells to be transplanted, and 2) it involves exposure to non-human cells, which raises concerns about xenopathogens. These considerations prompted the development of culture systems in which hESCs are grown in media conditioned by either MEFs[19] or human feeder alternatives.[20–22] Ultimately, the scaled production of hESCs for cell therapies would be greatly facilitated by the replacement of feeders and feeder-conditioned medium with recombinant or synthetic factors. Toward this end, the successful long-term culture of hESCs in chemically-defined media has been independently reported by multiple investigators.[23–26] Most of these culture systems involve media supplemented with high levels of basic fibroblast growth factor (bFGF), a critical molecule for hESC self-renewal.[26–29] Further refinements to hESC culture will likely result from ongoing small-molecule screens[30–32] and proteomic analyses of feeder-conditioned media.[33,34]

2.2. Induced Pluripotent Stem Cells

Somatic cells can be converted back to a pluripotent state by transferring their nuclei into oocytes[35,36] or following fusion with ESCs,[37] phenomena that indicated the latter two cell types have trans-activating cytoplasmic factors capable of mediating this "reprogramming." Yamanaka and co-workers built upon these observations and were the first to demonstrate directed reprogramming of cultured somatic cells via the forced expression of 24 ESC-associated factors.[38] To screen for successful reprogramming, they used transgenic MEFs in which β-galactosidase activity and neomycin resistance were driven by a promoter normally active only in ESCs. After transducing the MEFs with a panel of candidate reprogramming genes and then subjecting them to antibiotic selection, they observed the emergence of a β-galactosidase-positive, drug-resistant cell population with an ESC-like phenotype. The list of factors necessary but sufficient to convey this reprogramming activity was eventually reduced from 24 to just four factors: Oct4, Sox2, Klf4, and c-myc. The pluripotency of the resultant iPSCs was rigorously confirmed by multiple techniques, including trilineage differentiation within EBs, teratoma formation after transplantation in mice, and contribution to chimeric mice following injection into blastocysts.[38–41]

The discovery of iPSCs has produced tremendous excitement in the field, as iPSCs offer many or all of the advantages of ESCs while avoiding the ethical and political encumbrances associated with the latter cell type. Moreover, because iPSCs can be derived by reprogramming adult somatic cells, they also represent a potential source of patient-derived stem cells for use in disease modeling and autologous cell therapies. Given these enticements, it is unsurprising that the field has made extremely rapid progress since the Yamanaka group's original report.[38] Within 18 months following its publication, four groups independently reported the successful generation of human iPSCs by reprogramming human somatic cells via the overexpression of either Oct4, Sox2, Klf4, and c-Myc or Oct4, Sox2, Nanog, and LIN28.[42–45] Human iPSCs express typical hESC markers, show trilineage differentiation *in vitro*, and form teratomas after transplantation in immunodeficient mice. The field has also introduced several refinements that improve the efficiency of reprogramming and/or address safety concerns. For example, the oncogene c-Myc can be omitted from the reprogramming factors.[39] In fact, the list of required reprogramming factors has been successfully reduced to one (Oct-4) under certain conditions.[46] Moreover, while the reprogramming factors were initially delivered using integrating viral vectors, a number of presumably safer alternatives have been validated, including the use of membrane-permeant recombinant factors[47] and gene delivery via non-integrating adenoviral vectors,[48] plasmid transfection[49] or transposons.[50] Finally, the efficiency of reprogramming can be facilitated by small molecules,[51–53] which has raised hopes that gene-free, pharmacological approaches can be eventually developed.

3. PHENOTYPE OF CARDIOMYOCYTES DERIVED FROM HUMAN PLURIPOTENT STEM CELLS

Although invaluable information has come from work with murine ESC-CMs, we focus here on the phenotype and function of their human counterparts, as the latter are obviously more relevant to human infarct repair. While ESC-CMs from both species have many similarities, important differences in their rate of maturation and respective mechanisms of excitation-contraction coupling have been described elsewhere.[54–57]

hESC-CMs can be generated via spontaneous EB differentiation or using any of a number of guided differentiation approaches described below. Regardless of the method of derivation, hESC-CMs show an unambiguous, albeit immature cardiac phenotype. hESC-CMs express expected early cardiac-specific transcription factors including Nkx2.5, GATA4, myocyte enhancer factor 2C (MEF2c), Tbx-5, and Tbx-20.[54,58–62] They show spontaneous contractile activity and express sarcomeric proteins including α-actinin, cardiac troponins I and T, α- and β-myosin heavy chains (MHCs), atrial- and ventricular-myosin light chains (MLC2v and MLC2a), desmin, and tropomyosin.[54,60–64] They express connexins[60,63,65,66] and exhibit the corresponding intercellular electrical communication.[67–70]

By light microscopy, early hESC-CMs appear as small, fairly non-descript, mono-nucleated cells with variable spindled, rod-shaped, spherical, or triangular morphologies.[54,59,60,63] At the ultrastructural level, they show poorly organized sarcomeres and intercalated discs.[54,58,59] While prolonged duration in culture results in some degree of hypertrophy and improved sarcomeric organization and alignment,[59] no one has demonstrated their maturation to anything structurally resembling an adult ventricular myocyte. hESC-CMs show the robust proliferative activity characteristic of early chamber myocardium, with cell cycle activity slowly tapering off over several weeks of *in vitro* maturation.[58,66,71]

Recent studies suggest that the functional maturation of hESC-CMs *in vitro* may somewhat exceed their structural maturation.[56] Action potential (AP) recordings under current-clamp indicate their divergence into distinct nodal (pacemaker) and "working" (atrial and ventricular) cardiomyocytes.[63,64,72] Early hESC-CMs show immature AP properties (i.e. automaticity, a slow AP upstroke, and a depolarized maximum diastolic potential), but these improve somewhat with prolonged duration in culture.[64,73] In voltage-clamp studies, hESC-CMs exhibit most of the major cardiac ion currents, including fast sodium, L- and T-type calcium, pacemaker, as well as transient outward and inward rectifier potassium currents.[56,57,63,73] Depolarization in these cells is dominated by sodium influx via the $Na_V 1.5$ channel, and this current is at least partially responsible for their spontaneous electrical activity.[57] As in adult cardiomyocytes, depolarization activates L-type calcium channels in hESC-CMs, which results in a calcium influx that is amplified by release from sarcoplasmic reticulum calcium stores.[56,74,75] Recent data from our group indicates that this calcium-induced-calcium release process operates via a tight "local control" mechanism that is almost indistinguishable from that in adult myocardium.[56] hESC-CMs show the appropriate dose-dependent inotropic and chronotropic responses to β-adrenergic agonists, showing they have at least partial neurohormonal responsiveness.[54,60,64,76,77]

There has been less work to date examining the biomechanical properties of hESC-CMs. Using dynamic traction force microscopy, Mercola and co-workers estimated that individual hESC-CMs generated forces on the order of ~140 nN.[78] This value is in reasonable agreement with the estimated force produced by rat neonatal cardiomyocytes,[78] but it is only a small fraction of that reported for adult ventricular myocytes.[79] Interestingly, Hescheler and co-workers reported that hESC-CMs generated forces on the μN scale after co-culture on explanted sections of murine ventricle.[80] It is unclear whether the discrepancies in these two force measurements simply reflect the different modalities employed (dynamic traction force microscopy versus isometric force measurements by transducer) or if the hESC-CMs matured following electromechanical integration with adult myocardium.

Human iPSCs were described only recently,[42–45] so we have less information about the phenotype of hiPSC-derived cardiomyocytes (hiPSC-CMs). The early data suggests that they are quite similar to hESC-CMs, as would be expected (Figure 1). The Kamp group recently compared cardiomyocytes from hESCs and hiPSCs and found they had similar patterns of cardiac gene expression.[81] The two cell types also showed comparable electrophysiological properties, β-adrenergic responsiveness, and proliferative indices. Of note,

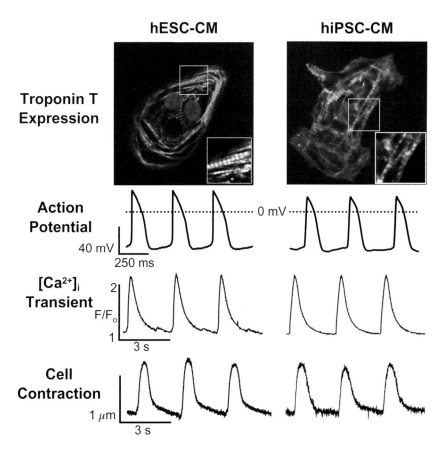

Figure 1.   The phenotype of hESC- and hiPSC-derived cardiomyocytes.
Both hESC- and hiPSC-derived cardiomyocytes (CMs) exhibit an unambiguous,
albeit immature cardiac phenotype. They express expected sarcomeric markers,
including troponin T, but show only patchy areas of sarcomeric organization and
alignment (inset). They also exhibit expected functional properties, including
spontaneous action potentials, calcium transients (measured using the fluorescent
indicator, fura-2), and contractile activity (measured using edge detection
videomicroscopy).

the hiPSC-CMs did show low-level residual expression of the reprogramming factors Oct-4 and Nanog, an observation that provides
further motivation for the development of reprogramming methods
that do not employ integrating viruses.

4. AVOIDING TUMOR FORMATION: APPROACHES TO DERIVING ENRICHED PREPARATIONS OF CARDIOMYOCYTES FROM PLURIPOTENT STEM CELLS

As noted above, one of the defining properties of pluripotent stem cells is their capacity to form teratomas after transplantation in an immunodeficient recipient. Although some investigators have suggested that the heart is a cardio-instructive environment that preferentially guides the differentiation of undifferentiated pluripotent stem cells into cardiomyocytes,[82–84] such observations have often not proven reproducible, and many investigators have demonstrated that the intra-cardiac transplantation of undifferentiated ESCs results in teratoma formation.[85–87] Hence, to overcome the risk of teratomas, the field must develop reliable methods for obtaining highly purified preparations of cardiomyocytes or restricted cardiovascular progenitors (Figure 2). In this section, we consider the currently available approaches for obtaining these cells, again emphasizing data from human pluripotent stem cells whenever possible.

4.1. Guided Differentiation into Cardiomyocytes

Most investigators have obtained cardiomyocytes from pluripotent stem cells using the methods first pioneered by Doetschman and co-workers with murine ESCs, i.e. differentiation within EBs in the presence of high concentrations of fetal calf serum.[11,88] Unfortunately, differentiation is relatively uncontrolled by this approach, and the cardiac purity of the resultant cell preparations is typically very low. In the case of hESCs, less than 10% of human EBs show spontaneous beating activity,[54] and cardiomyocytes are typically <1% of the total differentiated cell population.[78,89,90] The cardiac purity of such preparations can be somewhat improved by the mechanical dissection of spontaneously beating areas (to ~30%–70% cardiomyocytes[54,91]), but this method is tedious and not scalable to clinical applications.

The Mummery group was among the first to apply knowledge from embryology toward the development of a guided cardiac differentiation

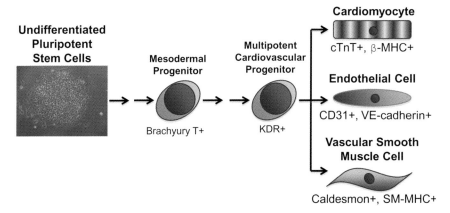

Figure 2.   The differentiation of pluripotent stem cells into useful cell types for cardiac repair. Pluripotent stem cells are highly expandable in the undifferentiated state, in which they grow as compact colonies of cells with a high nuclear/cytoplasmic ratio. As they differentiate into the cardiovascular lineages, there are intermediate progenitors that show increasingly restricted differentiation potential and characteristic patterns of marker expression. For example, KDR⁺ multipotent cardiovascular progenitors can differentiate into functional cardiomyocytes, endothelial cells, and vascular smooth muscle cells.[113,135] β-MHC: beta-myosin heavy chain, cTnT: cardiac troponin T, KDR: kinase insert domain receptor, SM-MHC: smooth muscle-myosin heavy chain, VE-Cadherin: vascular endothelial cadherin.

protocol for hESCs. Their approach, which attempts to recapitulate the cardio-inductive effects of anterior endoderm,[92–94] involves the co-culture of hESCs with END-2 cells, a visceral endoderm-like derivative of P19 cells.[63] When this co-culture was performed in the absence of serum, they obtained a yield of approximately one cardiomyocyte per 20 starting undifferentiated hESCs.[95] Similar results were obtained using END-2 cell conditioned medium, indicating that soluble factors mediated the cardiogenic activity. The efficacy of the conditioned medium was further improved by the elimination of insulin[96] and/or supplementation with a small molecule inhibitor of p38 MAP kinase,[97] which resulted in preparation of ~5%–10% cardiomyocytes without further enrichment. More recently, Davidson and co-workers attempted to identify the cardiogenic factors in END-2 conditioned medium by comparing the transcriptional profiles of END-2 cells and a related but non-cardiogenic cell type.[98] They found two enzymes involved in the synthesis of prostaglandin I2 were strongly expressed in the END-2 cells and that END-2 conditioned medium contained correspondingly high levels of the prostaglandin. When they treated hESCs with serum-and insulin-free medium supplemented with the p38 MAP kinase inhibitor and prostaglandin I2, this chemically-defined medium had cardio-inductive activity comparable to that of END-2 cell conditioned medium and resulted in preparations of ~10% cardiomyocytes (Figure 3A).[98]

Our group has recently reported a protocol for the guided differentiation of hESCs into cardiomyocytes that also involves defined factors (Figure 3B). Our protocol involves two members of the transforming growth factor-β (TGFβ) superfamily of signaling molecules, activin A and bone morphogenetic protein-4 (BMP-4).[89] Both factors were known to exert critical but time-dependent effects during cardiogenesis. Activation of activin/Nodal signaling promotes the induction of mesendoderm in the epiblast in model organisms[99–101] and ESC cultures.[102–106] Interestingly, this inductive effect is suppressed by insulin/phosphatidylinositol 3-kinase signaling,[107] which may relate to the previously discussed anti-cardiogenic effects of insulin on cells differentiated in the presence of END-2 cell conditioned medium.[96] At a slightly later point in development, the anterior endoderm and nearby ectodermal tissues release BMP-2 and BMP-4, signals that induce cardiogenesis in the precardiac mesoderm.[108,109] Our protocol attempts to recapitulate this sequence of signaling events by serially treating monolayer cultures of hESCs with activin and BMP-4 in serum-free medium. The growth factors are then removed, and the cells are maintained in serum-free medium in the absence of exogenous factors for an additional two to three weeks. This procedure typically yields preparations of 30%–60% cardiomyocytes.[89,110] Our directed differentiation protocol also works with hiPSCs.[42]

The Keller group has reported a third protocol (Figure 3C) that involves similar manipulation of TGFβ family signaling but also

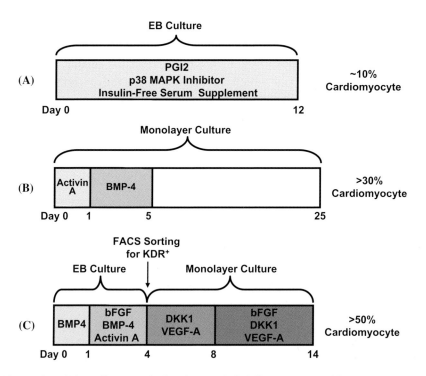

Figure 3. Selected protocols for the guided differentiation of hESCs into cardiomyocytes using chemically-defined factors. The estimated cardiac purity resulting from each protocol is indicated to the right. (A) Davidson and colleagues have described a protocol in which EBs cultured continuously in serum- and insulin-free medium that is supplemented with prostaglandin I2 and an inhibitor of p38 MAP kinase.[98] (B) Our group has reported a differentiation strategy in which hESCs are grown in monolayer culture, are serially pulsed with activin A and BMP-4 in serum-free medium, and then are grown in the absence of exogenous factors.[89] (C) The Keller group has reported a protocol that involves both EB and monolayer culture.[113] In brief, EBs are treated with BMP-4, bFGF, and activin A. KDR⁺ multipotent progenitors are then selected and replated as a monolayer. Thereafter, Dkk1 is added to prevent the inhibitory actions of Wnt signaling on cardiogenesis, VEGF to promote expansion of the KDR⁺ population, and bFGF to support the growth of the cardiovascular lineages.

incorporates information about the biphasic effects of Wnt signaling during cardiac development. Canonical Wnt signaling is required for early mesoderm induction, but it inhibits the cardiac induction of precardiac mesoderm at later timepoints.[111,112] Therefore, after differentiating hESCs into a primitive streak-like population by treatment with activin A, BMP-4, and bFGF, they apply the Wnt antagonist dickkopf homolog 1 (DKK1) to permit cardiac specification.[113] This protocol, which is discussed in more detail below, reportedly generates populations of ~40%–50% cardiomyocytes.

4.2. Enrichment Strategies

Of course, none of the preceding guided differentiation protocols result in a homogenous population of cardiomyocytes. It is not even clear whether guided differentiation to homogeneity is even possible, given the important function of non-cardiac tissues (e.g. anterior endoderm) in cardiac induction and patterning. With this in mind, the field continues to develop methods to enrich for cardiomyocytes.

The highest levels of cardiac purity have been obtained by genetic selection, whereby cardiomyocytes are isolated based on the expression of an antibiotic resistance gene or fluorescent reporter under the control of a cardiac- or muscle-specific promoter. This strategy was first pioneered by the Field group, who stably transfected murine ESCs with a transgene in which the cardiac-specific α-myosin heavy chain promoter drives expression of the aminoglycoside phosphotransferase gene (i.e. neomyocin resistance, which confers resistance to the antibiotic G418 in mammalian cells). EBs were formed using these transgenic mESCs, and then the differentiated EB outgrowths were subjected to antibiotic selection, which resulted in preparations of 99.6% pure ESC-CMs. This selection strategy was later scaled-up to purify large numbers of cardiomyocytes in bioreactors.[114,115] Similar genetic selection strategies have been used to generate preparations of >90% pure hESC-CMs, using transgenes in which reporter expression was driven by the human α-MHC,[78,116] troponin,[117] or MLC2v promoters.[118] Moreover, the Denning group obtained preparations of ~33% pure hESC-CMs using a negative selection strategy based on constitutive expression of the herpes simplex virus thymidine kinase/ganciclovir (HSVtk/GCV) suicide gene.[116] The latter preferentially kills rapidly proliferating cells, so the authors correctly reasoned that it would favor the elimination of the more rapidly cycling non-cardiac cell population.

However, for clinical applications, enrichment strategies that do not involve genetically modified cells would be preferable. One such approach is fluorescence- or magnetic activated cell sorting (FACS or MACS) based on a suitable cell surface marker. hESC-CMs have been sorted from mixed EB cultures using an antibody against CD166/ALCAM,[119] but there are a number of other potential cardiomyocyte surface markers that to our knowledge remain untested (e.g. CD56/NCAM[120]).

Enrichment strategies based on the unique physical properties of cardiomyocytes have also been pursued. For example, Doevendans and co-workers enzymatically dispersed differentiated cells from murine EBs and then used centrifugation on a discontinuous Percoll gradient to fractionate the cells by density.[121] The cardiomyocyte-enriched fraction was ~50% pure. This strategy was subsequently adapted to hESCs,[60] and in our hands, it results in a three- to seven-fold enrichment when applied to enzymatically dispersed hEBs[60,90] or hESC cultures differentiated with activin A and BMP-4.[89] Finally, Li and co-workers recently reported that non-invasive confocal Raman microspectroscopy was 66% accurate in identifying hESC-CMs within admixed cultures.[122] It remains to be seen whether this intriguing technique can be improved and scaled-up to permit the separation of large quantities of cardiomyocytes.

4.3. Useful Non-Myocyte Cell Types from Human Pluripotent Stem Cells

To overcome the risk of teratoma formation, the field must develop methods to eliminate undifferentiated pluripotent stem cells and undesirable differentiated cell types from preparations to be used in cell-based therapies. However, this does not mean that it is necessary or even desirable to remove *all* non-cardiac cell types. After all, while cardiomyocytes account for the vast majority of myocardial mass, they represent only approximately one-third of the adult heart by cell number.[123] The remaining two-thirds are comprised primarily of endothelial cells, smooth muscle cells, and fibroblasts. Correspondingly, if our goal is true regeneration of myocardial tissue rather than mere cardiomyocyte repopulation, it stands to reason that we may need to deliberately include these non-myocyte cell types (see also Chapter VII.). Indeed, early tissue engineering experience indicates that the organization, survival, and function of constructs seeded with a combination of hESC-CMs, endothelial and mesenchymal cells greatly exceed those of constructs seeded with hESC-CMs alone.[68,124] While most such studies to date have employed endothelial and mesenchymal cells from non-stem cell sources, one of the attractive features of human pluripotent stem cells is that they represent a single source for both cardiomyocytes and appropriate non-myocyte cell types. Indeed, efficient guided differentiation protocols have been reported for the generation of both endothelial cells[125–128] and smooth muscle-like cells.[125,128–130]

Moreover, multipotent cardiovascular progenitor cells are present during the intermediate stages of ESC differentiation, and these progenitors represent an attractive alternative to using a separate guided differentiation protocol for each cell type. Most of these progenitors have been identified using a genetic labeling strategy in which the promoter for a lineage-specific transcription factor (e.g. Nkx2.5,[131] Isl1,[132,133] or Mesp1[134]) drives expression of a fluorescent protein. Depending on the specific label and time point examined, the fluorescently-tagged progenitor can exhibit a variable capacity to differentiate into two or more of the four principal cell types present in myocardium (i.e. cardiomyocytes, endothelial cells, smooth muscle cells, and fibroblasts). However, because the progenitors identified by genetic labeling carry the intrinsic risks of gene therapy, they are obviously more appropriate for basic science studies rather than preclinical development.

By contrast, the Keller group has described an ESC-derived multipotent cardiovascular progenitor that can be identified based on its expression of an endogenous cell surface marker, KDR (also known as Flk1$^+$ or vascular endothelial growth factor receptor-2 (VEGFR-2)). In differentiating mESC cultures, the authors found two temporally distinct KDR$^+$ populations: an early hemangioblast population (i.e.

hematopoietic and endothelial progenitors) and a later cardiovascular progenitor population.[135,136] When the latter was isolated using the methylcellulose colony assay, it showed unambiguous capacity to differentiate into cardiomyocytes, endothelial cells, and vascular smooth muscle cells. A functionally equivalent tripotent progenitor can be identified in both developing mouse hearts[135] and hESC cultures differentiated using the previously discussed protocol from the Keller group.[113] In the human, the analogous progenitor population is characterized by low-level KDR expression and lack of expression of the stem cell marker c-Kit.

5. OTHER CHALLENGES TO THE DEVELOPMENT OF CARDIAC THERAPIES BASED ON PLURIPOTENT STEM CELLS

Even if the risk of teratoma formation can be eliminated, there are a number of other major hurdles to the development of safe, efficacious cardiac therapies based on pluripotent stem cells. Space limitations preclude a detailed discussion of all of these, but we consider here the three most important: immune rejection, graft cell death, and the risk of graft-related arrhythmias. We also describe potential approaches by which they each could be overcome.

5.1. Immune Rejection

Undifferentiated ESCs are immune-privileged,[137] but their immunogenicity increases with differentiation,[138] and intra-cardiac implants of ESC-derivatives are recognized and rejected by allogeneic recipients.[86,87,139] Hence, while hESC-CMs are less immunogenic than their adult counterparts,[140] their allotransplantation would likely necessitate some degree of pharmacological immunosuppression. Immunosuppressive drugs are currently employed in end-stage heart failure patients for whom the benefits of organ transplantation exceed the risks of immunosuppression, but the size of this patient population is small when compared to all those who could benefit from cell-based therapies. Moreover, while the levels of immunosuppressive therapy could likely be minimized by maintaining a bank of diverse HLA-type hESC lines, one analysis concluded that a bank with as many as 150 blood group compatible stem cell lines would be required to ensure a 58%–80% chance of an acceptable, albeit imperfect match in a typical patient population.[141] It is hard to imagine that an "off-the-shelf" hESC-CM therapy could ever be practical if it had to be maintained in ≥150 lines.

On the other hand, a number of alternative strategies to overcome immune rejection of hESC derivatives have been proposed, including the generation of autologous hESCs by somatic cell nuclear transfer (SCNT), hESCs homozygous for common HLA haplotypes by SCNT or parthenogenesis, or "universal donor" hESCs via genetic modification. (For a comprehensive review of these approaches, see Refs. 142 and 143.) It is also possible that immune tolerance to hESC derivatives could be induced by transplanting hESC-derived hematopoietic progenitors and establishing bone marrow chimerism. However, while all of the preceding approaches remain either conceptual or at an early stage of development, iPSCs represent a currently available source of potentially autologous pluripotent stem cells. It remains to be seen whether patient-derived iPSCs can be generated in a scalable, high-throughput fashion, but the rapid progress in this area gives reason for optimism.

5.2. Graft Cell Death

Even when immune rejection can be avoided, the vast majority of graft cells die shortly after intra-cardiac transplantation, a phenomenon that has been demonstrated with multiple candidate cell types for cell-based cardiac repair,[144–146] including hESC-CMs.[90,147] The causes of cell death following transplantation into infarcted myocardium are multi-factorial and include anoikis (cell death due to loss of attachment factors), ischemia, and inflammation. In the case of hESC-CM transplantation, graft cell death is at least partially counter-balanced by cardiomyocyte proliferation.[89–91] Moreover, our group has identified several interventions that improve the survival of hESC-CM grafts, including heat-shock[90] and erythropoietin[147] treatment of the hESC-CMs prior to transplantation, as well as delivery in extra-cellular matrix supplemented with a "cocktail" of anti-death factors.[89] Still, further progress will likely be required if these cells are ever to fully remuscularize a typical human infarct, which spans centimeters and involves the loss of approximately one billion cardiomyocytes.[148] In our opinion, tissue engineering approaches (e.g. gel-based carrier and scaffold systems seeded with hESC-derivatives[124,149,150]) offer the best hope for success. Such constructs can provide helpful anti-death and pro-angiogenic factors, while also enhancing structural organization and integration with host tissues (see Chapter VI.B.).

5.3. Electromechanical Integration and Arrhythmogenesis

Electromechanical coupling between graft and host cardiomyocytes is required for the graft cells to contract in synchrony with host myocardium and provide coordinated force generation. Rubart and co-workers have used elegant intravital imaging approaches to demonstrate that fetal cardiomyocytes do couple with host fibers following transplantation in intact and experimentally infarcted hearts.[4] One of the attractive features about cardiomyocytes derived from human pluripotent stem cells is that they also possess the cellular machinery (e.g. gap junctions) needed to undergo electromechanical integration. hESC-CMs have been shown to couple with co-cultured rat neonatal cardiomyocytes[68,69] and infarcted myocardium when co-cultured with an *ex vivo* slice preparation.[80] We also have strong evidence of their capacity for electromechanical integration following transplantation in uninfarcted recipients. The Gepstein group[68] transplanted hESC-CMs into the left ventricular apex of immuno-suppressed pigs in which the atrioventricular (AV) node had been previously ablated. Electrocardiography indicated ectopic pacemaker activity in the hESC-CM recipients that was not observed in control animals. The ectopic focus mapped to the site of cell injection, and points of host-graft contact with shared gap junctions were later found in these areas. Li and co-workers later demonstrated similar phenomena following transplantation in an AV node-ablated guinea pig heart model.[69] While these two studies prove hESC-CMs can undergo electromechanical integration *in vivo*, they involved uninfarcted recipients and short periods of follow-up (two weeks and three days post transplantation, respectively[68,69]). It remains to be seen whether these cells can undergo long-term integration in infarcted hearts, given the presence of insulating fibrosis and phenotypic changes in the border zone.

These issues are important, not only because host-graft coupling is required for the graft to contribute new systolic force, but also because they have major implications on the risk of graft-related arrhythmias. Concerns about arrhythmias apply to all cell-based cardiac therapies, but especially to those involving electrically excitable cell types, such as hESC- and hiPSC-CMs. Indeed, the latter could plausibly contribute to all three fundamental arrhythmia mechanisms: automaticity, triggered activity, and reentry. First, all hESC- and hiPSC-CMs show automaticity (spontaneous electrical activity),[64,81] and their capacity to function as an ectopic pacemaker was demonstrated by the aforementioned transplantation studies in AV node-ablated animals.[68,69] Moreover, both hESC- and hiPSC-CMs include a population of nodal cells,[63,64,72,81] which would presumably show sustained pacemaking activity following transplantation. Second, murine ESC-CMs seem to be particularly prone to early- and after-depolarizations,[151] triggered activity that is thought to underlie many episodes of ventricular tachycardia. Third, their transplantation may promote reentrant phenomena by increasing electrical heterogeneity in the recipient heart. hESC-CMs themselves are electrically heterogeneous, and they form irregularly shaped grafts that are generally isolated by scar tissue.[89,90] hESC-CMs also tend to be depolarized relative to adult ventricular cardiomyocytes.

While the preceding observations heighten concerns about the risk of graft-related arrhythmias, a recent study by Roell and co-workers[152] invites speculation that the transplantation of myogenic cell types might have the opposite effect and *reduce* the incidence of arrhythmias, a leading cause of death post infarction in humans. These investigators transplanted primary murine embryonic cardiomyocytes in a mouse cardiac cryoinjury model and observed a markedly decreased susceptibility to ventricular tachycardias in the cardiomyocyte recipients. Interestingly, this anti-arrythmogenic effect correlated with the expression of connexin-43 but not with the restoration of mechanical function.[152] While the mechanisms underlying this phenomenon remain undetermined, it invites speculation that the transplantation of cardiomyocytes from pluripotent stem cells could exert a similar anti-arrhythmic effect. In any event, because the latter cells could plausibly either increase or decrease the susceptibility to arrhythmias, transplantation studies with electrocardiographic endpoints are clearly warranted. These would be ideally performed in a large animal model with electrophysiological properties similar to humans.

6. PRECLINICAL EXPERIENCE WITH CARDIAC REPAIR USING PLURIPOTENT STEM CELLS

Although work in larger, more relevant animal models is reportedly commencing,[153] nearly all of the published studies examining the use of pluripotent stem cells, for infarct repair have been conducted in rodents. Early proof-of-concept for the use of pluripotent stem cells came from the Field group, who transplanted highly purified preparations of genetically-selected murine ESC-CMs into the uninfarcted hearts of adult dystrophic mice and demonstrated the formation of stable intra-cardiac grafts by anti-dystrophin immunostaining.[154] This early success motivated other investigators to transplant murine ESC-CMs in infarcted rodent hearts, in which the cells formed nascent myocardium and mediated beneficial effects on contractile function.[82,85,155–160]

In recent years, several laboratories have examined the capacity of hESC-CMs to mediate cardiac repair, so we will focus our discussion hereafter on this work with human cells (summarized in Table 1). Taken collectively, these studies support three conclusions: 1) hESC-CMs will engraft and form small amounts of nascent human myocardium, 2) if the input cell preparation is sufficiently homogenous, teratoma formation can likely be avoided, and 3) hESC-CM transplantation mediates beneficial effects on contractile function, at

Table 1. Preclinical Studies Examining the Transplantation of hESCs or hESC-Derived Cardiomyocytes in Rodent Infarct Models.

Study (Ref.)	Recipient	Cell Type (% CM)	Cell Dose (×10⁶)	Duration of Follow-Up	Major Findings
Cao et al.[169]	SCID-beige Mouse	hESC-CMs (~40%)	1		• Used Percoll-enriched hESC-CMs from EBs • Preserved FS in hESC-CM recipients by echo
Caspi et al.[91]	Rat (+ CyA)	hESC-CMs (71%)	1.5	≤ 9 wks	• Used hESC-CMs microdissected from spontaneously differentiating EBs • Preserved LV dimensions and FS in hESC-CM recipients by echo • No teratomas
Dai et al.[65]	Nude Rat	hESC-CMs (13%)	2	4 wks	• Used hESC-CMs generated by treating EBs with endoderm-conditioned medium • No teratomas (majority of the surviving graft was comprised of cardiomyocytes)
Kofidis et al.[170]	SCID-beige Mouse	hESC-CMs (15%)	1	3 wks	• Used Percoll-enriched hESC-CMs from EBs • Treatment of recipient with allopurinol and uricase enhanced hESC-CM graft survival • Preserved EF in hESC-CM recipients by MRI

(Continued)

Table 1. (*Continued*)

Study (Ref.)	Recipient	Cell Type (% CM)	Cell Dose (×10⁶)	Duration of Follow-Up	Major Findings
Laflamme et al.[89]	Nude Rat	hESC-CMs (83%)	10	4 wks	• Used Percoll-enriched hESC-CMs generated by directed differentiation with activin A and BMP-4 • Found that reliable engraftment required delivery in a pro-survival vehicle • No teratomas (surviving graft was >99% cardiomyocyte) • Preserved LV dimension and FS in hESC-CM recipients by echo; partially preserved regional wall motion by MRI
Leor et al.[171]	Nude Rat	hESC-CMs (ND)	0.5–1	2–4 wks	• Used hESC-CMs microdissected from spontaneously differentiating EBs • Observed limited cell engraftment, but found teratomas in 1 of 7 recipients • Improved FS and wall thinning in hESC-CM recipients, but these effects did not correlate with histological graft outcomes
Tomescot et al.[172]	Rat (+ CyA)	BMP-2-treated hESCs (ND)	3	8 wks	• Used hESCs treated for 48 hours with BMP-2 ± FGF inhibitor • Limited cell engraftment following transplantation in infarcted hearts by histology

(*Continued*)

Table 1.   (Continued)

Study (Ref.)	Recipient	Cell Type (% CM)	Cell Dose ($\times 10^6$)	Duration of Follow-Up	Major Findings
van Laake et al.[161,162,173]	NOD-SCID Mouse	hESC-CMs (~20%)	1–3	≤12 wks	• Used hESC-CMs generated by treating EBs with endoderm-conditioned medium • No teratomas (surviving graft was 80%–95% cardiomyocyte) • Preserved FS in hESC-CM recipients by MRI at 4 wks, but this effect was not sustained to the 12-wk timepoint • Beneficial functional effects at 4 wks correlated with vascularity but not graft size
Yang et al.[113]	NOD-SCID Mouse	KDR+ cardiac progenitors (~40%–50%)	0.1	2–11 wks	• Used KDR^{low}/c-Kit^{neg} trilineage cardiac progenitors generated by directed differentiation with activin A, BMP-4, FGF-2, VEGF, and DKK1 • Surviving graft included cardiac, endothelial, and vascular smooth muscle elements • Preserved FS in KDR+ progenitor recipients by MRI

least at early endpoints. To our knowledge, no one has yet reported the successful transplantation of hiPSC-CMs, but such experiments are undoubtedly in progress.

Our group has demonstrated that transplanted hESC-CMs can partially remuscularize infarcted rat hearts and mediate beneficial effects on regional and global cardiac function.[89] In this study, nude rats were infarcted by 60 minutes of ischemia followed by reperfusion. hESC-CMs of ~80% cardiomyocyte purity were generated by guided differentiation with activin/BMP-4 (Figure 3B), followed by enrichment by centrifugation on discontinuous Percoll gradients. At four days post infarction, 10×10^6 of these hESC-CMs were directly injected into the infarct zone via a left thoracotomy. Although engraftment was dismal when these cells were delivered in a saline vehicle, their survival was greatly enhanced when co-delivered with a pro-survival cocktail (PSC) composed of extracellular matrix (Matrigel) supplemented with anti-oncotic and anti-apoptotic factors. In the hearts receiving hESC-CMs in PSC, a small but significant portion of the infarct volume (up to 11%) was occupied by nascent human myocardium, as demonstrated by combining immunohistochemistry for cardiac markers and *in situ* hybridization with a human-specific pan-centromeric probe. No teratomas were observed and, in fact, >99% of the surviving graft cells were cardiomyocytes. Cardiac function was assessed at four weeks post transplantation by echocardiography and magnetic resonance imaging (MRI). Compared to controls receiving saline vehicle, PSC alone, or non-cardiac derivatives in PSC, the recipients of hESC-CMs in PSC showed significantly improved left ventricular dimensions, regional wall thickening, and global contractile function.

Qualitatively similar results have been independently reported by the Gepstein[91] and Mummery[161–163] groups, based on their transplantation studies in immunosuppressed rats and immunodeficient (NOD/SCID) mice, respectively. Of note, neither group found delivery in PSC necessary to ensure hESC-CM survival, but it appears likely that their grafts were substantially smaller in size than those we obtained by cell transplantation in PSC.

At seven to ten days post infarction, the Gepstein group transplanted 1.5×10^6 hESC-CMs of ~70% cardiomyocyte purity, obtained by microdissecting spontaneously beating EBs.[91] The recipient rat hearts were then evaluated at four- and eight-weeks post transplantation using histological and echocardiographic endpoints. As in our own study, no teratomas were found, and the graft consisted of small islands of human myocardium within the infarct and border zones. Moreover, the recipients of hESC-CMs showed greater preservation of left ventricular dimensions and global cardiac function than did controls as late as eight weeks post transplantation.

Recent hESC-CM transplantation studies by the Mummery group have included the longest duration of follow-up (up to 12 weeks).[161–163] These investigators used endoderm-conditioned medium to generate preparations of ~20% cardiomyocyte purity, which were then transplanted at doses of $1–3 \times 10^6$ cells into acutely infarcted mouse hearts. Interestingly, while the cardiac purity of their cell preparations was relatively low, the non-cardiac cells were gradually cleared from recipient hearts, such that the grafts were composed nearly exclusively of cardiomyocytes at later timepoints. We have observed a similar "clean-up" phenomenon when transplanting preparations of low cardiac purity into normal[90] but not infarcted[89] rat hearts. However, hESC-CM transplantation mediated only a transient beneficial effect on cardiac function in their model. In particular, while the recipients of hESC-CMs showed better cardiac function by MRI than did controls at four weeks post transplantation, no benefit was observed at 12 weeks post transplantation. While important differences in experimental design should be emphasized between Mummery's study and our own (e.g. mouse versus rat, acute versus subacute infarct, saline versus PSC, graft size, etc.), their findings are nonetheless troubling, especially in light of clinical trials suggesting that bone marrow-derived cells mediate similarly transient effects.[164] Long-term xenotransplantation experiments are challenging to perform, but the Mummery group's data underscore the need for additional preclinical testing with longer periods of follow-up.

7. SUMMARY AND CONCLUDING REMARKS

Pluripotent stem cells have a number of attractive properties for cell-based cardiac repair, including tremendous expandability in the undifferentiated state and the capacity to differentiate into multiple cell types relevant to cardiac repair. Cardiomyocytes have been historically generated from pluripotent stem cells using uncontrolled, EB-based approaches that are poorly cardiogenic, but more recently developed guided differentiation and enrichment protocols can produce large numbers of highly purified cardiomyocytes. This distinguishes ESCs and iPSCs from many adult cell types for whom the capacity to differentiate into significant numbers of unambiguous cardiomyocytes has been questioned.[165,166]

However, there are many unresolved issues that must be addressed before pluripotent stem cells can reach serious consideration for use in clinical cell-based therapies. While early preclinical transplantation studies in rodent models have been encouraging, the safety and efficacy of pluripotent stem cell-based therapies must be demonstrated in larger, more relevant animal models. Transplantation studies with longer endpoints should also be pursued, although these may be better modeled using cells from the same species as the recipient (e.g. porcine cardiomyocytes into infarcted pigs). Moreover, while the prospect of bypassing ethical objections and the need for immunosuppression with iPSC-based cardiac therapies is undeniably attractive, these cells are quite new, and more work is required to demonstrate their safety, phenotypic stability, and maturation potential. Finally, we need additional mechanistic information regarding the beneficial effects that have been observed in hESC-CM transplantation studies to date.

Given these uncertainties, we predict that cardiomyocytes from pluripotent stem cells will first find widespread use in a variety of *in vitro* applications. ESC-derived cardiomyocytes and cardiovascular progenitors have already been used as a model system for studying early human heart development. The availability of a renewable source of human cardiomyocytes has also caught the attention of the pharmaceutical industry, which is considering their use to screen novel compounds for efficacy, arrhythmogenic potential or other cardiotoxic effects.[167,168] We look forward to studies with iPSC-derived cardiomyocytes from patients with hereditable cardiac diseases (e.g. genetic cardiomyopathies, channelopathies, etc.), as these may provide long-sought connections between genotype and phenotype. More generally, the possibility of investigating patho-physiology and novel therapeutic approaches in patient-derived cardiomyocytes may bring us closer to the long-promised era of personalized medicine.

REFERENCES

1. Leor J, Patterson M, Quinones MJ, Kedes LH, *et al.* (1996) Transplantation of fetal myocardial tissue into the infarcted myocardium of rat. A potential method for repair of infarcted myocardium? *Circulation* 94: II332–336.
2. Li RK, Jia ZQ, Weisel RD, Mickle DA, *et al.* (1996) Cardiomyocyte transplantation improves heart function. *Ann Thorac Surg* 62: 654–660; discussion 660–651.
3. Scorsin M, Hagege AA, Marotte F, Mirochnik N, *et al.* (1997) Does transplantation of cardiomyocytes improve function of infarcted myocardium? *Circulation* 96: II-188–193.
4. Rubart M, Pasumarthi KB, Nakajima H, Soonpaa MH, *et al.* (2003) Physiological coupling of donor and host cardiomyocytes after cellu-lar transplantation. *Circ Res* 92: 1217–1224.
5. Zimmermann WH, Melnychenko I, Wasmeier G, Didie M, *et al.* (2006) Engineered heart tissue grafts improve systolic and diastolic function in infarcted rat hearts. *Nat Med* 12: 452–458.
6. Gnecchi M, Zhang Z, Ni A, Dzau VJ (2008) Paracrine mechanisms in adult stem cell signaling and therapy. *Circ Res* 103: 1204–1219.
7. Laflamme MA, Zbinden S, Epstein SE, Murry CE (2007) Cell-based therapy for myocardial ischemia and infarction: pathophysiological mechanisms. *Annu Rev Pathol* 2: 307–339.
8. Thomson JA, Itskovitz-Eldor J, Shapiro SS, Waknitz MA, *et al.* (1998) Embryonic stem cell lines derived from human blastocysts. *Science* 282: 1145–1147.

9. Evans MJ, Kaufman MH (1981) Establishment in culture of pluripotential cells from mouse embryos. *Nature* 292: 154–156.

10. Martin GR (1981) Isolation of a pluripotent cell line from early mouse embryos cultured in medium conditioned by teratocarcinoma stem cells. *Proc Natl Acad Sci USA* 78: 7634–7638.

11. Doetschman TC, Eistetter H, Katz M, Schmidt W, *et al.* (1985) The *in vitro* development of blastocyst-derived embryonic stem cell lines: formation of visceral yolk sac, blood islands and myocardium. *J Embryol Exp Morphol* 87: 27–45.

12. Bradley A, Evans M, Kaufman MH, Robertson E (1984) Formation of germ-line chimaeras from embryo-derived teratocarcinoma cell lines. *Nature* 309: 255–256.

13. Thomson JA, Marshall VS (1998) Primate embryonic stem cells. *Curr Top Dev Biol* 38: 133–165.

14. Ginis I, Luo Y, Miura T, Thies S, *et al.* (2004) Differences between human and mouse embryonic stem cells. *Dev Biol* 269: 360–380.

15. Ling V, Neben S (1997) *In vitro* differentiation of embryonic stem cells: immunophenotypic analysis of cultured embryoid bodies. *J Cell Physiol* 171: 104–115.

16. Kim J, Chu J, Shen X, Wang J, *et al.* (2008) An extended transcriptional network for pluripotency of embryonic stem cells. *Cell* 132: 1049–1061.

17. Sharma HW, Sokoloski JA, Perez JR, Maltese JY, *et al.* (1995) Differentiation of immortal cells inhibits telomerase activity. *Proc Natl Acad Sci USA* 92: 12343–12346.

18. Zeng X, Rao MS (2007) Human embryonic stem cells: long term stability, absence of senescence and a potential cell source for neural replacement. *Neuroscience* 145: 1348–1358.

19. Xu C, Inokuma MS, Denham J, Golds K, *et al.* (2001) Feeder-free growth of undifferentiated human embryonic stem cells. *Nat Biotechnol* 19: 971–974.

20. Inzunza J, Gertow K, Stromberg MA, Matilainen E, *et al.* (2005) Derivation of human embryonic stem cell lines in serum replacement medium using postnatal human fibroblasts as feeder cells. *Stem Cells* 23: 544–549.

21. Yoo SJ, Yoon BS, Kim JM, Song JM, *et al.* (2005) Efficient culture system for human embryonic stem cells using autologous human embryonic stem cell-derived feeder cells. *Exp Mol Med* 37: 399–407.

22. Xu C, Jiang J, Sottile V, McWhir J, *et al.* (2004) Immortalized fibroblast-like cells derived from human embryonic stem cells support undifferentiated cell growth. *Stem Cells* 22: 972–980.

23. Li Y, Powell S, Brunette E, Lebkowski J, *et al.* (2005) Expansion of human embryonic stem cells in defined serum-free medium devoid of animal-derived products. *Biotechnol Bioeng* 91: 688–698.

24. Lu J, Hou R, Booth CJ, Yang SH, *et al.* (2006) Defined culture conditions of human embryonic stem cells. *Proc Natl Acad Sci USA* 103: 5688–5693.

25. Ludwig TE, Levenstein ME, Jones JM, Berggren WT, *et al.* (2006) Derivation of human embryonic stem cells in defined conditions. *Nat Biotechnol* 24: 185–187.

26. Xu C, Rosler E, Jiang J, Lebkowski JS, *et al.* (2005) Basic fibroblast growth factor supports undifferentiated human embryonic stem cell growth without conditioned medium. *Stem Cells* 23: 315–323.

27. Bendall SC, Stewart MH, Menendez P, George D, *et al.* (2007) IGF and FGF cooperatively establish the regulatory stem cell niche of pluripotent human cells *in vitro*. *Nature* 448: 1015–1021.

28. Levenstein ME, Ludwig TE, Xu RH, Llanas RA, *et al.* (2006) Basic fibroblast growth factor support of human embryonic stem cell self-renewal. *Stem Cells* 24: 568–574.

29. Xu R-H, Peck RM, Li DS, Feng X, *et al.* (2005) Basic FGF and suppression of BMP signaling sustain undifferentiated proliferation of human ES cells. *Nature Methods* 3: 185–190.

30. Ware CB, Wang L, Mecham BH, Shen L, *et al.* (2009) Histone deacetylase inhibition elicits an evolutionarily conserved self-renewal program in embryonic stem cells. *Cell Stem Cell* 4: 359–369.

31. Desbordes SC, Placantonakis DG, Ciro A, Socci ND, *et al.* (2008) High-throughput screening assay for the identification of compounds regulating self-renewal and differentiation in human embryonic stem cells. *Cell Stem Cell* 2: 602–612.

32. Miyabayashi T, Teo JL, Yamamoto M, McMillan M, *et al.* (2007) Wnt/beta-catenin/CBP signaling maintains long-term murine

embryonic stem cell pluripotency. *Proc Natl Acad Sci USA* 104: 5668–5673.

33. Prowse AB, McQuade LR, Bryant KJ, Marcal H, *et al.* (2007) Identification of potential pluripotency determinants for human embryonic stem cells following proteomic analysis of human and mouse fibroblast conditioned media. *J Proteome Res* 6: 3796–3807.

34. Chin AC, Fong WJ, Goh LT, Philp R, *et al.* (2007) Identification of proteins from feeder conditioned medium that support human embryonic stem cells. *J Biotechnol* 130: 320–328.

35. Gurdon JB, Uehlinger V (1966) "Fertile" intestine nuclei. *Nature* 210: 1240–1241.

36. Wilmut I, Schnieke AE, McWhir J, Kind AJ, *et al.* (1997) Viable offspring derived from fetal and adult mammalian cells. *Nature* 385: 810–813.

37. Cowan CA, Atienza J, Melton DA, Eggan K (2005) Nuclear reprogramming of somatic cells after fusion with human embryonic stem cells. *Science* 309: 1369–1373.

38. Takahashi K, Yamanaka S (2006) Induction of pluripotent stem cells from mouse embryonic and adult fibroblast cultures by defined factors. *Cell* 126: 663–676.

39. Nakagawa M, Koyanagi M, Tanabe K, Takahashi K, *et al.* (2008) Generation of induced pluripotent stem cells without Myc from mouse and human fibroblasts. *Nat Biotechnol* 26: 101–106.

40. Okita K, Ichisaka T, Yamanaka S (2007) Generation of germline-competent induced pluripotent stem cells. *Nature* 448: 313–317.

41. Wernig M, Meissner A, Foreman R, Brambrink T, *et al.* (2007) *In vitro* reprogramming of fibroblasts into a pluripotent ES-cell-like state. *Nature* 448: 318–324.

42. Takahashi K, Tanabe K, Ohnuki M, Narita M, *et al.* (2007) Induction of pluripotent stem cells from adult human fibroblasts by defined factors. *Cell* 131: 861–872.

43. Yu J, Vodyanik MA, Smuga-Otto K, Antosiewicz-Bourget J, *et al.* (2007) Induced pluripotent stem cell lines derived from human somatic cells. *Science* 318: 1917–1920.

44. Park IH, Zhao R, West JA, Yabuuchi A, *et al.* (2007) Reprogramming of human somatic cells to pluripotency with defined factors. *Nature* 451: 141–146.

45. Lowry WE, Richter L, Yachechko R, Pyle AD, *et al.* (2008) Generation of human induced pluripotent stem cells from dermal fibroblasts. *Proc Natl Acad Sci USA* 105: 2883–2888.

46. Kim JB, Sebastiano V, Wu G, Arauzo-Bravo MJ, *et al.* (2009) Oct4-induced pluripotency in adult neural stem cells. *Cell* 136: 411–419.

47. Zhou H, Wu S, Joo JY, Zhu S, *et al.* (2009) Generation of induced pluripotent stem cells using recombinant proteins. *Cell Stem Cell* 4: 381–384.

48. Stadtfeld M, Nagaya M, Utikal J, Weir G, *et al.* (2008) Induced pluripotent stem cells generated without viral integration. *Science* 322: 945–949.

49. Okita K, Nakagawa M, Hyenjong H, Ichisaka T, *et al.* (2008) Generation of mouse induced pluripotent stem cells without viral vectors. *Science* 322: 949–953.

50. Woltjen K, Michael IP, Mohseni P, Desai R, *et al.* (2009) piggyBac transposition reprograms fibroblasts to induced pluripotent stem cells. *Nature* 458: 766–770.

51. Huangfu D, Maehr R, Guo W, Eijkelenboom A, *et al.* (2008) Induction of pluripotent stem cells by defined factors is greatly improved by small-molecule compounds. *Nat Biotechnol* 26: 795–797.

52. Huangfu D, Osafune K, Maehr R, Guo W, *et al.* (2008) Induction of pluripotent stem cells from primary human fibroblasts with only Oct4 and Sox2. *Nat Biotechnol* 26: 1269–1275.

53. Shi Y, Desponts C, Do JT, Hahm HS, *et al.* (2008) Induction of pluripotent stem cells from mouse embryonic fibroblasts by Oct4 and Klf4 with small-molecule compounds. *Cell Stem Cell* 3: 568–574.

54. Kehat I, Kenyagin-Karsenti D, Snir M, Segev H, *et al.* (2001) Human embryonic stem cells can differentiate into myocytes with structural and functional properties of cardiomyocytes. *J Clin Invest* 108: 407–414.

55. Itzhaki I, Schiller J, Beyar R, Satin J, *et al.* (2006) Calcium handling in embryonic stem cell-derived cardiac myocytes: of mice and men. *Ann N Y Acad Sci* 1080: 207–215.

56. Zhu WZ, Santana LF, Laflamme MA (2009) Local control of excitation-contraction coupling in human embryonic stem cell-derived cardiomyocytes. *PLoS ONE* 4: e5407.

57. Satin J, Kehat I, Caspi O, Huber I, *et al.* (2004) Mechanism of spontaneous excitability in human embryonic stem cell derived cardiomyocytes. *J Physiol* 559: 479–496.

58. Norstrom A, Akesson K, Hardarson T, Hamberger L, *et al.* (2006) Molecular and pharmacological properties of human embryonic stem cell-derived cardiomyocytes. *Exp Biol Med* 231: 1753–1762.

59. Snir M, Kehat I, Gepstein A, Coleman R, *et al.* (2003) Assessment of the ultrastructural and proliferative properties of human embryonic stem cell-derived cardiomyocytes. *Am J Physiol Heart Circ Physiol* 285: H2355–2363.

60. Xu C, Police S, Rao N, Carpenter MK (2002) Characterization and enrichment of cardiomyocytes derived from human embryonic stem cells. *Circ Res* 91: 501–508.

61. Lev S, Kehat I, Gepstein L (2005) Differentiation pathways in human embryonic stem cell-derived cardiomyocytes. *Ann N Y Acad Sci* 1047: 50–65.

62. Xu C, He JQ, Kamp TJ, Police S, *et al.* (2006) Human embryonic stem cell-derived cardiomyocytes can be maintained in defined medium without serum. *Stem Cells Dev* 15: 931–941.

63. Mummery C, Ward-van Oostwaard D, Doevendans P, Spijker R, *et al.* (2003) Differentiation of human embryonic stem cells to cardiomyocytes: role of coculture with visceral endoderm-like cells. *Circulation* 107: 2733–2740.

64. Bracken AP, Pasini D, Capra M, Prosperini E, *et al.* (2003) EZH2 is downstream of the pRB-E2F pathway, essential for proliferation and amplified in cancer. *EMBO J* 22: 5323–5335.

65. Dai W, Field LJ, Rubart M, Reuter S, *et al.* (2007) Survival and maturation of human embryonic stem cell-derived cardiomyocytes in rat hearts. *J Mol Cell Cardiol* 43: 504–516.

66. Cui L, Johkura K, Takei S, Ogiwara N, *et al.* (2007) Structural differentiation, proliferation, and association of human embryonic stem cell-derived cardiomyocytes *in vitro* and in their extracardiac tissues. *J Struct Biol* 158: 307–317.

67. Kehat I, Gepstein A, Spira A, Itskovitz-Eldor J, *et al.* (2002) High-resolution electrophysiological assessment of human embryonic stem cell-derived cardiomyocytes: a novel *in vitro* model for the study of conduction. *Circ Res* 91: 659–661.

68. Kehat I, Khimovich L, Caspi O, Gepstein A, *et al.* (2004) Electromechanical integration of cardiomyocytes derived from human embryonic stem cells. *Nat Biotechnol* 22: 1282–1289.

69. Xue T, Cho HC, Akar FG, Tsang SY, *et al.* (2005) Functional integration of electrically active cardiac derivatives from genetically engineered human embryonic stem cells with quiescent recipient ventricular cardiomyocytes: insights into the development of cell-based pacemakers. *Circulation* 111: 11–20.

70. Kehat I, Gepstein A, Spira A, Itskovitz-Eldor J, *et al.* (2002) High-resolution electrophysiological assessment of human embryonic stem cell-derived cardiomyocytes: a novel *in vitro* model for the study of conduction. *Circulation Research* 91: 659–661.

71. McDevitt TC, Laflamme MA, Murry CE (2005) Proliferation of cardiomyocytes derived from human embryonic stem cells is mediated via the IGF/PI 3-kinase/Akt signaling pathway. *J Mol Cell Cardiol* 39: 865–873.

72. Moore JC, Fu J, Chan YC, Lin D, *et al.* (2008) Distinct cardiogenic preferences of two human embryonic stem cell (hESC) lines are imprinted in their proteomes in the pluripotent state. *Biochem Biophys Res Commun* 372: 553–558.

73. Sartiani L, Bettiol E, Stillitano F, Mugelli A, *et al.* (2007) Developmental changes in cardiomyocytes differentiated from human embryonic stem cells: a molecular and electrophysiological approach. *Stem Cells* 25: 1136–1144.

74. Liu J, Fu JD, Siu CW, Li RA (2007) Functional sarcoplasmic reticulum for calcium-handling of human embryonic stem cell-derived cardiomyocytes: insights for driven maturation. *Stem Cells* 25:3038–3044.

75. Satin J, Itzhaki I, Rapoport S, Schroder EA, *et al.* (2008) Calcium handling in human embryonic stem cell-derived cardiomyocytes. *Stem Cells* 26: 1961–1972.

76. Friedrich G, Soriano P (1991) Promoter traps in embryonic stem cells: a genetic screen to identify and mutate developmental genes in mice. *Genes Dev* 5: 1513–1523.

77. Yokoo N, Baba S, Kaichi S, Niwa A, *et al.* (2009) The effects of cardioactive drugs on cardiomyocytes derived from human induced pluripotent stem cells. *Biochem Biophys Res Commun* 387: 482–488.

78. Kita-Matsuo H, Barcova M, Prigozhina N, Salomonis N, *et al.* (2009) Lentiviral vectors and protocols for creation of stable hESC lines for fluorescent tracking and drug resistance selection of cardiomyocytes. *PLoS One* 4: e5046.
79. Bluhm WF, McCulloch AD, Lew WY (1995) Active force in rabbit ventricular myocytes. *J Biomech* 28: 1119–1122.
80. Pillekamp F, Reppel M, Rubenchyk O, Pfannkuche K, *et al.* (2007) Force measurements of human embryonic stem cell-derived cardiomyocytes in an *in vitro* transplantation model. *Stem Cells* 25: 174–180.
81. Zhang J, Wilson GF, Soerens AG, Koonce CH, *et al.* (2009) Functional cardiomyocytes derived from human induced pluripotent stem cells. *Circ Res* 104: e30–41.
82. Hodgson DM, Behfar A, Zingman LV, Kane GC, *et al.* (2004) Stable benefit of embryonic stem cell therapy in myocardial infarction. *Am J Physiol Heart Circ Physiol* 287: H471–479.
83. Behfar A, Zingman LV, Hodgson DM, Rauzier JM, *et al.* (2002) Stem cell differentiation requires a paracrine pathway in the heart. *FASEB J* 16: 1558–1566.
84. Nelson TJ, Martinez-Fernandez A, Yamada S, Perez-Terzic C, *et al.* (2009) Repair of acute myocardial infarction by human stemness factors induced pluripotent stem cells. *Circulation* 120: 408–416.
85. Kolossov E, Bostani T, Roell W, Breitbach M, *et al.* (2006) Engraftment of engineered ES cell-derived cardiomyocytes but not BM cells restores contractile function to the infarcted myocardium. *J Exp Med* 203: 2315–2327.
86. Nussbaum J, Minami E, Laflamme MA, Virag JA, *et al.* (2007) Transplantation of undifferentiated murine embryonic stem cells in the heart: teratoma formation and immune response. *FASEB J* 21: 1345–1357.
87. Swijnenburg RJ, Tanaka M, Vogel H, Baker J, *et al.* (2005) Embryonic stem cell immunogenicity increases upon differentiation after transplantation into ischemic myocardium. *Circulation* 112: 1166–172.
88. Robbins J, Gulick J, Sanchez A, Howles P, *et al.* (1990) Mouse embryonic stem cells express the cardiac myosin heavy chain genes during development *in vitro*. *J Biol Chem* 265: 11905–11909.

89. Laflamme MA, Chen KY, Naumova AV, Muskheli V, *et al.* (2007) Cardiomyocytes derived from human embryonic stem cells in pro-survival factors enhance function of infarcted rat hearts. *Nat Biotechnol* 25: 1015–1024.

90. Laflamme MA, Gold J, Xu C, Hassanipour M, *et al.* (2005) Formation of human myocardium in the rat heart from human embryonic stem cells. *Am J Pathol* 167: 663–671.

91. Caspi O, Huber I, Kehat I, Habib M, *et al.* (2007) Transplantation of human embryonic stem cell-derived cardiomyocytes improves myocardial performance in infarcted rat hearts. *J Am Coll Cardiol* 50: 1884–1893.

92. Fullilove SL (1970) Heart induction: distribution of active factors in newt endoderm. *J Exp Zool* 175: 323–326.

93. Schultheiss TM, Xydas S, Lassar AB (1995) Induction of avian cardiac myogenesis by anterior endoderm. *Development* 121: 4203–4214.

94. Sugi Y, Lough J (1994) Anterior endoderm is a specific effector of terminal cardiac myocyte differentiation of cells from the embryonic heart forming region. *Dev Dyn* 200: 155–162.

95. Passier R, Oostwaard DW, Snapper J, Kloots J, *et al.* (2005) Increased cardiomyocyte differentiation from human embryonic stem cells in serum-free cultures. *Stem Cells* 23: 772–780.

96. Freund C, Ward-van Oostwaard D, Monshouwer-Kloots J, van den Brink S, *et al.* (2008) Insulin redirects differentiation from cardiogenic mesoderm and endoderm to neuroectoderm in differentiating human embryonic stem cells. *Stem Cells* 26: 724–733.

97. Graichen R, Xu X, Braam SR, Balakrishnan T, *et al.* (2008) Enhanced cardiomyogenesis of human embryonic stem cells by a small molecular inhibitor of p38 MAPK. *Differentiation* 76: 357–370.

98. Xu XQ, Graichen R, Soo SY, Balakrishnan T, *et al.* (2008) Chemically defined medium supporting cardiomyocyte differentiation of human embryonic stem cells. *Differentiation* 76: 958–970.

99. Yatskievych TA, Ladd AN, Antin PB (1997) Induction of cardiac myogenesis in avian pregastrula epiblast: the role of the hypoblast and activin. *Development* 124: 2561–2570.

100. Conlon FL, Lyons KM, Takaesu N, Barth KS, *et al.* (1994) A primary requirement for nodal in the formation and maintenance of the primitive streak in the mouse. *Development* 120: 1919–1928.

101. Smith JC, Price BM, Van Nimmen K, Huylebroeck D (1990) Identification of a potent Xenopus mesoderm-inducing factor as a homologue of activin A. *Nature* 345: 729–731.
102. D'Amour KA, Agulnick AD, Eliazer S, Kelly OG, *et al.* (2005) Efficient differentiation of human embryonic stem cells to definitive endoderm. *Nat Biotechnol* 23: 1534–1541.
103. Vallier L, Reynolds D, Pedersen RA (2004) Nodal inhibits differentiation of human embryonic stem cells along the neuroectodermal default pathway. *Dev Biol* 275: 403–421.
104. Vallier L, Touboul T, Chng Z, Brimpari M, *et al.* (2009) Early cell fate decisions of human embryonic stem cells and mouse epiblast stem cells are controlled by the same signalling pathways. *PLoS One* 4: e6082.
105. Golob JL, Paige SL, Muskheli V, Pabon L, *et al.* (2008) Chromatin remodeling during mouse and human embryonic stem cell differentiation. *Dev Dyn* 237: 1389–1398.
106. Tada S, Era T, Furusawa C, Sakurai H, *et al.* (2005) Characterization of mesendoderm: a diverging point of the definitive endoderm and mesoderm in embryonic stem cell differentiation culture. *Development* 132: 4363–4374.
107. McLean AB, D'Amour KA, Jones KL, Krishnamoorthy M, *et al.* (2007) Activin a efficiently specifies definitive endoderm from human embryonic stem cells only when phosphatidylinositol 3-kinase signaling is suppressed. *Stem Cells* 25: 29–38.
108. Schultheiss TM, Burch JB, Lassar AB (1997) A role for bone morphogenetic proteins in the induction of cardiac myogenesis. *Genes Dev* 11: 451–462.
109. Ladd AN, Yatskievych TA, Antin PB (1998) Regulation of avian cardiac myogenesis by activin/TGFbeta and bone morphogenetic proteins. *Dev Biol* 204: 407–419.
110. Zhu WZ, Hauch KD, Xu C, Laflamme MA (2009) Human embryonic stem cells and cardiac repair. *Transplant Rev* 23: 53–68.
111. Naito AT, Shiojima I, Akazawa H, Hidaka K, *et al.* (2006) Developmental stage-specific biphasic roles of Wnt/beta-catenin signaling in cardiomyogenesis and hematopoiesis. *Proc Natl Acad Sci USA* 103: 19812–19817.
112. Ueno S, Weidinger G, Osugi T, Kohn AD, *et al.* (2007) Biphasic role for Wnt/beta-catenin signaling in cardiac specification in

zebrafish and embryonic stem cells. *Proc Natl Acad Sci USA* 104: 9685–9690.

113. Yang L, Soonpaa MH, Adler ED, Roepke TK, *et al.* (2008) Human cardiovascular progenitor cells develop from a KDR(+) embryonic-stem-cell-derived population. *Nature* 453: 524–528.

114. Zandstra PW, Bauwens C, Yin T, Liu Q, *et al.* (2003) Scalable production of embryonic stem cell-derived cardiomyocytes. *Tissue Eng* 9: 767–778.

115. Zweigerdt R, Burg M, Willbold E, Abts H, *et al.* (2003) Generation of confluent cardiomyocyte monolayers derived from embryonic stem cells in suspension: a cell source for new therapies and screening strategies. *Cytotherapy* 5: 399–413.

116. Anderson D, Self T, Mellor IR, Goh G, *et al.* (2007) Transgenic enrichment of cardiomyocytes from human embryonic stem cells. *Mol Ther* 15: 2027–2036.

117. Gallo P, Grimaldi S, Latronico MV, Bonci D, *et al.* (2008) A lentiviral vector with a short troponin-I promoter for tracking cardiomyocyte differentiation of human embryonic stem cells. *Gene Ther* 15: 161–170.

118. Huber I, Itzhaki I, Caspi O, Arbel G, *et al.* (2007) Identification and selection of cardiomyocytes during human embryonic stem cell differentiation. *FASEB J* 21: 2551–2563.

119. Rust W, Balakrishnan T, Zweigerdt R (2009) Cardiomyocyte enrichment from human embryonic stem cell cultures by selection of ALCAM surface expression. *Regen Med* 4: 225–237.

120. Blau HM, Webster C, Pavlath GK (1990) Purification and proliferation of human myoblasts isolated with fluorescence activated cell sorting. *Adv Exp Med Biol* 280: 97–100.

121. Doevendans PA, Kubalak SW, An RH, Becker DK, *et al.* (2000) Differentiation of cardiomyocytes in floating embryoid bodies is comparable to fetal cardiomyocytes. *J Mol Cell Cardiol* 32: 839–851.

122. Chan JW, Lieu DK, Huser T, Li RA (2009) Label-free separation of human embryonic stem cells and their cardiac derivatives using Raman spectroscopy. *Anal Chem* 81: 1324–1331.

123. Weber KT, Brilla CG (1991) Pathological hypertrophy and cardiac interstitium. Fibrosis and renin-angiotensin-aldosterone system. *Circulation* 83: 1849–1865.

124. Caspi O, Lesman A, Basevitch Y, Gepstein A, *et al.* (2007) Tissue engineering of vascularized cardiac muscle from human embryonic stem cells. *Circ Res* 100: 263–272.

125. Ferreira LS, Gerecht S, Shieh HF, Watson N, *et al.* (2007) Vascular progenitor cells isolated from human embryonic stem cells give rise to endothelial and smooth muscle like cells and form vascular networks *in vivo. Circ Res* 101: 286–294.

126. Levenberg S, Golub JS, Amit M, Itskovitz-Eldor J, *et al.* (2002) Endothelial cells derived from human embryonic stem cells. *Proc Natl Acad Sci USA* 99: 4391–4396.

127. Goldman O, Feraud O, Boyer-Di Ponio J, Driancourt C, *et al.* (2009) A boost of BMP4 accelerates the commitment of human embryonic stem cells to the endothelial lineage. *Stem Cells* 27: 1750–1759.

128. Sone M, Itoh H, Yamahara K, Yamashita JK, *et al.* (2007) Pathway for differentiation of human embryonic stem cells to vascular cell components and their potential for vascular regeneration. *Arterioscler Thromb Vasc Biol* 27: 2127–2134.

129. Huang H, Zhao X, Chen L, Xu C, *et al.* (2006) Differentiation of human embryonic stem cells into smooth muscle cells in adherent monolayer culture. *Biochem Biophys Res Commun* 351: 321–327.

130. Xie CQ, Zhang J, Villacorta L, Cui T, *et al.* (2007) A highly efficient method to differentiate smooth muscle cells from human embryonic stem cells. *Arterioscler Thromb Vasc Biol* 27: e311–312.

131. Wu SM, Fujiwara Y, Cibulsky SM, Clapham DE, *et al.* (2006) Developmental origin of a bipotential myocardial and smooth muscle cell precursor in the mammalian heart. *Cell* 127: 1137–1150.

132. Moretti A, Caron L, Nakano A, Lam JT, *et al.* (2006) Multipotent embryonic isl1+ progenitor cells lead to cardiac, smooth muscle, and endothelial cell diversification. *Cell* 127: 1151–1165.

133. Bu L, Jiang X, Martin-Puig S, Caron L, *et al.* (2009) Human ISL1 heart progenitors generate diverse multipotent cardiovascular cell lineages. *Nature* 460: 113–117.

134. Bondue A, Lapouge G, Paulissen C, Semeraro C, *et al.* (2008) Mesp1 acts as a master regulator of multipotent cardiovascular progenitor specification. *Cell Stem Cell* 3: 69–84.

135. Kattman SJ, Huber TL, Keller GM (2006) Multipotent flk-1+ cardio-vascular progenitor cells give rise to the cardiomyocyte, endothelial, and vascular smooth muscle lineages. *Dev Cell* 11: 723–732.
136. Kattman SJ, Adler ED, Keller GM (2007) Specification of multipotential cardiovascular progenitor cells during embryonic stem cell differentiation and embryonic development. *Trends Cardiovasc Med* 17: 240–246.
137. Li L, Baroja ML, Majumdar A, Chadwick K, *et al.* (2004) Human embryonic stem cells possess immune-privileged properties. *Stem Cells* 22: 448–456.
138. Drukker M, Katz G, Urbach A, Schuldiner M, *et al.* (2002) Characterization of the expression of MHC proteins in human embryonic stem cells. *Proc Natl Acad Sci USA* 99: 9864–9869.
139. Swijnenburg RJ, Schrepfer S, Govaert JA, Cao F, *et al.* (2008) Immunosuppressive therapy mitigates immunological rejection of human embryonic stem cell xenografts. *Proc Natl Acad Sci USA* 105: 12991–12996.
140. Drukker M, Katchman H, Katz G, Even-Tov Friedman S, *et al.* (2006) Human embryonic stem cells and their differentiated derivatives are less susceptible to immune rejection than adult cells. *Stem Cells* 24: 221–229.
141. Taylor CJ, Bolton EM, Pocock S, Sharples LD, *et al.* (2005) Banking on human embryonic stem cells: estimating the number of donor cell lines needed for HLA matching. *Lancet* 366: 2019–2025.
142. Odorico JS, Kaufman DS, Thomson JA (2001) Multilineage differentiation from human embryonic stem cell lines. *Stem Cells* 19: 193–204.
143. Drukker M (2008) Recent advancements towards the derivation of immune-compatible patient-specific human embryonic stem cell lines. *Semin Immunol* 20: 123–129.
144. Zhang M, Methot D, Poppa V, Fujio Y, *et al.* (2001) Cardiomyocyte grafting for cardiac repair: graft cell death and anti-death strategies. *J Mol Cell Cardiol* 33: 907–921.
145. Suzuki K, Smolenski RT, Jayakumar J, Murtuza B, *et al.* (2000) Heat shock treatment enhances graft cell survival in skeletal myoblast transplantation to the heart. *Circulation* 102: III216–221.

146. Muller-Ehmsen J, Whittaker P, Kloner RA, Dow JS, *et al.* (2002) Survival and development of neonatal rat cardiomyocytes transplanted into adult myocardium. *J Mol Cell Cardiol* 34: 107–116.

147. Robey TE, Saiget MK, Reinecke H, Murry CE (2008) Systems approaches to preventing transplanted cell death in cardiac repair. *J Mol Cell Cardiol* 45: 567–581.

148. Beltrami CA, Finato N, Rocco M, Feruglio GA, *et al.* (1994) Structural basis of end-stage failure in ischemic cardiomyopathy in humans. *Circulation* 89: 151–163.

149. Lesman A, Habib M, Caspi O, Gepstein A, *et al.* (2009) Transplantation of a tissue-engineered human vascularized cardiac muscle. *Tissue Eng Part A* 16(1): 115–125.

150. Shapira-Schweitzer K, Habib M, Gepstein L, Seliktar D (2009) A photopolymerizable hydrogel for 3-D culture of human embryonic stem cell-derived cardiomyocytes and rat neonatal cardiac cells. *J Mol Cell Cardiol* 46: 213–224.

151. Zhang YM, Hartzell C, Narlow M, Dudley SC, Jr. (2002) Stem cell-derived cardiomyocytes demonstrate arrhythmic potential. *Circulation* 106: 1294–1299.

152. Roell W, Lewalter T, Sasse P, Tallini YN, *et al.* (2007) Engraftment of connexin 43-expressing cells prevents post-infarct arrhythmia. *Nature* 450: 819–824.

153. Zweigerdt R (2007) The art of cobbling a running pump — will human embryonic stem cells mend broken hearts? *Semin Cell Dev Biol* 18: 794–804.

154. Klug MG, Soonpaa MH, Koh GY, Field LJ (1996) Genetically selected cardiomyocytes from differentiating embronic stem cells form stable intracardiac grafts. *J Clin Invest* 98: 216–224.

155. Min JY, Yang Y, Converso KL, Liu L, *et al.* (2002) Transplantation of embryonic stem cells improves cardiac function in postinfarcted rats. *J Appl Physiol* 92: 288–296.

156. Min JY, Yang Y, Sullivan MF, Ke Q, *et al.* (2003) Long-term improvement of cardiac function in rats after infarction by transplantation of embryonic stem cells. *J Thorac Cardiovasc Surg* 125: 361–369.

157. Menard C, Hagege AA, Agbulut O, Barro M, *et al.* (2005) Transplantation of cardiac-committed mouse embryonic stem cells to

infarcted sheep myocardium: a preclinical study. *Lancet* 366: 1005–1012.

158. Naito H, Nishizaki K, Yoshikawa M, Yamada T, *et al.* (2004) Xenogeneic embryonic stem cell-derived cardiomyocyte transplantation. *Transplant Proc* 36: 2507–2508.

159. Cho SW, Gwak SJ, Kim IK, Cho MC, *et al.* (2006) Granulocyte colony-stimulating factor treatment enhances the efficacy of cellular cardiomyoplasty with transplantation of embryonic stem cell-derived cardiomyocytes in infarcted myocardium. *Biochem Biophys Res Commun* 340: 573–582.

160. Cai J, Yi FF, Yang XC, Lin GS, *et al.* (2007) Transplantation of embryonic stem cell-derived cardiomyocytes improves cardiac function in infarcted rat hearts. *Cytotherapy* 9: 283–291.

161. van Laake LW, Passier R, den Ouden K, Schreurs C, *et al.* (2009) Improvement of mouse cardiac function by hESC-derived cardiomyocytes correlates with vascularity but not graft size. *Stem Cell Res* 3: 106–112.

162. van Laake LW, Passier R, Doevendans PA, Mummery CL (2008) Human embryonic stem cell-derived cardiomyocytes and cardiac repair in rodents. *Circ Res* 102: 1008–1010.

163. van Laake LW, Passier R, Monshouwer-Kloots J, Verkleij AJ, *et al.* (2007) Human embryonic stem cell-derived cardiomyocytes survive and mature in the mouse heart and transiently improve function after myocardial infarction. *Stem Cell Res* 1: 9–24.

164. Meyer GP, Wollert KC, Lotz J, Steffens J, *et al.* (2006) Intracoronary bone marrow cell transfer after myocardial infarction: eighteen months' follow-up data from the randomized, controlled BOOST (BOne marrOw transfer to enhance ST-elevation infarct regeneration) trial. *Circulation* 113: 1287–1294.

165. Laflamme MA, Murry CE (2005) Regenerating the heart. *Nat Biotechnol* 23: 845–856.

166. Reinecke H, Minami E, Zhu WZ, Laflamme MA (2008) Cardiogenic differentiation and transdifferentiation of progenitor cells. *Circ Res* 103: 1058–1071.

167. Sartipy P, Bjorquist P, Strehl R, Hyllner J (2007) The application of human embryonic stem cell technologies to drug discovery. *Drug Discov Today* 12: 688–699.

168. Caspi O, Itzhaki I, Arbel G, Kehat I, *et al.* (2009) *In vitro* electrophysiological drug testing using human embryonic stem cell derived cardiomyocytes. *Stem Cells Dev* 18: 161–172.

V. What is the Paracrine Effect of Stem Cells?

A. Stem Cell Therapy for Heart Repair: The Paracrine Paradigm

Massimiliano Gnecchi, Maria Mirotsou
and Victor J. Dzau

1. INTRODUCTION

Ischemic heart disease and congestive heart failure represent major causes of morbidity and mortality in Western countries. Moreover, cardiovascular diseases are expected to become an extremely relevant health and economic burden in Asia over the next decade.[1]

In most cases, heart failure is the result of pathological changes in the myocardium in response to acute myocardial infarction (AMI). Following AMI, cardiomyocytes begin to die and, if blood supply is not timely restored, all the cardiac tissue served by the infarcted related artery undergoes necrosis or apoptosis. Currently, despite the variety of treatment options, there are no therapeutic interventions that can replace necrotic cardiac tissue with new functional myocytes and/or develop a vascular network supporting those myocytes and the surrounding ischemic tissue. Recently, stem cell therapy has been proposed as innovative approach to treat heart disease.[2] For example, over the last decade the use of adult stem cells (ASCs) to treat post AMI cardiac damage has been tested in several preclinical models. Different adult stem and progenitor cell types, including unfractionated bone marrow cells (BMCs) and mononuclear cells (BMMNCs), mesenchymal stem cells (MSCs), hematopoietic stem cells (HSCs), endothelial progenitor cells (EPCs), skeletal myoblasts (SMs), and cardiac stem cells (CSCs), have been tested in experimental animal models (see Chapters IV. B. to IV. D. and Ref. 2). Overall, the results are extremely promising. Importantly, the first clinical trials also suggest that the administration of ASCs to patients suffering AMI is safe and lead to a small but significant improvement in cardiac function (see Chapter VIII. and Ref. 3).

However, many questions still remain unanswered, in particular the choice of appropriate cell type, dose, method and timing of delivery. Moreover, the biological mechanisms mediating the therapeutic effects of stem cell therapy remain to be elucidated.

Originally, cardiac and vascular regeneration have been proposed as main mechanisms involved.[4,5] However, adult stem cell plasticity has been harshly questioned and the magnitude of cardiac regeneration derived from stem cell transdifferentiation described by some investigators has not been confirmed by others.[6,7] Overall, the number of newly generated cardiomyocytes seems too low to justify the important functional recovery reported in most of the experimental studies in which stem cells were transplanted into infarcted hearts. Likewise, the increased number of vessels observed in stem cell-treated hearts seems not to be the result of neovasculogenesis but rather of angiogenesis and arteriogenesis.[8,9] Furthermore, it has been reported that stem cells exert their beneficial effects acutely, within 72 hours; clearly, differentiation into contracting cardiomyocytes cannot occur in such a short time.[10,11]

Accordingly, we and others have proposed a third mechanism of stem cell action, namely the paracrine effect. In this chapter we will expose the growing body of evidence supporting the hypothesis that paracrine mechanisms mediated by stem cell-released factors play an essential role in the reparative process observed after ASC therapy for AMI.

2. PARACRINE MECHANISMS

It has been shown that ASCs, particularly MSCs, produce and secrete a broad variety of cytokines, chemokines, and growth factors that may potentially be involved in cardiac repair (Table 1).[9,12] Furthermore, hypoxic stress increases the production of several of these factors and tissue concentrations of proteins such as vascular endothelial growth factor (VEGF), basic fibroblast growth factor (b-FGF), hepatocyte growth factor (HGF), insulin growth factor I (IGF-I) and adreno-medullin, just to name some, are significantly increased in ischemic hearts treated with MSCs or multipotent human BM stem cells (hBMSCs).[8,13,14]

The most convincing evidences in favor of paracrine mechanisms come from experimental studies where the administration of condi-tioned medium (CM) from ASCs is able to recapitulate the beneficial effects observed after stem cell therapy. For example, we have demon-strated that CM from MSCs, particularly from genetically modified MSCs overexpressing Akt-1 (Akt-MSCs), exerts cardiomyocyte pro-tection.[10,11] Takahashi and co-workers injected CM from BM-MNCs into acutely infarcted hearts and observed increased capillary density, decreased infarct size, and improved cardiac function compared with controls.[15]

The paracrine factors may influence adjacent cells and exert their actions via several mechanisms. Myocardial protection and neovascu-larization are the most extensively studied. Furthermore, the post

Table 1.   Putative Paracrine Factors Secreted by Adult Stem Cells.

Putative Secreted Factor	Abbreviation	Proposed Function
Adrenomedullin	ADM	Cytoprotection
Angio-associated migratory protein	AAMP	Angiogenesis
Angiogenin	ANG	Angiogenesis, cell proliferation
Angiopoietin-1	AGPT1	Cell migration, vessel stabilization
Bone morphogenetic protein-2	BMP2	Development
Bone morphogenetic protein-6	BMP6	Cell differentiation, growth
Connective tissue growth factor	CTGF	Angiogenesis, cell growth
Endothelin-1	EDN1	Cytoprotection, cell proliferation
Fibroblast growth factor-2	FGF2	Cell proliferation and migration
Fibroblast growth factor-7	FGF7	Cell proliferation and stabilization
Hepatocyte growth factor	HGF	Cytoprotection, angiogenesis, cell migration
Insulin-like growth factor-1	IGF-1	Cytoprotection, cell migration, contractility
Interleukin-1	IL-1	VEGF induction
Interleukin-6	IL-6	VEGF induction
Kit ligand/Stem cell factor	KITLG (SCF)	Cell proliferation and migration

(*Continued*)

Table 1. (*Continued*)

Putative Secreted Factor	Abbreviation	Proposed Function
Leukemia inhibitory factor	LIF	Cell proliferation, cytoprotection
Macrophage migration inhibitory factor	MIF	Cell proliferation, inflammatory response
Matrix metalloproteinase-1	MMP1	Loosens matrix, tubule formation
Matrix metalloproteinase-2	MMP2	Loosens matrix, tubule formation
Matrix metalloproteinase-9	MMP9	Loosens matrix
Macrophage-specific colony-stimulating factor	M-CSF	Monocyte proliferation/migration
Plasminogen activator	PA	Degrading matrix molecules
Platelet-derived growth factor	PDGF	Cell proliferation and migration
Pleiotrophin	PTN	Cell proliferation
Secreted frizzled-related protein-1	SFRP1	Development
Secreted frizzled-related protein-2	SFRP2	Development
Stem cell-derived factor-1	SDF-1	Progenitor cell homing
Thymosin-β4	TMSB4	Cell migration, cytoprotection
Tissue inhibitor of metalloproteinase-1	TIMP-1	Cell migration
Tissue inhibitor of metalloproteinase-2	TIMP-2	Cell migration
Transforming growth factor-β	TGF-β	Vessel maturation, cell proliferation
Tumor necrosis factor-α	TNF-α	Degrade matrix molecules, cell proliferation
Vascular endothelial growth factor	VEGF	Cytoprotection, proliferation, migration, angiogenesis

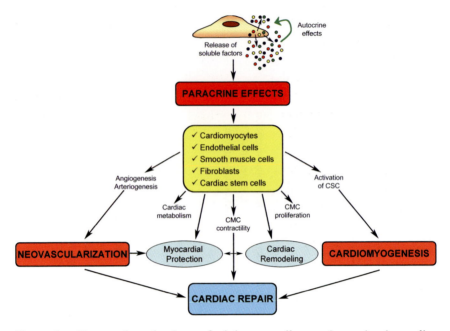

Figure 1.   Proposed mechanisms of adult stem cell paracrine action in cardiac repair.   Paracrine effects can positively influence many processes, among them cardiomyogenesis and neovascularization (see text for details). Paracrine effects, cardiomyogenesis, and vasculogenesis lead to cardiac repair. The released factors may also exert autocrine actions modulating the biology of stem cells including self-renewal and proliferation.

infarction inflammatory and fibrogenic processes, cardiac metabolism, cardiac contractility, and/or endogenous cardiac regeneration may also be positively influenced in a paracrine fashion (Figure 1). In addition, the released factors may exert autocrine actions on the biology of stem cells themselves (Figure 1). Thus, the paracrine/autocrine hypothesis extends the concept of tissue regeneration to include the influence of stem cell released factors on cell niches and tissue repair.

2.1. Myocardial Protection

An immediate paracrine effect of stem cells in an ischemic environment is the release of cytoprotective molecules that increase cardiomyocyte survival. Besides necrosis, apoptotic mechanisms have been progressively linked with both the acute and chronic deleterious effects of cardiac ischemia.[16]

Our group characterized the spectrum of BM-derived stem cell paracrine actions and demonstrated that MSCs exert direct cyto-protective action on ischemic cardiomyocytes and that these properties are enhanced in Akt-MSCs. In particular, we showed that CM from hypoxic Akt-MSCs reduces apoptosis and necrosis of isolated rat cardiomyocytes exposed to low oxygen tension.[10] To further validate the protective properties of the MSCs, we studied the effect of the CM *in vivo*, using a rat experimental model of coronary occlusion and demonstrated a dramatic limita-tion of infarct size and cardiac apoptosis after administration of CM from Akt-MSCs.[10] These data showed that MSCs exert cardiac protective effects through the release of paracrine factors and that the activation of the Akt pathway markedly enhances the production/release of these putative molecules. In a follow-up study we confirmed our previous results and documented how the limitation of the infarct size was matched by preservation of cardiac function.[11] We then used qPCR to verify whether Akt upregulates the expression of putative secreted factors in MSCs. Our data showed that various known cytoprotective molecules such as VEGF, bFGF, HGF, IGF-I, and thymosin ß4 (TMSB4) were significantly upregulated in the Akt-MSCs at baseline nor-moxia compared with native MSCs and increased further after exposure to hypoxia.[11]

Recently, our original findings in rodents have been successfully replicated by others in a large animal model: intramyocardial injection of Akt-MSCs into pig infarcted hearts led to limitation of infarct size and preservation of heart function.[17] Furthermore, numerous other studies have confirmed that ASCs exert paracrine cytoprotective effects on ischemic cardiomyocytes.[14,15,18]

For example, it has been reported that the rate of apoptotic mouse cardiomyocytes exposed to hypoxia is reduced if they are co-cultured with BM-MSCs.[18] The same group also performed an *in vivo* experiment in which MSCs or hypoxia-preconditioned MSCs were injected into infarcted mouse hearts. The GFP-labeled MSCs engrafted mostly at the border area but the absolute number of newly generated cardiomyocytes was extremely low in both groups. However, infarcted area and ventricular remodeling were significantly prevented, particularly in the hypoxia-preconditioned MSC treated animals. TUNEL staining showed that the number of apoptotic cardiomyocytes was significantly reduced in both groups compared with saline controls, but the reduction was more pronounced when hypoxia-preconditioned cells were administered.

It has been documented that also rat BM-MNCs can release various cytoprotective factors, including VEGF, PDGF, IL-1β and IGF-1.[15] Administration of CM from BM-MNCs inhibited apoptosis *in vitro* and limited infarct size *in vivo*. Importantly, also human BM-MNCs exert potent cytoprotective effects.[19]

Interestingly, it has been reported that TMSB4, a powerful cytoprotective protein, is secreted not only from MSCs but also from other stem cell types such as embryonic-derived endothelial progenitor cell (eEPCs). In particular it has been shown that eEPCs mediate cardiac protection through paracrine effects and that if TMSB4 is knocked-down by shRNA the cytoprotective effects exerted by eEPCs are attenuated both *in vitro* and *in vivo*.[20]

2.2. Neovascularization

Stem cell administration into infarcted hearts also leads to neovascularization and to improved tissue blood flow. However, despite evidences that BM-derived stem or progenitor cells incorporate into vascular structures, several studies suggest that only a small number of vessels contain donor cells.[6,21] Accordingly, it has been proposed that angiogenesis and arteriogenesis, stimulated by paracrine factors released by the ASCs, may be the main mechanism involved. The molecular processes leading to angiogenesis and arteriogenesis are mediated by proteins such as VEGF, bFGF, HGF, angiopoietins and others. These molecules lead to EC and VSMC migration, proliferation, vessel enlargement and maturation, and synthesis of extracellular matrix. As mentioned, it has been shown that BM stem cells express several key pro-angiogenic and pro-arteriogenic mediators.[8,9,12]

For example, it has been reported that the injection of BM-MNCs into the infarct zone in a swine model of myocardial ischemia resulted in a significant increase in regional blood flow and capillary density at three weeks.[22] The BM-MNCs expressed bFGF, VEGF, and angiopoietin-1 and the levels of these angiogenic ligands were significantly increased in cell treated hearts compared with controls. Furthermore, the cardiac levels of other two pro-angiogenic factors, interleukin-1β (IL-1β) and tumor necrosis factor-α (TNF-α) were markedly increased after BM-MNC administration. The authors concluded that the released factors likely contributed to stimulate angiogenesis in BM-MNC treated animals. The pro-angiogenic paracrine action of BM-MNCs has been validated by the documentation of increased microvessel density in ischemic hearts injected solely with CM from MNCs.[15]

The MSCs also secrete high amounts of pro-angiogenic and pro-arteriogenic factors and their administration results in increased neovascularization mainly via paracrine effects as shown by our group and other investigators.[9,12] For example, Epstein and collaborators injected 1×10^6 MSCs in the adductor muscle of mice 24 hours after femoral artery ligation and showed that, compared with animals treated with control medium or mature ECs, distal limb perfusion improved and conductance vessels increased in number and total cross-sectional area.[9] The MSCs did not incorporate into mature collaterals but stained positive for VEGF and bFGF. Quantification of VEGF and bFGF tissue levels documented a significant increase of these proteins in the muscle of MSC-treated animals compared with controls. Based on these results the authors concluded that MSCs, in their model, contributed to collateral remodeling through paracrine mechanisms. The same authors performed a gene expression profiling of MSCs grown under normal conditions or under hypoxia stimulation and documented that these cells express a wide range of arteriogenic cytokines at baseline and that several of them are upregulated by hypoxia.[8] ELISA assays and immunoblotting of the CM from MSCs (MSC-CM) confirmed at protein level the transcriptomic analysis. Furthermore, MSC-CM promoted the proliferation and the migration of ECs and VSMCs in a dose-dependent manner *in vitro* and enhanced collateral flow recovery in a model of hind limb ischemia *in vivo*. Other studies, testing MSC transplantation in experimental infarcted hearts, reported an increase in capillary density in treated animals compared with controls, despite the presence of few ECs of donor origin.[23] In these cases, even though not directly proven, a pro-angiogenic paracrine action seems the most reasonable explanation to the effects observed.

Finally, it has also been shown that EPCs enhance angiogenesis trough paracrine mechanisms. In particular, VEGF and stromal derived factor-1 (SDF-1) released in the CM by EPCs promote the migration of mature ECs and the formation of capillaries via differentiation-independent mechanisms.[24] In addition, several reports showed that transplantation of EPCs into ischemic hearts promotes angiogenesis and arteriogenesis *in vivo*.[25] Taken together, all these reports show that ASCs improve neovascularization of ischemic tissues mainly through pro-angiogenic and pro-arteriogenic paracrine effects.

2.3. Cardiac Remodeling

Beside cytoprotection and neovascularization, paracrine factors released by transplanted stem cell may alter the extracellular matrix and favorably influence the post infarction remodeling of the heart chambers.

It has been shown in animal models that MSC transplantation decreases fibrosis in the heart and other organs such as lung, liver and kidney.[26] Since MSCs express a number of molecules involved in the biogenesis of extracellular matrix such as collagenase, metalloproteinases (MMPs), serine proteases and serine protease inhibitors, it has been suggested that transplanted MSCs can inhibit the fibrosis through paracrine actions. Xu and co-workers evaluated the effects of grafting MSCs on extracellular matrix in infarcted myocardium.[27] They showed that MSC transplantation significantly attenuated the increased cardiac expression of collagen types I and III, tissue inhibitor of metalloproteinase-1 (TIMP-1), and tumor growth factor-beta (TGF-β) observed in infarcted control hearts. Interestingly, it has also been reported that MSC-CM significantly attenuates cardiac fibroblast proliferation and inhibits type I and type III collagen expression in cardiac fibroblasts.[28]

Another study has shown that direct injection of hMSCs into ischemic rat hearts decreased fibrosis, apoptosis and left ventricular dilatation while increasing myocardial thickness; this resulted in the prevention of systolic and diastolic cardiac dysfunction without evidences of myocardial regeneration.[29]

It has also been suggested that soluble factors released by ASCs may directly affect cardiac remodeling by releasing cardiotrophic factors.[30] Intracoronary injection of CM from EPCs in a porcine model of myocardial infarction resulted with time in increased cardiomyocyte size. Further *in vitro* experiments showed that the CM from EPCs increased cell mass of cultured cardiomyocytes and that these effects were partly mediated by TGF-β1.

Finally, stem cells may also affect remodeling by releasing local signaling molecules that limit local inflammation when injected into injured tissues. Ohnishi and co-workers[31] demonstrated that MSC transplantation into a rat model of acute myocarditis attenuated the increase in CD68$^+$ inflammatory cells and monocyte chemoattractant protein-1 (MCP-1) expression in the myocardium. Isolated adult rat cardiomyocytes (ARVCs) cultured in standard conditions were injured by MCP-1, which may play an important role in myocarditis; in contrast, in the presence of MSC-CM the MCP-1-induced injury was significantly attenuated.

2.4. Cardiac Contractility

Some evidences suggest that the administration of ASCs positively influences cardiac contractility. For example, we reported that the spontaneous contractile activity of ARVCs exposed to prolonged hypoxia was markedly increased in the presence of the CM from hypoxic Akt-MSCs compared with controls.[11] The contractile activity was vigorous and synchronized, very different from the typical spontaneous contractility observed when the ARVCs were exposed to hypoxic stress in standard growth medium. Based on this observation, we hypothesized that the Akt-MSC-CM contains inotropic factors. Preliminary experiments on freshly isolated ARVCs exposed to standard growth medium or to different concentrations of MSC-CM or Akt-MSC-CM seem to confirm that MSCs do release inotropic factors that positively modulate cell contractility. Using an edge-to-edge detection system we observed that cell shortening, maximal rate of relengthening (+dL/dt), and maximal rate of shortening (−dL/dt) are significantly improved in the presence of CM from hypoxic MSCs, particularly Akt-MSCs, compared with standard conditions (unpublished data). Results from other groups support the paracrine action of BM-derived stem cells on cardiomyocyte contractility. For example, Takahashi and collaborators showed that CM from BM-MNCs preserves the contractility capacity of ARVCs.[15] Fractional shortening and ± dL/dt of ARVCs cultivated for 72 hours in standard growth medium were significantly lower compared with those recorded at baseline. Conversely, cell contractility parameters did not decrease significantly when ARVCs were cultured in the presence of CM from BM-MNCs pre-exposed to normoxia or hypoxia. Of note, the supernatant from hypoxic BM-MNCs was more effective compared with CM from normoxic cells in preserving ARVC contractility. This observation further suggests that both the production and the release of the factors mediating inotropic paracrine effects are increased under hypoxia. Currently, the nature and identity of putative molecules can only be hypothesized and further experiments are required to elucidate the pathways and the mediators involved.

Interestingly, Dhein and collaborators have reported that administration of BM-derived stem cells in doxorubicine-induced decompensated rabbit hearts significantly limited the down regulation of β-adrenoceptor normally present in this heart failure model and significantly increased cardiac contractility measured *in vivo*.[32] The positive and significant correlation between β-adrenoceptor density and contractility described would support the hypothesis of a relationship between these two parameters. Moreover, in failing hearts receiving BM cell transplantation β-adrenoceptor density was nearly normalized in the septum while still below the normal level in the left ventricle. Considering that the cells were injected in the free wall and that the number of engrafted cells was low, the authors proposed that the effects observed might be due to paracrine action. Experiments on isolated cardiomyocytes co-cultured with BM-derived stem cells or exposed to CM may help to clarify this mechanism.

2.5. Cardiac Metabolism

Profound changes in cardiac metabolism occur in the acute and subacute phase after AMI. These changes greatly influence infarct size and ventricular structural remodeling. Therapeutic interventions preventing or limiting cardiac metabolic alterations may be beneficial. Recently, Feygin and co-workers have confirmed that the border zone of infarcted pig hearts is affected by significant bioenergetic abnormalities and that MSC administration can partially attenuate this metabolic remodeling.[33] Because of the low cell engraftment observed in their model, the authors hypothesized that MSCs did not provide a structural contribution to the damaged heart and the observed beneficial effects were likely the result of paracrine mechanisms. Our group has reported that Akt-MSCs significantly prevent metabolic remodeling in infarcted rat hearts.[34] Specifically, treatment with Akt-MSCs spared phosphocreatine stores and limited the increase in 2-DG uptake in the residual intact myocardium compared with the saline or the MSCs control animals. Furthermore, Akt-MSCs treated hearts had normal pH, whereas low pH was measured in the saline and MSC groups.

2.6. Cardiac Regeneration

Accumulating evidences suggest that stem cell paracrine mechanisms may influence cardiomyocyte regeneration by activating endogenous stem cells such as resident CSCs. It is indeed conceivable that some of the factors released by transplanted stem cells may enhance proliferation, mobilization, differentiation, survival, and function of endogenous CSCs.

For instance, *in vitro* studies testing CM from rat MSCs have shown that paracrine factors released from those cells induce adult rat cKit+ CSCs proliferation and differentiation toward the cardiac line-age.[35] Similar effects have also been obtained with EPCs.[24]

Furthermore, indirect evidences have been provided by experiments in which intramyocardial administration of HGF and IGF-1 at the infarct border zone induced CSC migration, proliferation, and differentiation.[36] Since MSCs release both HGF and IGF-1, particularly under hypoxic stimulation,[11] it is reasonable to hypothesize that MSC injected into ischemic hearts may attract and activate resident CSCs. In support of this hypothesis, it has been shown that MSC injection into infarcted pig hearts resulted in newly formed cardiomyocytes, some of which stained positive for c-Kit, a marker expressed by CSCs, and others for Ki67, a marker of cell proliferation.[37] The authors concluded that endogenous cardiac regeneration was present. However, the true origin of those replicating cardiomyocytes remains to be determined since the co-staining for c-Kit and Ki67 was not performed.

Finally, many recent studies have identified stem cell-derived molecules acting in a paracrine fashion as potential regulators of resident CSC proliferation and differentiation, namely fibronectin, heparin-binding EGF-like growth factor as well as various ligands of the Notch and Wnt signaling pathways.[38,39]

3. FUTURE DIRECTION IN STEM CELL-MEDIATED PARACRINE MECHANISMS

3.1. Enhancing Beneficial Paracrine Effects

The demonstration that stem cells secrete therapeutic factors provides a potential breakthrough. Characterization of those paracrine mediators may lead to the possibility of replacing stem cell-based therapy with soluble factor-based therapy in which a single or a "mixture" of molecules are administered to patients. This strategy, compared with cell therapy, may result technically easier to translate to the bedside.

Nevertheless, transplantation of stem cells for their paracrine effects still represents a reasonable strategy because the beneficial factors may remain partly unidentified and because multiple factors might be functioning synergistically. For this reason, it is important to discover new strategies to improve the paracrine properties of stem cells. One possible approach is to modify the cells *ex vivo* with viral vectors. By combining gene with cell therapy, it is possible to enhance stem cell function and viability. Furthermore, the cells may be used as vehicle for gene therapy to deliver secreted gene products able to exert paracrine or endocrine actions. Our group was the first to conceptualize this approach and showed that genetically modified EPCs have greater reparative potential than unmodified cells[12] and that MSCs overexpressing the pro-survival gene Akt-1 exert more powerful functional benefits than wild-type cells.[40] Conceived initially as a strategy to increase cell viability, the overexpression of Akt also enhanced the production of secreted factors capable of cardioprotection both *in vitro* and *in vivo*. The salutary effects of this approach have been further validated and extended by several other groups. For instance, BCL-2 engineered MSCs were shown to increase the survival of MSCs transplanted into infarcted rat hearts and lead to higher expression level of VEGF, increased capillary density and smaller infarct size relative to controls.[41] Tang and collaborators showed that MSCs overexpressing hypoxia regulated heme oxygenase-1 (HO-1-MSCs) had higher survival rate compared with control MSCs once injected into ischemic hearts and led to significant cardiac recovery.[42] The beneficial effects obtained with the HO-1-MSCs appeared primarily due to paracrine protective action since the low number of newly generated cardiomyocytes could not alone explain the structural and functional improvements reported. Thus, as in the case of the Akt-MSCs, the original intervention, intended to increase stem cell viability, triggered strong paracrine mechanisms. Similarly HGF-, bFGF-, adrenomedullin- and VEGF-engineered MSCs were shown to survive better in ischemic tissues and to exert more powerful cytoprotection and/or pro-neovascularization effects compared with control MSCs.[26]

An alternative approach to improve paracrine effects is represented by cell preconditioning. As mentioned earlier in this chapter, preconditioning of MSCs with hypoxia before administration results in improved cell survival and stimulation of paracrine mechanisms for heart repair.[18] It has been recently reported that MSCs cultured under hypoxia become more resistant to apoptosis by activating the Akt signaling pathway. Injection of hypoxic pre-conditioned MSCs into ischemic hind limbs restored blood flow via HGF overexpression compared with MSCs grown under standard conditions.[43]

Other groups have used growth factor or other molecules for pre-conditioning and showed that this strategy enhance cell survival and differentiation and stimulate the release of soluble factors.[44] For instance, Ventura and collaborators have developed hyaluronan mixed esters of butyric and retinoic acid (HBR), eliciting a remarkable increase in the transcription of cardiac lineage-promoting genes and cardiac differentiation in embryonic and mesenchymal stem cells.[45] In the same cells, HBR treatment also enhanced the gene expression of VEGF and HGF. Transplantation of MSCs preconditioned *ex vivo* with HBR into infarcted rat hearts led to complete normalization of myocardial performance and to a dramatic reduction in scar formation via regeneration and paracrine mechanisms.

3.2. Discovery of Therapeutic Molecules

The demonstration that stem cells express and release biologically active mediators has important implications in the understanding of stem cell biology and action. As discussed earlier, stem cells express and secrete powerful paracrine soluble factors that enhance cell survival and activate mechanisms of endogenous heart repair and regeneration. Specific studies should be carried out to better understand how stem cells mediate their favorable effects. In particular, the factors and the pathways involved need to be defined. Studying in detail the temporal and spatial expression of paracrine factors in the heart after stem cell administration will increase our understanding of the complex process of heart protection, remodeling, repair and regeneration. Furthermore, identifying the complete scope and nature of paracrine factors involved in stem cell-mediated cardiac repair is a demanding task, although extremely relevant and worth pursuing.

It appears reasonable to speculate that either proteins or small peptides are among the stem cell-derived paracrine mediators. For this reason, a possible approach is represented by microarray transcript profiling followed by confirmation at protein level. Our group used this strategy to analyze Akt-MSCs *vs.* native MSCs[46] and reported that approximately 650 transcripts were differentially regulated between Akt-MSCs and control MSCs. A subanalysis of the data, focused on potential cytoprotective genes, revealed 62 transcripts encoding for 51 unique genes potentially contributing to the paracrine effects of Akt-MSCs. Among those genes, secreted frizzled related protein 2 (Sfrp2) was one of the most dramatically up-regulated. We then showed that Sfrp2 exerts multiple biologic effects including cardiac protection. Indeed, we have demonstrated that administration of Sfrp2 alone protects ischemic cardiomyocytes and that the pro-survival effect of Akt-MSCs was markedly attenuated upon knockdown of Sfrp2 with siRNA. Furthermore, we have shown that the mechanism of action of Sfrp2 is through modulation of the Wnt signaling, a pathway involved in cell proliferation and differentiation. Recent data suggest that Sfrp2 also influences the remodeling process by regulating type 1 collagen synthesis and processing via interaction with the bone morphogenetic protein 1 (a TLD-like metalloproteinase).[47] Furthermore, current *in vitro* studies support the hypothesis that Sfrp2 is also involved in promoting angiogenesis by activating the non-canonical Wnt/calcineurin/NFAT pathway.[48] Taken together, these evidences strongly support the hypothesis that Sfrp2 plays a pivotal role in stem cell-mediated paracrine mechanisms (Figure 2).

Using the same microarray approach, we have also discovered several novel factors that are expressed in the Akt activated stem cells; the nature and relative contribution to heart repair of each of these factors are currently under intense investigation.

Other groups have proposed to use proteomics to identify the putative paracrine factors. However, the analysis of secreted proteins represents a challenge for current proteomic techniques. Proteins are usually secreted at low concentrations in culture media, which makes their recovery very difficult. In addition, culture media are rich in salts and other compounds that interfere

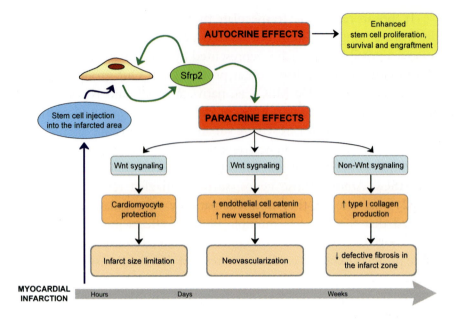

Figure 2.   Sfrp2 and myocardial repair. Myocardial injury triggers dynamic molecular and cellular changes leading to ventricular scarring, remodeling and fibrosis. During the acute phase Sfrp2 inhibits Wnt signaling resulting in repression of apoptosis in adult cardiomyocytes. At a later stage Sfrp2 may contribute to neovascularization by promoting the Wnt canonical signaling in endothelial cells and or endothelial precursors. Finally during the maturation phase Sfrp2 can modulate the deposition of type 1 collagen via Wnt independent mechanism (see text for details).

with most proteomics techniques, making selective precipitation of proteins almost mandatory for a correct analysis. In addition, the presence of serum proteins may dramatically influence the dynamic range of the sample and thereby identification of secreted proteins, as well. Thus, reliable methods for profiling secretory proteins are highly desirable. Efficient pipelines should be developed for the enrichment and analysis of the secretome of different cell lines, free of essential contaminants. These new methods should encompass the optimization of cell incubation conditions in serum-free medium, the subfractionation of the CM with appropriate chromatographic techniques, the establishment of biochemical assays to monitor the paracrine effects of the isolated protein, and the use of the CM in heterologous cell systems for biological assays.

Despite these limitations, several interesting results have already been obtained with proteomics. For instance, Estrada and co-workers have used high-resolution, two-dimensional liquid chromatography tandem mass spectrometry (LC-MS/MS) to profile the proteome of murine MSCs and identified 258 proteins specifically expressed by mMSCs, 54 of which were classified as secreted proteins.[49] More recently, CM from EPC cultures was analyzed using a matrix-assisted laser desorption/ionization tandem time-of-flight mass spectrometer combined with offline peptide separation by nanoflow liquid chromatography.[50] This extensive proteomic analysis revealed the presence of the pro-angiogenic factor thymidine phosphorylase (TP). Functional experiments demonstrated that inhibition of TP results in a significant increase in basal and oxidative stress-induced apoptosis, whereas supplementation with 2-deoxy-D-ribose-1-phosphate (dRP), the enzymatic product of TP, abrogates this effect. Moreover, dRP produced in EPC cultures stimulates endothelial cell migration in a paracrine manner, as demonstrated by gene-silencing experiments in transmigration and wound repair assays. Finally, the effect of TP on angiogenesis was investigated by implantation of Matrigel plugs in mice. In these *in vivo* experiments, dRP strongly promoted neovascularization. These data support the concept that EPCs exert their pro-angiogenic activity in a paracrine manner and demonstrate a key role of TP activity in their survival and pro-angiogenic potential. A comprehensive characterization of the paracrine factors and their pathways will allow the identification of possible novel drugs and/or therapeutical targets.

Finally, it has been hypothesized that stem cell-released factors may also exert autocrine effects. For example, it has been shown that the Sfrp2 protein, beside its regulatory action in cardiac remodeling and cytoprotection, exert an important role in MSC autoregulation. In particular, MSCs isolated from MRL/MpJ, a peculiar mouse strain characterized by uncommon autoregenerative capacity, express higher levels of Sfrp2 compared with the MSCs isolated from wild-type Bl/6 strain, and that knockdown of Sfrp2 significantly affected MSC expansion potential, impaired differentiation potential towards the vascular lineage and reduced cell engraftment.[51] The identification and characterization of autocrine factors may improve our understanding of how stem cells regulate their own fate, behavior, stemness, lineage commitment and many aspects of their biology.

3.3. Potential Hurdles for Protein Therapy

Administration of soluble factors instead of stem cells may be more easily translated into the clinical arena since it has several advantages. However, there are also technical problems related with protein/peptide administration. The most obvious limitation of protein therapy is represented by the need to maintain therapeutic concentrations in order to induce the desired effect. Establishing the threshold concentration remains to be determined and represents a difficult task. Different actions may require different concentrations and timing. Substantial differences between animal models and humans further complicate the scenario. For example, it has been shown that a single dose of specific growth factors is effective in enhancing neovascularization in animals but not in humans.[52]

Protein stability and pharmacokinetic represent other possible hurdles. To overcome these problems a variety of strategies have emerged for manipulating protein properties, stability, specificity, immunogenicity and pharmacokinetic.[53] Mechanisms for altering these properties include manipulation of primary structure, incorporation of chemical and post translation modifications and utilization of fusion partner. The protein and peptide therapeutics have already become an important class of drugs due to advancement in molecular biology and recombinant technology. Currently, most of therapeutic proteins are administered by the parenteral route, which has many drawbacks. Various delivery strategies have evolved over the past few years to improve delivery of proteins and peptides, including the use of biopolymers and nanomaterials for controlled release of proteins (see Chapter V and Ref. 54) and delivery via non-invasive routes such as subcutaneous release or dermal patches. The non-invasive approach remains challenging due to poor absorption and enzymatic instability, pharmacokinetics and pharmacodynamics of protein therapeutics. Development of an oral dosage form for protein therapeutics is still the most desirable one but with greater challenge. Even though the road to reach optimal protein therapy is full of hurdles, we anticipate that the constant development and application of rational protein design technology will enable significant improvements in the efficacy and safety of existing protein therapeutics, as well as allow the generation of entirely novel classes of proteins and mode of action. In this case, curing AMI with a single protein or, most likely, with a cocktail of proteins may become a reality.

4. CONCLUSION

The demonstration of paracrine/autocrine mechanisms improves our understanding of stem cell biology and action in tissue repair and regeneration. Although stem cell therapy holds promise in the future treatment of heart disease such as AMI, chronic ischemic heart disease, and congestive heart failure, its current use is significantly hampered by biologic and technologic challenges. Genetic modification of stem cells represents an important advancement because this approach may enhance cell function and overcome the issue of cell viability, scalability, and immune tolerance. Furthermore, the recent demonstration that genetically modified cells may secrete novel therapeutic factors provides a potential strategy in that, rather than administering cells, one may be able to administer specific soluble factors produced by these cells for cardiac therapy.

ACKNOWLEDGMENTS

This work was supported by the National Heart, Lung, and Blood Institute grants RO1 HL35610, HL81744, HL72010, and HL73219 (to V.J.D.); the Edna Mandel Foundation; and the Foundation Leducq (to V.J.D.). M.G. is supported by the Fondazione IRCCS Policlinico San Matteo Pavia, Italy; the Ministero Italiano dell'Università e della Ricerca (MIUR); the Ministero Italiano della

Sanità; the Fondazione Cariplo; and the Fondazione Banca del Monte di Lombardia.

REFERENCES

1. Sasayama S (2008) Heart disease in Asia. *Circulation* 118: 2669–2671.
2. Laflamme MA, Murry CE (2005) Regenerating the heart. *Nat Biotechnol* 23: 845–856.
3. Martin-Rendon E, Brunskill SJ, Hyde CJ, Stanworth SJ, *et al.* (2008) Autologous bone marrow stem cells to treat acute myocardial infarction: a systematic review. *Eur Heart J* 29: 1807–1818.
4. Asahara T, Murohara T, Sullivan A, Silver M, *et al.* (1997) Isolation of putative progenitor endothelial cells for angiogenesis. *Science* 275: 964–967.
5. Orlic D, Kajstura J, Chimenti S, Jakoniuk I, *et al.* (2001) Bone marrow cells regenerate infarcted myocardium. *Nature* 410: 701–705.
6. Balsam LB, Wagers AJ, Christensen JL, Kofidis T, *et al.* (2004) Haematopoietic stem cells adopt mature haematopoietic fates in ischaemic myocardium. *Nature* 428: 668–673.
7. Murry CE, Soonpaa MH, Reinecke H, Nakajima H, *et al.* (2004) Haematopoietic stem cells do not transdifferentiate into cardiac myocytes in myocardial infarcts. *Nature* 428: 664–668.
8. Kinnaird T, Stabile E, Burnett MS, Lee CW, *et al.* (2004) Marrow-derived stromal cells express genes encoding a broad spectrum of arteriogenic cytokines and promote *in vitro* and *in vivo* arteriogenesis through paracrine mechanisms. *Circ Res* 94: 678–685.
9. Kinnaird T, Stabile E, Burnett MS, Shou M, *et al.* (2004) Local delivery of marrow-derived stromal cells augments collateral perfusion through paracrine mechanisms. *Circulation* 109: 1543–1549.
10. Gnecchi M, He H, Liang OD, Melo LG, *et al.* (2005) Paracrine action accounts for marked protection of ischemic heart by Akt-modified mesenchymal stem cells. *Nat Med* 11: 367–368.
11. Gnecchi M, He H, Noiseux N, Liang OD, *et al.* (2006) Evidence supporting paracrine hypothesis for Akt-modified mesenchymal stem cell-mediated cardiac protection and functional improvement. *FASEB J* 20: 661–669.

12. Dzau VJ, Gnecchi M, Pachori AS, Morello F, *et al.* (2005) Therapeutic potential of endothelial progenitor cells in cardiovascular diseases. *Hypertension* 46: 7–18.

13. Nagaya N, Kangawa K, Itoh T, Iwase T, *et al.* (2005) Transplantation of mesenchymal stem cells improves cardiac function in a rat model of dilated cardiomyopathy. *Circulation* 112: 1128–1135.

14. Yoon YS, Wecker A, Heyd L, Park JS, *et al.* (2005) Clonally expanded novel multipotent stem cells from human bone marrow regenerate myocardium after myocardial infarction. *J Clin Invest* 115: 326–338.

15. Takahashi M, Li TS, Suzuki R, Kobayashi T, *et al.* (2006) Cytokines produced by bone marrow cells can contribute to functional improvement of the infarcted heart by protecting cardiomyocytes from ischemic injury. *Am J Physiol Heart Circ Physiol* 291: H886–893.

16. Kajstura J, Cheng W, Reiss K, Clark WA, *et al.* (1996) Apoptotic and necrotic myocyte cell deaths are independent contributing variables of infarct size in rats. *Lab Invest* 74: 86–107.

17. Lim SY, Kim YS, Ahn Y, Jeong MH, *et al.* (2006) The effects of mesenchymal stem cells transduced with Akt in a porcine myocardial infarction model. *Cardiovasc Res* 70: 530–542.

18. Uemura R, Xu M, Ahmad N, Ashraf M (2006) Bone marrow stem cells prevent left ventricular remodeling of ischemic heart through paracrine signaling. *Circ Res* 98: 1414–1421.

19. Kubal C, Sheth K, Nadal-Ginard B, Galinanes M (2006) Bone marrow cells have a potent anti-ischemic effect against myocardial cell death in humans. *J Thorac Cardiovasc Surg* 132: 1112–1118.

20. Hinkel R, El-Aouni C, Olson T, Horstkotte J, *et al.* (2008) Thymosin beta4 is an essential paracrine factor of embryonic endothelial progenitor cell-mediated cardioprotection. *Circulation* 117: 2232–2240.

21. Rehman J, Li J, Orschell CM, March KL (2003) Peripheral blood "endothelial progenitor cells" are derived from monocyte/macrophages and secrete angiogenic growth factors. *Circulation* 107: 1164–1169.

22. Kamihata H, Matsubara H, Nishiue T, Fujiyama S, *et al.* (2001) Implantation of bone marrow mononuclear cells into ischemic myocardium enhances collateral perfusion and regional function via side

supply of angioblasts, angiogenic ligands, and cytokines. *Circulation* 104: 1046–1052.

23. Nagaya N, Fujii T, Iwase T, Ohgushi H, *et al.* (2004) Intravenous administration of mesenchymal stem cells improves cardiac function in rats with acute myocardial infarction through angiogenesis and myogenesis. *Am J Physiol Heart Circ Physiol* 287: H2670–2676.

24. Urbich C, Aicher A, Heeschen C, Dernbach E, *et al.* (2005) Soluble factors released by endothelial progenitor cells promote migration of endothelial cells and cardiac resident progenitor cells. *J Mol Cell Cardiol* 39: 733–742.

25. Kawamoto A, Tkebuchava T, Yamaguchi J, Nishimura H, *et al.* (2003) Intramyocardial transplantation of autologous endothelial progenitor cells for therapeutic neovascularization of myocardial ischemia. *Circulation* 107: 461–468.

26. Gnecchi M, Zhang Z, Ni A, Dzau VJ (2008) Paracrine mechanisms in adult stem cell signaling and therapy. *Circ Res* 103: 1204–1219.

27. Xu X, Xu Z, Xu Y, Cui G (2005) Effects of mesenchymal stem cell transplantation on extracellular matrix after myocardial infarction in rats. *Coron Artery Dis* 16: 245–255.

28. Ohnishi S, Sumiyoshi H, Kitamura S, Nagaya N (2007) Mesenchymal stem cells attenuate cardiac fibroblast proliferation and collagen synthesis through paracrine actions. *FEBS Lett* 581: 3961–3966.

29. Berry MF, Engler AJ, Woo YJ, Pirolli TJ, *et al.* (2006) Mesenchymal stem cell injection after myocardial infarction improves myocardial compliance. *Am J Physiol Heart Circ Physiol* 290: H2196–2203.

30. Doyle B, Sorajja P, Hynes B, Kumar AH, *et al.* (2008) Progenitor cell therapy in a porcine acute myocardial infarction model induces cardiac hypertrophy, mediated by paracrine secretion of cardiotrophic factors including TGFbeta1. *Stem Cells Dev* 17: 941–951.

31. Ohnishi S, Yanagawa B, Tanaka K, Miyahara Y, *et al.* (2007) Transplantation of mesenchymal stem cells attenuates myocardial injury and dysfunction in a rat model of acute myocarditis. *J Mol Cell Cardiol* 42: 88–97.

32. Dhein S, Garbade J, Rouabah D, Abraham G, *et al.* (2006) Effects of autologous bone marrow stem cell transplantation on beta-adrenoceptor density and electrical activation pattern in a rabbit model of non-ischemic heart failure. *J Cardiothorac Surg* 1: 17.

33. Feygin J, Mansoor A, Eckman P, Swingen C, *et al.* (2007) Functional and bioenergetic modulations in the infarct border zone following autologous mesenchymal stem cell transplantation. *Am J Physiol Heart Circ Physiol* 293: H1772–1780.

34. Gnecchi M, He H, Melo LG, Noiseaux N, *et al.* (2009) Early beneficial effects of bone marrow-derived mesenchymal stem cells overexpressing Akt on cardiac metabolism after myocardial infarction. *Stem Cells* 27: 971–979.

35. Yu XY, Geng YJ, Li XH, Lin QX, *et al.* (2009) The effects of mesenchymal stem cells on c-kit up-regulation and cell-cycle re-entry of neonatal cardiomyocytes are mediated by activation of insulin-like growth factor 1 receptor. *Mol Cell Biochem* 332: 25–32.

36. Linke A, Muller P, Nurzynska D, Casarsa C, *et al.* (2005) Stem cells in the dog heart are self-renewing, clonogenic, and multipotent and regenerate infarcted myocardium, improving cardiac function. *Proc Natl Acad Sci USA* 102: 8966–8971.

37. Amado LC, Saliaris AP, Schuleri KH, St John M, *et al.* (2005) Cardiac repair with intramyocardial injection of allogeneic mesenchymal stem cells after myocardial infarction. *Proc Natl Acad Sci USA* 102: 11474–11479.

38. Boni A, Urbanek K, Nascimbene A, Hosoda T, *et al.* (2008) Notch1 regulates the fate of cardiac progenitor cells. *Proc Natl Acad Sci USA* 105: 15529–15534.

39. Koyanagi M, Haendeler J, Badorff C, Brandes RP, *et al.* (2005) Non-canonical Wnt signaling enhances differentiation of human circulating progenitor cells to cardiomyogenic cells. *J Biol Chem* 280: 16838–16842.

40. Mangi AA, Noiseux N, Kong D, He H, *et al.* (2003) Mesenchymal stem cells modified with Akt prevent remodeling and restore performance of infarcted hearts. *Nat Med* 9: 1195–1201.

41. Li W, Ma N, Ong LL, Nesselmann C, *et al.* (2007) Bcl-2 engineered MSCs inhibited apoptosis and improved heart function. *Stem Cells* 25: 2118–2127.

42. Tang YL, Tang Y, Zhang YC, Qian K, *et al.* (2005) Improved graft mesenchymal stem cell survival in ischemic heart with a hypoxia-regulated heme oxygenase-1 vector. *J Am Coll Cardiol* 46: 1339–1350.

43. Rosova I, Dao M, Capoccia B, Link D, *et al.* (2008) Hypoxic preconditioning results in increased motility and improved therapeutic potential of human mesenchymal stem cells. *Stem Cells* 26: 2173–2182.

44. Haider H, Ashraf M (2008) Strategies to promote donor cell survival: combining preconditioning approach with stem cell transplantation. *J Mol Cell Cardiol* 45: 554–566.

45. Ventura C, Cantoni S, Bianchi F, Lionetti V, *et al.* (2007) Hyaluronan mixed esters of butyric and retinoic Acid drive cardiac and endothelial fate in term placenta human mesenchymal stem cells and enhance cardiac repair in infarcted rat hearts. *J Biol Chem* 282: 14243–14252.

46. Mirotsou M, Zhang Z, Deb A, Zhang L, *et al.* (2007) Secreted frizzled related protein 2 (Sfrp2) is the key Akt-mesenchymal stem cell-released paracrine factor mediating myocardial survival and repair. *Proc Natl Acad Sci USA* 104: 1643–1648.

47. Kobayashi K, Luo M, Zhang Y, Wilkes DC, *et al.* (2009) Secreted Frizzled-related protein 2 is a procollagen C proteinase enhancer with a role in fibrosis associated with myocardial infarction. *Nat Cell Biol* 11: 46–55.

48. Courtwright A, Siamakpour-Reihani S, Arbiser JL, Banet N, *et al.* (2009) Secreted frizzle-related protein 2 stimulates angiogenesis via a calcineurin/NFAT signaling pathway. *Cancer Res* 69: 4621–4628.

49. Estrada R, Li N, Sarojini H, An J, *et al.* (2009) Secretome from mesenchymal stem cells induces angiogenesis via Cyr61. *J Cell Physiol* 219: 563–571.

50. Pula G, Mayr U, Evans C, Prokopi M, *et al.* (2009) Proteomics identifies thymidine phosphorylase as a key regulator of the angiogenic potential of colony-forming units and endothelial progenitor cell cultures. *Circ Res* 104: 32–40.

51. Alfaro MP, Pagni M, Vincent A, Atkinson J, *et al.* (2008) The Wnt modulator sFRP2 enhances mesenchymal stem cell engraftment, granulation tissue formation and myocardial repair. *Proc Natl Acad Sci USA* 105: 18366–18371.

52. Post MJ, Laham R, Sellke FW, Simons M (2001) Therapeutic angiogenesis in cardiology using protein formulations. *Cardiovasc Res* 49: 522–531.

53. Malik DK, Baboota S, Ahuja A, Hasan S, *et al.* (2007) Recent advances in protein and peptide drug delivery systems. *Curr Drug Deliv* 4: 141–151.

54. Segers VF, Tokunou T, Higgins LJ, MacGillivray C, *et al.* (2007) Local delivery of protease-resistant stromal cell derived factor-1 for stem cell recruitment after myocardial infarction. *Circulation* 116: 1683–1692.

V. What is the Paracrine Effect of Stem Cells?

B. Delivery Systems to the Heart

Kathryn M. Rafferty and Michael E. Davis

1. INTRODUCTION

The human heart does not exhibit the capacity for *de novo* regeneration after injury, despite the presence of resident and circulating stem cells in the myocardium. Therapeutic cell and drug implantation strategies to promote endogenous cell turnover and regeneration of lost myocardium have met many challenges, including poor drug retention, and suboptimal engraftment and survival of the implanted cells. Tissue engineering approaches are in rapid development to address these limitations with delivery vehicles to engraft cells, and deliver and retain drugs and/or genes in extracellular matrix (ECM)-mimetic scaffolds. Upon delivery to the myocardium, biomaterial scaffolding creates localized therapeutic levels of the embedded cargo, potentially creating a reservoir for therapeutic agent release. Ideal repair systems for adult post myocardial infarction (MI) ischemic tissue temporarily restore mechanical function, direct biological responses, and promote healing.[1] In contrast, clinical repair of congenital heart defects requires dynamic systems that grow and remodel during development. Perhaps the greatest challenge to advancement of therapeutic regeneration is our limited understanding of all the mechanisms underlying cardiac repair. In cell-based therapy, it is unclear whether the implanted cells differentiate into functional cardiomyocytes and endothelial cells within the heart, and further debatable as to whether the benefit of cell therapy is due to donor cell engraftment or to secreted paracrine factors (see Chapters IV. B. to IV. C. and V. A. and Ref. 2). Furthermore, there are some reports that introduction of empty biomaterial scaffolds promotes functional improvements equal to those induced by growth factor-conjugated scaffolds.[3,4] Thus, a milieu of challenges confronts cardiac regeneration technologies, but these must be satisfied to meet escalating clinical needs for cardiac repair.

The issue of maintaining a local, therapeutically efficacious concentration of proteins in the heart has spurred the development of cardiac vehicles to deliver cells, paracrine factors, and genetic modulators to the injured myocardium. These vehicles leverage biocompatible natural and synthetic polymers to deliver therapeutics over a designed time frame, which reduces the systemic concentration of drug or biologic needed, and allows more patient-friendly dosing regimens in the clinic. We will discuss both current and emerging tissue engineered cardiac delivery systems, describe the key features of each, and summarize how they are currently used to promote cardiac tissue regeneration. Efforts will be made to present the state-of-the-art in biomaterial development and discuss their current and/or future application in cardiac repair.

2. CELL-BASED DELIVERY SYSTEMS

2.1. Cells and Microparticles

Microencapsulation is a widely used delivery system for biologic tissue regeneration, and accommodates cell, protein, or gene entrapment. Base polymer for microspheres includes poly(α-hydroxy acids), such as poly(lactic-co-glycolic acid) (PLGA), alginate, gelatin, and most recently, cyclohexane 1,4-diylacetone dimethylene ketal (PCADK). Many microsphere formulations and encapsulation protocols are commercially available, including Gliadel (Eisai), Lupron Depot (Abbott Labs), Zoladex (Astra Zeneca), and CultiSphere-S (Percell Biolytica). Microparticle synthesis typically occurs in a layered protocol, where the primary microbead is surrounded by the addition of hydrophilic and hydrophobic layers. Therapeutic drugs are mixed with the polymer during synthesis to be preferentially partitioned within the layered microsphere. The biomaterial microparticle, thus, provides a multi-drug controlled release system to sequentially deliver therapeutics at a rate tunable to the degradation kinetics of each layer.

Microparticle technology was originally used to control the release of growth factors, but has more recently been adapted for cell encapsulation. Encapsulation of vascular endothelial growth factor (VEGF)-overexpressing Chinese Hamster Ovary (CHO) cells in the core of a trilayered alginate-poly-L-lysine-alginate microcapsule successfully improves angiogenesis in rat ischemic myocardium.[5] In this system, the cells remain encapsulated throughout treatment, providing controlled cellular biosynthesis and release of VEGF into the myocardium. The encapsulation also limits the titer of anti-CHO cell antibodies in the blood, suggesting microencapsulation as a strategy to blunt immune reactions against xenogenic cell implantation. The survival and functionality of embryonic stem cell-derived cardiomyocytes (EB-CMs) entrapped in CultiSphere-S microcarriers has also been validated *in vitro*.[6] Entrapped EB-CMs maintain functionality similar to control cultured embryoid bodies and eventually exhibit a phenotype typical of muscle myofilaments and Z-discs in culture.

Inflammatory responses toward the implant, leading to death of the embedded cells, are a major clinical obstacle to cell-based therapy.

One strategy to bypass xenogenic rejection is to augment cellular encapsulation with anti-inflammatory drugs. PLGA microspheres dually loaded with cells and the nonsteroidal anti-inflammatory agent Ibuprofen improve *in vivo* biocompatibility after injection when compared to microparticles loaded with cells only.[7] Polymer composition is also intimately connected to inflammatory rejection. The novel polymer formulation, PCADK, can be engineered into microspheres that degrade into non-acidic by-products of non-reactive acetone and a diol compound (discussed in detail in paracrine delivery section),[8] and may thus negate the need for use of adjuvants to improve biocompatibility.

2.2. Cells and Hydrogels

Biodegradable hydrogels provide another modality to deliver cells in a dynamic scaffold for myocardial regeneration and repair. Hydrogels are generally described as loosely crosslinked networks of polymers swollen with aqueous media, typically greater than 90% water by weight. Proteins are incorporated into the hydrogel network by swelling the crosslinked matrix in a solution of the therapeutic. Ideally, the chemical composition of hydrogel scaffolds is optimized so that polymer degradation timing coordinates with cell secretion of ECM and neoangiogenesis within the scaffold.[9] Ideal hydrogel grafts eventually become indistinguishable from the native myocardium. Polymer degradation properties provide the capacity for growth and remodeling, creating an array of hydrogel-based cardiac delivery systems that are adaptable for mild to severe heart disease.

Hydrogel-based cell delivery systems are often 3D grafts applied epicardially to ischemic myocardium. After experimental MI in rats, collagen scaffolds seeded with human mesenchymal stem cells (MSCs) and applied to the ischemic region demonstrated significant improvements in cardiac function, as measured by improved fractional shortening and increased anterior wall thickness.[10] Improved systolic function is also obtained upon application of collagen patches seeded with skeletal muscle cells.[11] Collagen patch technology has recently been enhanced by functionalization with peptides containing RGD motifs to anchor and increase the mechanical functionality of seeded rat neonatal cardiomyocytes (nrCMs).[12] Contractile performance and striated organization of nrCMs were improved *in vitro* when seeded on the functionalized constructs. Spontaneous contraction of nrCMs on biomaterials was also observed when seeded on to polyethylene glycol (PEGylated)-fibrinogen hydrogel scaffolds.[13] In this study the chemical composition of the hydrogel was tuned proportionally by increasing or decreasing PEG content. Likewise, the study shows that embedded nrCM synchronous contraction is directly linked to material stiffness. Videos of soft versus stiff beating nrCM-embedded hydrogels can be viewed in the authors' supplementary data.[13] While PEGylated hydrogel patches have not currently been applied *in vivo*, PEGylated-fibrin patches have successfully delivered MSCs to rat myocardium *ex vivo*.[14] Application of the MSC-seeded PEGylated-fibrin patch to healthy rat epicardium shows increased MSC viability compared to injection of MSCs alone. Taken together, these "cardiac patch" delivery systems show clearly that cell implantation imparts functional improvement. However, the variety of cells delivered suggests that successful cell-based therapeutics may ultimately be independent of the cell type.

Injectable hydrogels have also emerged as a promising mode for cell delivery, as localizing therapeutic cellular scaffolds within the damaged left ventricular free wall intramyocardial space may more readily promote neoangiogenesis and cell engraftment when compared to epicardial delivery. Injectable systems also bypass concerns that cardiac patches distort ventricular geometry,[15] because the gels acquire the geometry of the inner heart wall before solidifying at body temperature. Mouse embryonic stem cells (ESCs) injected into the intramyocardial space in both collagen type I and Matrigel scaffolds demonstrate significant improvement in heart function in rat and mouse models of experimental MI.[16–18] Similar results are demonstrated using ESC-laden temperature-sensitive chitosan-based hydrogels.[19] The cell-laden hydrogel improves ejection fraction by 5% and left ventricular shortening by 3%.[20] This ESC-seeded chitosan scaffold has a degradation time of four to six weeks and allows differentiation of ESCs toward the cardiac lineage, as suggested by the presence of cardiac troponin T-positive cells in the border zone. Similarly, injection of skeletal myoblast-laden[21] and catheter-assisted percutaneous injection of MSC-laden fibrin glue[22] into the rat heart post MI improves cell retention and survival. Fibrin glue-based delivery systems have been subsequently adapted for dual cell and growth factor therapy. PEGylated fibrin glue seeded with both bone marrow mononuclear cells (BMNC) and hepatocyte growth factor (HGF) more effectively improves left ventricular function and apoptosis in a mouse MI model, when compared to naked, BMNC-only, or HGF-only PEGylated-fibrin glue injection.[23] As dual cell and drug therapy exhibits positive therapeutic effects on the heart, so also does application of dual delivery modes. In a mouse model of MI, Cortes-Morichetti and co-workers applied myocardial injections of human umbilical cord blood mononuclear cells (HUCBCs) followed by application of a HUCBC-seeded collagen patch to the ischemic region.[24] This two-hit strategy blunts ventricular dilation better than either therapy alone, defining a novel system to return mechanical function to the injured heart.

In order to improve growth potential of the hydrogel system, natural polymers may be constructed into a reinforced graft with porous fabric mesh of poly-L-lactic acid (PLLA) or PLGA. Efficacy of a hybrid construct is demonstrated in tissue engineered blood vessels seeded with human bone marrow cells (BMCs) and implanted in both human and canine cardiac surgery.[25] Eight weeks after implantation, the human construct retains patency and exhibits functional improvement. As postimplantation remodeling cannot be longitudinally followed in humans, histologic analysis in the canine model demonstrates that BMCs expressing endothelial cell lineage markers, such as CD34, CD31, Flk-1, and Tie-2, adhere to the scaffold. Recently, a hybrid 3D fibrous patch (3DFC, Theregen, Inc. San Francisco, CA) was seeded with fibroblasts and applied to the rat heart at the time of experimental MI.[26] At three weeks, the cell-seeded patch improved ejection fraction and myocardial blood flow compared to the no-patch control. The 3DFC patch also supports *in vitro* survival of nrCMs, which exhibit spontaneous, synchronized beating when adhered to the patch.[27] Another recent advancement in hybrid cell-seeded biomaterial patches demonstrates cardiac differentiation of MSCs *in vitro*. Silk fibroin patches functionalized with either chitosan or hyaluronic acid increase MSC cardiogenic differentiation, as measured by the expression of cardiac transcription factors Nkx2.5 and GATA4.[28]

Another subset of liquid hydrogels are peptide hydrogels, also called nanofibers, that can be designed *de novo*, allowing researchers to tailor peptide properties through choice of amino acid sequences. The enhanced ability of nanofiber-based hydrogels to cross the cell membrane provides an intracellular drug delivery system.[29] Nanofiber gels also alter the microenvironment for implanted cells, which is protective for both the implanted and endogenous cells. Davis and co-workers used a novel technique coined the "biotin sandwich," to tether insulin-like growth factor (IGF-1), a factor that promotes cell survival and growth, to self-assembling nanofibers via biotin linkages.[30] This strategy provides sustained delivery of IGF-1 to ischemic myocardium for up to 28 days, while the free protein was cleared rapidly as it is quite small. Delivery of nrCM in tandem with the IGF-1 nanofiber hydrogel after experimental MI improved implanted cell growth and survival, eventually leading to an improvement in systolic function and decreased ventricular volume. This similar technique proved successful in a separate system, improving the growth and efficacy of adult-derived cardiac stem cell (CSC) therapy following MI.[31] Because CSC growth and survival is enhanced via the IGF-1 receptor system, this cell-based therapy offers the potential to mobilize other endogenous CSCs toward the ischemic tissue.

2.3. Cells in Monolayers

Cellular monolayers, or cell sheets, are a promising strategy to deliver cells and paracrine factors to the heart without a biomaterial carrier. Monolayers are single layers of cells with intact cell-cell contacts and detached basement membranes. Currently they are obtained under two protocols. One is to culture cells on temperature sensitive poly-N-isopropylacrylamide (PIPAAm) coated dishes,[32] which release the cell sheet at temperatures lower then 32°C. Grafted PIPAAm is hydrophobic at temperatures above 32°C and hydrophilic at temperatures lower than 32°C. Because hydrophobic surfaces promote cell attachment better than hydrophilic surfaces, lowering the culture dish temperature causes the entire cell monolayer to switch from attached to detached. A second method is to grow cells to confluence on a dish grafted with a matrix protein that is readily degraded by cell secreted enzymes.[33] Upon confluence, proteases secreted at the cell apical membrane cleave the fibrin anchors so that the sheet can be released by gentle scraping. CM monolayers obtained with this method exhibit spontaneous beating up to 60 beats per minute after six days in culture.[33]

This cell delivery method has demonstrated *in vivo* improvement in cardiac function, a elegant review of which is provided by Masuda and co-workers.[34] Adipose tissue derived-MSC monolayers made with the PIPAAm method and applied epicardially to the left ventricle after MI in rat increase diastolic thickness and left ventricle fractional shortening.[35] Secretion of angiogenic and antiapoptosis factors, such as VEGF and HGF, was also detected. Myoblast cell sheets significantly reduced ventricular dilation in a hamster model of dilated cardiomyopathy.[36] When autologous myoblast sheets were applied to rat heart post MI, similar improvements were detected.[37] Interestingly, the implanted myoblasts did not transdifferentiate into cardiac cells, indicating that the observed angiogenic and antifibrotic benefits may be through paracrine mechanisms. *In vitro* and *in vivo* studies with the fibrinolytic release technique show that overlapping cell sheets beat in synchrony and action potential is conserved without delay.[33] Epicardially ablated rat heart covered with a bilayered CM cell sheet *in vivo* maintained pacing, while the bare ablated heart exhibited delayed action potential propagation. In addition, the monolayer patch remained free of necrosis and apoptosis, and it is suggested that the conserved cell-cell contacts within the monolayer sheet provide an anti-inflammatory property, which is absent in free injected cells.

3. PARACRINE FACTOR-BASED DELIVERY SYSTEMS

Protein therapeutics hold great promise for cardiac repair and a variety of angiogenic, pro-survival, and chemotaxis-related proteins have been examined for their potential in cardiac regeneration and repair. However, drug retention within the myocardium has proved the limiting obstacle to this therapy, e.g. 24 hours after intravenous or intracoronary injection of free bFGF in a porcine MI model, less than 1% of administered protein accumulated in the myocardium.[38] One major reason for this rapid clearance is the vascular nature of the myocardium. Each myocyte is surrounded by a network of capillaries that act as a sink, pulling factors away by a variety of potential mechanisms. To overcome this, subsequent protein therapeutic studies have injected repeated doses of proteins to achieve relevant therapeutic concentrations, resulting in improvements in cardiac function; however, the unknown or potentially toxic effects of repeated injections or systemic delivery make development of local delivery vehicles critical.

4. MICROPARTICLE-BASED DELIVERY

Hydrophilic gelatin-based drug delivery vehicles have been long studied and can be processed into a number of different geometries for other applications including microspheres[39] and nanofibers.[40] Many studies present therapeutic benefit upon injection of intracoronary gelatin and alginate microsphere-conjugated basic fibroblast growth factor (bFGF) in both rodent and large animal models of MI.[41-44] In an acute study, it was found that 32% of bFGF-microspheres are retained in infarcted rat myocardium compared to under 2% of freely injected bFGF.[42] The authors also demonstrated the chronic effects of bFGF-microsphere injection into porcine infarcted myocardium, reporting improved neovascularization and left ventricular function compared to animals receiving empty microspheres. Microspheres have also been used to "prevascularize" ischemic myocardium for cell implantation. Microspheres conjugated with bFGF improved retention of freely injected nrCM cells,[45] which demonstrated subsequent improvement in left ventricular function[46] four weeks after infarction. The optimal mode of injection has also been investigated in a microsphere-based study. Hoshino and co-workers delivered bFGF-microspheres to pig heart via anterograde injection (AI) in coronary arteries, retrograde injection (RI) in coronary veins, or direct myocardial injection (DI).[47] Although retention after AI was highest, only RI and DI induced functional improvements post MI. Between these two methods, therapeutic angiogenesis was only detected after DI, indicating that the method of microsphere delivery is intimately connected to the therapeutic benefit.

Controlled release of large proteins, such as bFGF, is effective in gelatin microspheres because of the relatively large pore size in the particles. However, sustained delivery of hydrophobic drugs, such as small molecule inhibitors, requires a tightly crosslinked polymer scaffold with small pores. Encapsulation of the p38-inhibitor SB239063 (p38$_i$) in PCADK is a novel controlled-release system that results in significantly improved cardiac function after MI in rats.[8] PCADK polymer provides a unique biocompatible delivery system because its hydrolysis generates neutral degradation products, compounds 1,4-cyclohexanedimethanol and acetone. Injection of p38$_i$-PCADK microspheres sustains delivery of the small molecule inhibitor in the myocardium for up to six days and compared to PGLA microspheres, this delivery system did not generate an inflammatory response. Microspheres are also utilized in hybrid delivery systems. Fabrication protocol of a dual protein-encapsulated microsphere/nanofiber delivery system has recently been developed to enhance the growth and survival of seeded MSCs.[48] In a one-step mechanism, IGF-1 encapsulated PGLA microspheres were grafted onto nanofibrous ECM-mimetic scaffolds. IGF-1 bioactivity within the scaffold supported MSC attachment, growth, and survival compared to scaffolds without IGF-1 incorporation.

5. HYDROGEL-BASED DELIVERY

As discussed in cell-based systems, immobilizing growth factors in hydrogel grafts offers the added benefit of temporary myocardial support and mechanical stabilization. A study by Kobayashi and co-workers used a crosslinked gelatin hydrogel sheet in order to deliver erythropoietin to a rabbit model of myocardial infarction.[49] A freeze-dried gelatin patch was reconstituted in a solution of erythropoietin and placed on the surface of the heart following a 30-minute period of ischemia before reperfusion. *In vivo* release studies using radiolabeled erythropoietin indicated a rapid burst release with 80% of encapsulated erythropoietin released in the first 24-hours postimplantation. This was followed by a steady zero-order release of approximately 46 IU/day for the following two weeks. Sustained, local delivery of erythropoietin was able to improve cardiac function, as well as elevate biochemical markers of cell survival and angiogenesis. bFGF immobilized in chitosan hydrogel was successfully grafted on the ischemic rabbit heart via ultraviolet light polymerization. After four weeks, the systolic pressure of the left ventricle improved compared to application of the empty hydrogel. Histologic analysis also demonstrates that controlled-release of bFGF retains myocardial viability and increases CD31 staining of blood vessels. Positive effects of sustained platelet-derived growth factor (PDGF) to ischemic myocardium have also been demonstrated in a nanofiber-based delivery system. Sustained delivery of PDGF in infarcted rat myocardium for up to 14 days was achieved by loading the large protein within self-assembling nanofibers.[50] This study demonstrated that controlled release of PDGF improved endogenous cardiomyocyte survival, decreased infarct size, and increased cardiac function more effectively than naked nanofibers or PDGF alone. Important for efforts to define the underlying biology of repair, the study proposes a mechanism for cardiac functional improvement, suggesting that enhancing PDGF signaling in infarcted myocardium improves critical communication between endothelial cells and cardiomyocytes. In a separate study, Engel and co-workers utilized self-assembling nanofibers to deliver FGF-1 and combined it with the delivery of a p38 inhibitor via repeated intraperitoneal injection.[51] This strategy produced a synergistic effect as FGF-1 improved angiogenesis and p38 inhibition induced cardiomyocyte division.

Hydrogels are also utilized for dual growth factor delivery. Hao and co-workers developed an injectable alginate hydrogel which sequentially delivers VEGF and PDGF-BB to the injured myocardium.[52] In this system, VEGF is released first to stimulate immature vessel growth and PDGF-BB release follows to enhance vessel maturation via recruitment of smooth muscle cells. As expected, the biphasic delivery system increases vessel maturation within the injured myocardium.

6. NON-VIRAL GENE DELIVERY SYSTEMS

There is a paucity of investigation on gene delivery to ischemic myocardium, as to-date the majority of efforts have focused on localized cell and drug delivery. Myocardial gene delivery systems are also challenged because necrotic heart tissue is resistant to transfection.[53] However, introducing genes and genetically modified cells into ischemic tissue offers the potential to create a reservoir of therapeutic factor biosynthesis *in situ*. Tissue engineered, non-viral delivery systems are promising vectors to release DNA cargo into the endosomal space with improved efficiency and limited cytotoxicity.[54] Non-viral vectors are typically positively charged polymers that complex with negatively charged DNA to condense and protect the oligonucleotide during transfection.[55]

Early studies support non-viral vectors as a strategy for cellular genetic modification. Endothelial progenitor cells (EPCs) readily internalize green fluorescent protein (GFP) DNA condensed in a gelatin vector by phagocytosis.[55] These cells were then used to administer adrenomedullin (AM) DNA intravenously into rats with monocrotaline-induced pulmonary hypertrophy. The modified EPCs engraft into the pulmonary tissue sufficiently and cells transfected with AM DNA significantly lower pulmonary arterial pressure and improve vascular resistance. More recently, post MI delivery of MSCs internalized with the non-viral vector cationized dextran, condensed with AM, also offer improvement to post MI cardiac function.[56] Also demonstrated in a rat MI model, myocardial injection of pleiotrophin plasmid in fibrin glue biopolymer improves neovasculogenesis compared to injection of naked plasmid alone.[57] Cationic poly (b-aminoesters) (PDMA) are other non-viral vectors with potential applications in cardiac delivery. PDMA vectors carrying a GFP plasmid improve transfection efficiency into both neonatal mouse cardiomyocytes and human umbilical vein endothelial cells (HUVECs) compared to naked DNA.[58,59] Water-soluble lipopolymer (WSLP) is another emerging non-viral vector that shows efficient release of cargo from the aortic smooth muscle cell endosome.[60] Injection of WSLP loaded with the VEGF gene into rabbit myocardium induced 50% greater expression of VEGF protein than injection of naked DNA alone. Multi-block co-polymers (MBCPs) offer yet another strategy to transfect the myocardium. MBCPs complexed with plasmid DNA (pDNA) show robust localization and retention with healthy rat myocardium.[61] It is proposed that this smart polymeric system releases pDNA-encapsulated micelles that enter the cell through endosomal uptake when activated by body temperature.

Despite the advantages of hydrogel-based cell delivery, a number of disadvantages must be overcome to prepare these systems for clinical use. Controlling the chemical composition of biodegradable hydrogels is pivotal since degradation prior to sufficient graft remodeling by endogenous and seeded cells can lead to deleterious myocardial weakening.[9] While many hydrogel chemistries are amenable to tunable degradation, the high diffusivity of proteins in the largely aqueous matrix can result in release rates largely dictated by protein transport out of the hydrogel rather than hydrolysis of the polymer network. This significantly reduces the timeframe over which a hydrogel can release physiologic concentrations of proteins. On the same token, prolonged hydrogel residence time can distort ventricular geometry and affect mechanical operation of the heart muscle.[62] In addition, insufficient nutrient diffusion into the 3D construct and collagenous encapsulation[1] are also addressed by optimization of the hydrogel polymer composition. Collagenous encapsulation may be followed by secondary angiogenesis of the implant, generating concern that the therapeutic benefit of the hydrogel may be attributed to pathologic rejection of the implant rather than to the therapy itself. Although nanofiber technology overcomes the major immunogenic limitations, the nanoscale comes at a loss to drug loading availability.[29] Moreover, nanoscale materials may be susceptible to endocytosis and enzyme degradation compared to other hydrogel scaffolds, thereby limiting their *in vivo* residence time and efficacy. Lastly, the low modulus of soft hydrogels may inadequately stabilize weakened myocardial tissue.[29]

7. FUTURE FOR CARDIAC DELIVERY USING BIOMATERIAL SCAFFOLDS

Biomaterials chemistry and engineering is fast growing with many applications outside cardiac and vascular repair. Initial application of new technology is often pioneered in the areas of bone and cartilage regeneration and may require some extensive modification for cardiac use. The future of tissue engineering may lie in scaffold patterning, wherein the biomaterial is functionalized to direct cellular deposition of ECM and organization of regenerated tissue. Rat adrenal pheochromocytoma cells have recently been differentiated on hydrogelating self-assembling fibers (hSAFs), which constitute a novel nanofiber hydrogel with a greater control of self-assembly and, thus, greater control of cellular biology postimplantation.[63] Using a novel technique to tether small-molecular chemical functional groups to 3D scaffolds, MSCs were directed toward osteogenic and adipogenic lineage.[64] Non-toxic click chemistry has recently been engineered to create cell-embedded hydrogels that are patterned with biological functionalities.[65] This technology is promising because it offers chemical tools to study cellular behavior within materials patterned in closer likeness to biological tissue. Other technologies, including cellular encapsulation in nanoporous microspheres to increase oxygenation,[66] are also being explored. Finally, as alluded to by Davis and co-workers in a recent review, many groups are moving toward "smart materials" that respond to their environment.[67] Segers and co-workers put this idea to the test by engineering a self-assembling peptide/stromal cell-derived factor fusion that contained a matrix metalloproteinase (MMP) cleavage sequence in between the peptide and protein.[68] In the setting of acute MI, MMP levels are upregulated and thus should progressively induce release of the protein. While this idea has been used in potential tumor treatments,[69] finding more potential smart mechanisms for MI treatment is a rapidly growing area.

In closing, the cardiac delivery systems presented clearly demonstrate the utility of therapeutic cell, protein, and gene therapy. The various successful strategies and biologics employed also underscore our deficiency in understanding the mechanisms underlying cardiac repair. The functional benefits derived in these studies, however, support further optimization of biomaterial and therapeutic combinations that best integrate and direct myocardial regeneration within the injured heart. As more is learned about the exact mechanisms that prevent successful endogenous and exogenous repair, biomaterials can be generated to better suit the needs of the damaged tissue.

REFERENCES

1. Ratner BD, Bryant SJ (2004) BIOMATERIALS: where we have been and where we are going. *Annu Rev Biomed Eng* 6: 41–75.
2. Tang YL, Zhao Q, Qin X, Shen L, *et al.* (2005) Paracrine action enhances the effects of autologous mesenchymal stem cell transplantation on vascular regeneration in rat model of myocardial infarction. *Ann Thorac Surg* 80: 229–237.
3. Singelyn JM, Dequach JA, Seif-Naraghi SB, Littlefield RB, *et al.* (2009) Naturally derived myocardial matrix as an injectable scaffold for cardiac tissue engineering. *Biomaterials* 30: 5409–5416.
4. Landa N, Miller L, Feinberg MS, Holbova R, *et al.* (2008) Effect of injectable alginate implant on cardiac remodeling and function after recent and old infarcts in rat. *Circulation* 117: 1388–1396.
5. Zhang H, Zhu SJ, Wang W, Wei YJ, *et al.* (2008) Transplantation of microencapsulated genetically modified xenogeneic cells augments angiogenesis and improves heart function. *Gene Ther* 15: 40–48.
6. Akasha AA, Sotiriadou I, Doss MX, Halbach M, *et al.* (2008) Entrapment of embryonic stem cells-derived cardiomyocytes in macroporous biodegradable microspheres: preparation and characterization. *Cell Physiol Biochem* 22: 665–672.
7. Baruch L, Benny O, Gilert A, Ukobnik M, *et al.* (2009) Alginate-PLL cell encapsulation system co-entrapping PLGA-microspheres for the continuous release of anti-inflammatory drugs. *Biomed Microdevices* 11: 1103–1113.
8. Sy JC, Seshadri G, Yang SC, Brown M, *et al.* (2008) Sustained release of a p38 inhibitor from non-inflammatory microspheres inhibits cardiac dysfunction. *Nat Mater* 7: 863–868.

9. Isenberg BC, Williams C, Tranquillo RT (2006) Small-diameter artificial arteries engineered *in vitro. Circ Res* 98: 25–35.

10. Simpson D, Liu H, Fan TH, Nerem R, *et al.* (2007) A tissue engineering approach to progenitor cell delivery results in significant cell engraftment and improved myocardial remodeling. *Stem Cells* 25: 2350–2357.

11. Giraud M-N, Ayuni E, Cook S, Siepe M, *et al.* (2008) Hydrogel-based engineered skeletal muscle grafts normalize heart function early after myocardial infarction. *Artif Organs* 32: 692–700.

12. Schussler O, Coirault C, Louis-Tisserand M, Al-Chare W, *et al.* (2009) Use of arginine-glycine-aspartic acid adhesion peptides coupled with a new collagen scaffold to engineer a myocardium-like tissue graft. *Nat Clin Pract Cardiovasc Med* 6: 240–249.

13. Shapira-Schweitzer K, Seliktar D (2007) Matrix stiffness affects spontaneous contraction of cardiomyocytes cultured within a PEGylated fibrinogen biomaterial. *Acta Biomater* 3: 33–41.

14. Zhang G, Wang X, Wang Z, Zhang J, *et al.* (2006) A PEGylated fibrin patch for mesenchymal stem cell delivery. *Tissue Eng* 12: 9–19.

15. Huang NF, Yu J, Sievers R, Li S, *et al.* (2005) Injectable biopolymers enhance angiogenesis after myocardial infarction. *Tissue Eng* 11: 1860–1866.

16. Kofidis T, de Bruin JL, Hoyt G, Lebl DR, *et al.* (2004) Injectable bioartificial myocardial tissue for large-scale intramural cell transfer and functional recovery of injured heart muscle. *J Thorac Cardiovasc Surg* 128: 571–578.

17. Kofidis T, Lebl DR, Martinez EC, Hoyt G, *et al.* (2005) Novel injectable bioartificial tissue facilitates targeted, less invasive, large-scale tissue restoration on the beating heart after myocardial injury. *Circulation* 112: I-173–177.

18. Kofidis T, de Bruin JL, Hoyt G, Ho Y, *et al.* (2005) Myocardial restoration with embryonic stem cell bioartificial tissue transplantation. *J Heart Lung Transplant* 24: 737–744.

19. Hoemann CD, Chenite A, Sun J, Hurtig M, *et al.* (2007) Cytocompatible gel formation of chitosan-glycerol phosphate solutions

supplemented with hydroxyl ethyl cellulose is due to the presence of glyoxal. *J Biomed Mater Res A* 83: 521–529.

20. Lu WN, Lu SH, Wang HB, Li DX, *et al.* (2009) Functional improvement of infarcted heart by co-injection of embryonic stem cells with temperature-responsive chitosan hydrogel. *Tissue Eng Part A* 15: 1437–1447.

21. Christman KL, Vardanian AJ, Fang Q, Sievers RE, *et al.* (2004) Injectable fibrin scaffold improves cell transplant survival, reduces infarct expansion, and induces neovasculature formation in ischemic myocardium. *J Am Coll Cardiol* 44: 654–660.

22. Martens T, Godier A, Parks J, Wan L, *et al.* (2009) Percutaneous cell delivery into the heart using hydrogels polymerizing *in situ*. *Cell Transplant* 18: 297–304.

23. Zhang G, Hu Q, Braunlin EA, Suggs LJ, *et al.* (2008) Enhancing efficacy of stem cell transplantation to the heart with a PEGylated fibrin biomatrix. *Tissue Eng Part A* 4: 1025–1036.

24. Cortes-Morichetti M, Frati G, Schussler O, Van Huyen J, *et al.* (2007) Association between a cell-seeded collagen matrix and cellular cariomyoplasty for myocardial support and regeneration. *Tissue Eng* 13: 2681–2687.

25. Matsumura G, Hibino N, Ikada Y, Kurosawa H, *et al.* (2003) Successful application of tissue engineered vascular autografts: clinical experience. *Biomaterials* 24: 2303–2308.

26. Thai H, Juneman E, Lancaster J, Hagerty T, *et al.* (2009) Implantation of a three-dimensional fibroblast matrix improves left ventricular function and blood flow after acute myocardial infarction. *Cell Transplant* 18: 283–295.

27. Lancaster JJ, Johnson NM, Juneman E, Thai HM, *et al.* (2009) Construction of a spontaneously contracting biologically active cardiomyocyte scaffold. In: *Basic Cardiovascular Sciences Conference*, Lake Las Vegas, Nevada.

28. Yang M-C, Wang S-S, Chou N-K, Chi N-H, *et al.* (2009) The cardiomyogenic differentiation of rat mesenchymal stem cells on silk fibroin-polysaccharide cardiac patches *in vitro*. *Biomaterials* 30: 3757–3765.

29. Goldberg M, Langer R, Xinqiao J (2007) Nanostructured materials for applications in drug delivery and tissue engineering. *J Biomater Sci Polym Ed* 3: 241–268.

30. Davis ME, Hsieh PC, Takahashi T, Song Q, *et al.* (2006) Local myocardial insulin-like growth factor 1 (IGF-1) delivery with biotinylated peptide nanofibers improves cell therapy for myocardial infarction. *Proc Natl Acad Sci USA* 103: 8155–8160.

31. Padin-Iruegas ME, Misao Y, Davis ME, Segers VFM, *et al.* (2009) Cardiac progenitor cells and biotinylated insulin-like growth factor-1 nanofibers improve endogenous and exogenous myocardial regeneration after infarction. *Circulation* 120: 876–887.

32. Yamada N, Okano T, Sakai H, Karikusa F, *et al.* (1990) Thermo-responsive polymeric surfaces; control of attachment and detachment of cultured cells. *Makromol Chem Rapid Commun* 11: 571–576.

33. Itabashi Y, Miyoshi S, Kawaguchi H, Yuasa S, *et al.* (2004) A new method for manufacturing cardiac cell sheets using fibrin-coated dishes and its electrophysiological studies by optical mapping. *Artific Organs* 29: 95–103.

34. Masuda S, Shimizu T, Yamato M, Okano T (2008) Cell sheet engineering for heart tissue repair. *Adv Drug Deliv Rev* 60: 277–285.

35. Miyahara Y, Nagaya N, Kataoka M, Yanagawa B, *et al.* (2006) Monolayered mesenchymal stem cells repair scarred myocardium after myocardial infarction. *Nat Med* 12: 459–465.

36. Kondoh H, Sawa Y, Miyagawa S, Sakakida-Kitagawa S, *et al.* (2006) Longer preservation of cardiac performance by sheet-shaped myoblast implantation in dilated cardiomyopathic hamsters. *Cardiovasc Res* 69: 466–475.

37. Sawa Y, Memon IA, Matsuda H. (2005) Myocardial regeneration therapy with tissue implantation of autologous myoblast sheets for severe impaired heart failure. In: Mori H, Matsuda H (eds.) *Cardiovascular Regeneration Therapies Using Tissue Engineering Approaches.* Springer, Tokyo, pp. 53–65.

38. Laham RJ, Rezaee M, Post M, Sellke FW, *et al.* (1999) Intracoronary and intravenous administration of basic fibroblast growth factor: myocardial and tissue distribution. *Drug Metab Dispos* 27: 821–826.

39. Madan PL (1978) Effect of intensity of agitation on disintegration time of tablets. *J Pharm Sci* 67: 210–214.

40. Sisson K, Zhang C, Farach-Carson MC, Chase DB, *et al.* (2009) Evaluation of cross-linking methods for electrospun gelatin on cell growth and viability. *Biomacromolecules* 10: 1675–1680.

41. Yamamoto T, Suto N, Okubo T, Mikuniya A, *et al.* (2001) Intramyocardial delivery of basic fibroblast growth factor-impregnated gelatin hydrogel microspheres enhances collateral circulation to infarcted canine myocardium. *Jpn Circ J* 65: 439–444.

42. Sakakibara Y, Tambara K, Sakaguchi G, Lu F, *et al.* (2003) Toward surgical angiogenesis using slow-released basic fibroblast growth factor. *Eur J Cardiothorac Surg* 24: 105–112.

43. Iwakura A, Fujita M, Kataoka K, Tambara K, *et al.* (2003) Intramyocardial sustained delivery of basic fibroblast growth factor improves angiogenesis and ventricular function in a rat infarct model. *Heart Vessels* 18: 93–99.

44. Liu Y, Sun L, Huan Y, Zhao H, *et al.* (2006) Application of bFGF and BDNF to improve angiogenesis and cardiac function. *J Surg Res* 136: 85–91.

45. Sakakibara Y, Nishimura K, Tambara K, Yamamoto M, *et al.* (2002) Prevascularization with gelatin microspheres containing basic fibroblast growth factor enhances the benefits of cardiomyocyte transplantation. *J Thorac Cardiovasc Surg* 124: 50–56.

46. Yamamoto M, Sakakibara Y, Nishimura K, Komeda M, *et al.* (2003) Improved therapeutic efficacy in cardiomyocyte transplantation for myocardial infarction with release system of basic fibroblast growth factor. *Artif Organs* 27: 181–184.

47. Hoshino K, Kimura T, De Grand AM, Yoneyama R, *et al.* (2006) Three catheter-based strategies for cardiac delivery of therapeutic gelatin microspheres. *Gene Ther* 13: 1320–1327.

48. Wang F, Li Z, Tamama K, Sen CK, *et al.* (2009) Fabrication and characterization of prosurvival growth factor releasing, anisotropic scaffolds for enhanced mesenchymal stem cell survival/growth and orientation. *Biomacromolecules* 10: 2609–2618.

49. Kobayashi H, Minatoguchi S, Yasuda S, Bao N, *et al.* (2008) Post-infarct treatment with an erythropoietin-gelatin hydrogel drug delivery system for cardiac repair. *Cardiovasc Res* 79: 611–620.

50. Hsieh PC, Davis ME, Gannon J, MacGillivray C, *et al.* (2006) Controlled delivery of PDGF-BB for myocardial protection using injectable self-assembling peptide nanofibers. *J Clin Invest* 116: 237–248.

51. Engel FB, Hsieh PC, Lee RT, Keating MT (2006) FGF1/p38 MAP kinase inhibitor therapy induces cardiomyocyte mitosis, reduces scarring,

and rescues function after myocardial infarction. *Proc Natl Acad Sci USA* 103: 15546–15551.

52. Hao X, Silva EA, Mansson-Broberg A, Grinnemo K-H, *et al.* (2007) Angiogenic effects of sequential release of VEGF-A165 and PDGF-BB with alginate hydrogels after myocardial infarction. *Cardiovasc Res* 75: 178–185.

53. Murry C, Kay M, Bartosek T, Hauschka S, *et al.* (1996) Muscle differentiation during repair of myocardial necrosis in rats via gene transfer with MyoD. *J Clin Invest* 98: 2209–2217.

54. Yockman JW, Kastenmeier A, Erickson HM, Brumbach JG, *et al.* (2008) Novel polymer carriers and gene constructs for treatment of myocardial ischemia and infarction. *J Control Release* 132: 260–266.

55. Nagaya N, Kangawa K, Kanda M, Uematsu M, *et al.* (2003) Hybrid cell-gene therapy for pulmonary hypertension based on phagocytosing action of endothelial progenitor cells. *Circulation* 108: 889–895.

56. Jo J, Nagaya N, Miyahara Y, Kataoka M, *et al.* (2007) Transplantation of genetically engineered mesenchymal stem cells improves cardiac function in rats with myocardial infarction: benefit of a novel nonviral vector, cationized dextran. *Tissue Eng* 13: 313–322.

57. Christman KL, Fang Q, Yee MS, Johnson KR, *et al.* (2005) Enhanced neovasculature formation in ischemic myocardium following delivery of pleiotrophin plasmid in a biopolymer. *Biomaterials* 26: 1139–1144.

58. Xu P, Li S-Y, Li Q, Ren J, *et al.* (2006) Biodegradable cationic polyester as an efficient carrier for gene delivery to neonatal cardiomyocytes. *Biotechnol Bioeng* 95: 893–903.

59. Green JJ, Shi J, Chiu E, Leshchiner ES, *et al.* (2006) Biodegradable polymeric vectors for gene delivery to human endothelial cells. *Bioconjug Chem* 17: 1162–1169.

60. Lee M, Rentz J, Han SO, Bull DA, *et al.* Water-soluble lipopolymer as an efficient carrier for gene delivery to myocardium. *Gene Ther* 10: 585–593.

61. Namgung R, Nam S, Kim SK, Son S, *et al.* (2009) An acid-labile temperature-responsive sol-gel reversible polymer for enhanced gene delivery to the myocardium and skeletal muscle cells. *Biomaterials* 30: 5225–5233.

62. Shapira K, Dikovsky D, Habib M, Gepstein L, *et al.* (2008) Hydrogels for cardiac tissue regeneration. *Biomed Mater Eng* 18: 309–314.

63. Banwell EF, Abelardo ES, Adams DJ, Birchall MA, *et al.* (2009) Rational design and application of responsive [alpha]-helical peptide hydrogels. *Nat Mater* 8: 596–600.

64. Benoit DSW, Schwartz MP, Durney AR, Anseth KS (2008) Small functional groups for controlled differentiation of hydrogel-encapsulated human mesenchymal stem cells. *Nat Mater* 7: 816–823.

65. DeForest CA, Polizzotti BD, Anseth KS (2009) Sequential click reactions for synthesizing and patterning three-dimensional cell microenvironments. *Nat Mater* 8: 659–664.

66. Gimi B, Kwon J, Su Y, Nemani K, *et al.* (2009) Cell encapsulation and oxygenation in nanoporous microcontainers. *Biomed Microdevices* 11(6): 1205–1212.

67. Davis ME, Hsieh PC, Grodzinsky AJ, Lee RT (2005) Custom design of the cardiac microenvironment with biomaterials. *Circ Res* 97: 8–15.

68. Segers VFM, Tokunou T, Higgins LJ, MacGillivray C, *et al.* (2007) Local delivery of protease-resistant stromal cell derived factor-1 for stem cell recruitment after myocardial infarction. *Circulation* 116: 1683–1692.

69. Chau Y, Padera RF, Dang NM, Langer R (2006) Antitumor efficacy of a novel polymer-peptide-drug conjugate in human tumor xenograft models. *Int J Cancer* 118: 1519–1526.

VI. How to Create an Artificial Heart *Ex Vivo*?

A. The Mechanical Heart

JoAnn Lindenfeld and Patricia A. Kelly

1. INTRODUCTION

Heart failure (HF) is an important public health problem, affecting more than five million people in both the United States and Europe. Despite advances in medical and device therapy, mortality from advanced HF remains high, with a five-year survival rate of only 20%.[1] The finite pool of available donor hearts limits cardiac transplantation to all but a small number of patients, leaving many without options. Accordingly, there has been growing interest in the use of mechanical hearts as an alternative therapy for these patients. Although they were initially used as a short term support to allow myocardial recovery after an acute illness or surgery, or to allow time for a donor heart to become available for transplantation (bridge to transplantation, BTT), improvements in size, thrombogenicity, ease of implantation, and durability have allowed studies using these devices as the final or destination therapy (DT). With continued improvements, these devices are likely to be implanted earlier in the course of HF and may soon rival transplantation as the preferred therapy for patients with advanced HF.

2. HISTORY

In the 1950s, the advent of cardiopulmonary bypass to maintain oxygenation and organ perfusion during cardiac surgical procedures spawned interest in the use of mechanical devices for more prolonged circulatory support. Spencer and colleagues reported in 1965 the successful application of roller-pump technology to temporarily support a patient who could not be weaned from cardiopulmonary bypass.[2] Due to a number of technical limitations roller pumps were impractical for use as true assist devices and their application was not widely accepted. In 1966, DeBakey successfully utilized a pneumatically driven diaphragm pump to support a 37-year-old woman following aortic and mitral valve replacements. For ten days, circulatory support was maintained via a circuit from the left atrium to the right axillary artery. The patient was successfully weaned and recovered.[3] Still, it was nearly two decades before the development of widespread interest in ventricular assist devices and the first report of successful BTT using a left ventricular assist device.[4] Today, there are a number of approved and investigational devices being used for both short- and long-term support.

Initial progress with the development of a total artificial heart (TAH) also dates back to the 1950s, with the report by Akutsu and Kolff of the implantation of a TAH in a dog who subsequently survived for 90 minutes.[5] In 1969, Cooley implanted a device in a man who could not be weaned from bypass following a left ventricular aneurysm repair. The patient survived 64 hours until cardiac transplantation.[6] It was not until 1982, however, with the widely publicized case of Dr. Barney Clark, that the idea of a TAH received worldwide attention. Dr. William DeVries implanted the Jarvik-7 TAH into Clark, a 61-year-old retired dentist, who survived for 112 days.[7] A total of five patients underwent implantation of the Jarvik-7 device during this initial evaluation, with the longest survival being 620 days.[8] Several other models have been developed worldwide and used in relatively small numbers of patients, but widespread utilization has been limited by complications such as infection, thrombosis and stroke.

3. SITUATIONS IN WHICH TO CONSIDER A MECHANICAL HEART

Mechanical hearts may be used for temporary or long-term cardiac support (Figure 1). Temporary devices may provide cardiac support for high-risk procedures, to provide time for myocardial recovery such as in patients with post-cardiotomy shock or acute myocarditis, or as a "bridge to decision." Patients with uncertain neurological function following cardiac arrest may benefit from temporary support to allow time to assess recovery and make a decision about placement of a long-term device. If CNS function improves, a chronic device may be placed as BTT or as "DT" (DT) — thus the temporary device serves as a "bridge to bridge." Long-term or chronic devices are most

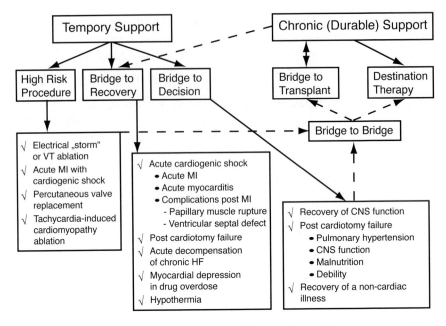

Figure 1. Situations to consider mechanical heart.

often placed to bridge critically ill patients to cardiac transplantation (BTT) or as DT for those who are not candidates for transplantation. Long-term devices may also be used for "bridge to recovery" in patients with long-standing HF and systolic dysfunction where recovery requires months and as a "bridge to decision" in clinical situations such as pulmonary hypertension where sufficient improvement to permit successful cardiac transplantation may also take months.

4. MECHANICAL HEARTS: TYPES OF DEVICES

Mechanical hearts may be categorized by the expected duration of use (temporary or chronic); the method of implantation (percutaneous or surgical); type of support (counterpulsation, cardiopulmonary support (CPS), ventricular assist device (VAD), or TAH); by type of blood flow (pulsatile *vs.* continuous); and by the chamber(s) supported (left ventricle, right ventricle, or biventricular).

4.1. Temporary (Short-Term) *vs.* Chronic (Long-Term)

Temporary devices include counterpulsation devices, cardiopulmonary bypass, and VADs. Intra-aortic balloon pumps may increase cardiac output but are not considered mechanical hearts. Cardiopulmonary bypass is utilized in the very short term — either in the catheterization laboratory or during cardiac surgery. Use of extracorporeal membrane oxygenation (ECMO) is generally used for short-term support.[9] While neither cardiopulmonary bypass nor ECMO are considered mechanical hearts, they are currently the only available methods of support for patients who have both primary respiratory failure and cardiac failure.

Chronic or long-term devices were originally designed to support the circulation for months or years. There are both extra- and intracorporeal devices. Most of these devices are equipped with portable consoles to allow the recipient a reasonable quality of life living outside the hospital as well as with batteries to allow several hours unattached to the console.

4.2. Percutaneous *vs.* Surgical

Devices that can be placed percutaneously are generally short-term devices while both short- and long-term devices are available for surgical implantation.

4.2.1. *Pulsatile vs. continuous (non-pulsatile) flow*

Pulsatile devices deliver a stroke volume very similar to a normal stroke volume generating both a systolic and diastolic pressure. Continuous flow devices are often not truly non-pulsatile, as there may be a detectable pulse as the left ventricle augments flow in systole. However, these devices generate a markedly diminished pulse pressure, which is often not detectable with a sphygmomanometer and thus they are termed "non-pulsatile." Initially it was uncertain whether non-pulsatile flow would provide the same quality of organ perfusion but that notion has been largely dispelled as these devices do seem as effective as pulsatile devices in restoring blood flow and vital organ function as well as improving exercise capacity, permitting myocardial recovery, and enhancing survival to and following cardiac transplantation.[10–12] However, peripheral vascular reactivity as assessed by brachial artery flow-mediated dilation is greater with pulsatile than non-pusatile flow.[13] Moreover, continuous or non-pulsatile devices have been reported to have unanticipated side effects such as gastrointestinal (GI) bleeding, discussed later in this text.[14]

4.2.2. *Chamber supported*

Devices may support the left ventricle, the right ventricle, or both ventricles (bi-ventricular).

5. MECHANICAL HEARTS: SPECIFIC DEVICES

5.1. Temporary Ventricular Assist Devices

A number of temporary devices placed either percutaneously, such as the Impella 2.5 or 5.0 (Abiomed-Impella, CardioSystems GmbH, Aachen, Germany) and TandemHeart (CardiacAssist, Inc., Pittsburgh, PA) or implanted surgically, such as the, Abiomed BVS 5000 (Abiomed, Danvers, MA) or Bio-Medicus (Medtronic, Inc. Minneapolis, MN) are available but are beyond the scope of this discussion.

5.2. Chronic Ventricular Assist Devices

5.2.1. *First generation devices*

First generation devices are pulsatile volume displacement pumps. Volume displacement is accomplished via a pusher plate activated either pneumatically or electrically. These devices deliver a stroke volume similar to that provided by a normal heart generating both a systolic and diastolic blood pressure. All these devices are relatively large because they must have a reservoir for the blood accumulation prior to ejection. The large size, when implanted intracorporeally, requires more extensive surgical dissection resulting in an increased incidence of bleeding, pocket hematomas, and infection. The requirement for an air vent channel also increases the size of the percutaneous drive line. The pusher plate action and bearings are noisy in the first generation devices and the moving parts and valves result in limited durability.

Thoratec PVAD and IVAD: The Thoratec paracorporeal VAD (PVAD) has a 65 ml pumping chamber and two mechanical valves — one in each of the inflow and outflow Dacron conduits. The PVAD is activated pneumatically delivering flows up to 7.2 L/min.[15] The inflow and outflow conduits are implanted in the heart and great vessels, externalized through the abdominal wall and connected to the pump, which sits on the abdominal wall. This paracorporeal position allows the device to be implanted in smaller patients with a recommended body surface area of >0.73 m^2. The PVAD may be utilized for left, right, or for biventricular assistance. Full anticoagulation with warfarin, in addition to aspirin, is recommended. The Thoratec PVAD was approved by the FDA as a bridge to transplant in 1995 and for postcardiotomy support in 1998. More than 3000 PVADs have been implanted worldwide. Survival to recovery or transplantation for patients receiving the PVAD was 64.8% for LVAD only, 56.6% for BiVAD, and 31.2% for RVADs.[16] The smaller and more portable TLC (Thoratec Laboratories Corp.)-II console for out of hospital use was approved in 2003.

The Thoratec intracorporeal VAD (IVAD) has the same conduits, pumping mechanism, and mechanical valves as the PVAD but has been reconfigured for intracorporeal implantation. Changes include a titanium housing for the pump to allow an intracorporeal position, a modestly reduced weight and volume, and a reduction in the drive line from 20 to 9 mm.[17] The IVAD, like the PVAD, can be utilized for single or biventricular support. The pump is implanted in either a pre- or intraperitoneal position. The consoles for the PVAD and IVAD are identical, as are the anticoagulation recommendations. The IVAD version of the Thoratec PVAD was FDA approved in 2004.

Berlin Heart EXCOR: The Berlin Heart EXCOR (Berlin Heart AG, Berlin, Germany) is a paracorporeal, pneumatically-driven device that may be utilized for short- or mid-term single or biventricular support. As with the Thoratec PVAD the cannulae are tunneled beneath the abdominal wall to exit and attach to the pump(s), which sit on the abdomen.[18] The EXCOR is available in sizes for both adult and pediatric use. In July 2011 an advisory panel to the FDA Division of Cardiovascular Devices unanimously recommended approval of the Berlin Heart EXCOR under a humanitarian devices exemption.

Thoratec Heartmate I: The Thoratec HeartMate I is a pulsatile device that requires surgical implantation, and is currently configured for left ventricular support only. It is constructed of titanium with a polyurethane diaphragm and a pusher plate actuator. Porcine valves are suspended within the Dacron inflow and outflow cannulae. The inlet and outlet cannulae are implanted in the left ventricular apex and the ascending aorta, respectively. An important feature of the HeartMate I is the proprietary titanium microsphere coating of the surface of the blood pumping chamber that allows formation of a "pseudointima" that resists thrombus formation. It is the only device that requires only antiplatelet therapy with aspirin and dipyridamole while maintaining a low thromboembolic rate. Important modifications to the HeartMate I incorporated into the current HeartMate XVE (Figure 2) include the conversion from pneumatic to vented electric, an outflow graft revised to provide "bend relief" that prevents kinking and abrasion, modification of the porcine valves to prevent incompetence and commissural dehiscence, a reduction in the pressure of the pumping chamber to reduce stress on the diaphragm, motor and bearings, repositioning of the diaphragm with software changes that have reduced buckling of the diaphragm and techniques to immobilize the drive line to allow tissue ingrowth, thus decreasing drive line infections. This device has been used in more than 5000 patients worldwide. Its large size precludes implantation in people with a BSA of <1.5 M^2. The major limiting factor has been device longevity. Bearing wear often occurs at 18–24 months, ultimately leading to device failure requiring

(A)   (B)   (C)

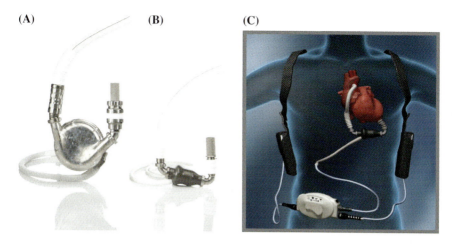

Figure 2.   HeartMate XVE (A) and HeartMate II (B) and (C) HeartMate II implanted. Reprinted with permission from Thoratec Corporation.

surgical placement of a new device. The HeartMate XVE durability is improved compared to the VE with a six and 12 months freedom from failure of 97% and 82% compared to 92% and 73% for the VE.[19,20] When used as a BTT or for myocardial recovery, overall survival until device removal is 60%–70% with a perioperative mortality of 15%–20%.[21-23]

Novocor Left Ventricular Assist Device: The Novocor Left Ventricular Assist Device (World Heart Corp., Oakland, CA) was the first chronic device used successfully as a bridge to transplant in 1984. Despite its excellent long-term durability, it is no longer available for several reasons including its size and a high incidence of thromboembolic events, due in part to the need for full anticoagulation.[22-25]

LionHeart VAD: The LionHeart (Arrow International Inc., Reading, PA) has a pulsatile blood-sac design.[26] It has a transcutaneous energy transmission system (TETS) and thus requires no drive line. Unidirectional blood flow is maintained via two Delrin disk monostrut valves. The inlet cannula is implanted in the left ventricular apex and the outlet cannula is implanted as an end-to-side anastomosis to the ascending aorta. Maximum pump outflow is approximately 8 L/min, with a stroke volume of 64 mL.[26] The LionHeart Device is no longer commercially available but a smaller, similar device is being developed (www.worldheart.com).

5.2.2. *Second generation devices*

The large size and durability problems noted with first generation devices improved considerably with the development of axial, continuous (non-pulsatile) flow devices. These devices have only one moving part and no valves and thus are more durable than the HeartMate I. They are also considerably smaller, allowing insertion in smaller patients and a less extensive surgical procedure with less perioperative bleeding. The smaller size has nearly eliminated problems of abdominal compression, early satiety, etc. associated with first generation devices. A smaller drive line appears to reduce drive line and device infections. Full anticoagulation with warfarin and aspirin is required for all these devices.

HeartMate II (Thoratec Corp, Pleasanton, CA.): The Heartmate II (Figure 2) consists of a titanium tube 12 mm in diameter. Blood flows into the tube over an inlet "stator" that has three guide veins that direct blood flow over a rotor that has three curved blades that turn rapidly (6000 to 15,000 rpm) and impart kinetic energy to the blood (Figure 3). As the blood flows over the outflow stator and its veins, the radial velocity of the blood is converted to axial flow.[27] Although flow is continuous, pulsatility increases as left ventricular end-diastolic pressure increases. The pump weighs only 350 gm (one-fourth of the HeartMate I), is 7 by 4 cm and is able to deliver up to 10 L/min (Figure 2). Because there is no need for a blood sac and thus no venting, the percutaneous drive line is considerably smaller than in first generation devices. The HeartMate II has been approved by the FDA for use as BTT in 2008 based on a study of 133 patients with a median follow-up of 126 days and survival of 75% and 68% at six and 12 months (see below Clinical Trials and

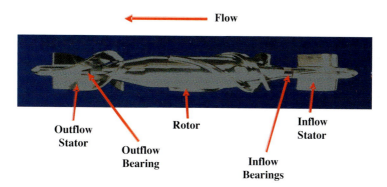

Figure 3.   HeartMate II axial flow pump: Internal view. Reprinted with permission from Thoratec Corporation.

Ref. 28). The HeartMate II has been used in more than 1200 patients worldwide.

Jarvik 2000: The Jarvik 2000 (Jarvik Heart Inc., New York, NY) is an electrically powered, axial flow pump with the pump placed directly in the LV cavity. No pocket dissection is required to place the pump, essentially eliminating the risk of pocket infection. No inflow cannula is required and the outflow cannula is generally placed in the descending aorta rather than the ascending aorta. The pump is quite small (85 g and 2.4 × 5.5 cm) with a titanium housing and a single moving part — the rotating vaned impeller. All blood contact surfaces are made of titanium. Flows as high as 8 L/min can be generated. Although device failure has not been a problem, thrombus may form in the left ventricle at the base of the pump and interfere with pump filling. This can be prevented by intermittent decreases in the pump speed which allows increased filling and washout of the ventricle.[29] An intermittent speed controller is available in Europe but not yet approved in the United States. The driveline may be externalized at the base of the skull in patients for chronic use. Results to date are encouraging with survival similar to the HeartMate II although thromboembolic rates appear to be higher.[30,31] Infection rates have been quite low.

MicroMed DeBakey and Berlin INCOR: The MicroMed-D (Micro-Med Technology, Inc. Houston, Texas) and the Berlin Heart INCOR (Berlin Heart AG, Berlin, Germany) are both axial flow pumps currently undergoing clinical investigation.[32,33]

5.2.3. *Third generation devices*

These devices provide either axial or centrifugal continuous (non-pulsatile) flow. The VentrAssist (Ventracor LTD., Chatswood New South Wales, Australia) and DuraHeart (Terumo, Somerset, New Jersey, USA) are magnetically levitated and are being evaluated in Phase II trials. The HeartWare (Thoratec, Pleasanton, CA) (Figure 4) device is also magnetically levitated and is in phase III clinical trials and is particularly intriguing as it is very small and has no bearings, which is likely to provide very long-term durability. It is implanted in the pericardial space, reducing the risk of pocket infections. The device weighs only 145 g but can provide up to 10 L of flow. A Phase III trial for BTT with the HeartWare device is currently enrolling subjects in the US (www.clinicaltrials.gov; NCT00751972).[34] Even smaller devices, such as the Synergy Micropump (Circulite, Inc., Saddle Brook, N.J.), which is the size of a AA battery providing partial ventricular support of up to 3 L/min, are being developed.[35] The device withdraws blood from the left atrium (instead of the left ventricle) and pumps the blood though a superficially placed micro-pump (instead of a pump placed

(A)   **(B)**   **(C)**

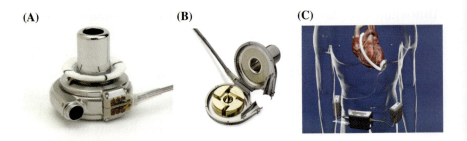

Figure 4.   HeartWare centrifugal flow pump. (A) External and **(B)** internal views. **(C)** Implanted position. Reprinted with Permission from HeartWare, Inc.

deep in the chest cavity) to the subclavian artery (instead of the aorta). All of these devices require full anticoagulation.

5.3. Total Artificial Heart (TAH)

Although associated with significant survival benefits, successful use of LVADs may be limited by right HF, low flow, valvular regurgitation and arrhythmias. These problems have stimulated continued interest in the development of a total artificial heart (TAH). In 2004, the US FDA approved the CardioWest TAH (SynCardia, Inc., Tucson, AZ) for use as BTT. The pneumatic device, based on the original Jarvik-7 prototype, is attached to the recipient's atria and completely replaces both ventricles and all four cardiac valves. Because the device is powered by a large external console the patient is confined to the hospital. In Europe, a portable driver is available and patients can be discharged home to await transplantation. This feature is not yet approved for use in the United States. Recipients must be fully anticoagulated while the device is implanted.

As of June 2009, 800 patients in the United States and Europe had been treated with the CardioWest TAH, for a total of more than 150 patient years of therapy. Copeland and colleagues published one of the earliest reports, which included 81 of these patients, in 2004. When compared to a group of historical controls who received no mechanical circulatory support, the CardioWest TAH patients demonstrated a significant survival benefit. Seventy-nine percent of patients in the treatment group survived to transplantation *vs.* 46% of controls (p < 0.001). Moreover, one year overall survival was 70% in the device group *vs.* 31% in the control group (p < 0.001). Adverse events were not dissimilar to those encountered with other mechanical circulatory assist devices and included bleeding (62%), infection (77%), neurologic events (27%), peripheral thromboembolism (14%) and device malfunction.[36]

Abiocor: The AbioCor TAH (Abiomed, Inc., Danvers, MA) was FDA approved for a Humanitarian Device Exemption (HDE) for selected patients in late 2006. Criteria for implantation include endstage biventricular HF, a life expectancy of <30 days, and ineligibility for heart transplantation. This grapefruit-sized, electrically powered device is the only U.S. FDA-approved circulatory assist device that does not require percutaneous lines. Two artificial ventricles are attached to the recipient atria, aorta and pulmonary artery. The internal battery can power the unit independently for up to 20 minutes. An external transcutaneous energy transfer (TET) coil, connected to either a bedside console or a portable unit, transmits energy through the skin to drive the device and recharge the internal battery.

The first Abiocor device was implanted in a 58-year-old man in 2001. Dowling and co-workers subsequently published a small multicenter series of seven transplant-ineligible patients with biventricular failure who underwent Abiocor implantation.[37] There were two intraoperative deaths, three fatal strokes and one death due to multi-organ failure. Two patients were ultimately discharged to a transitional care unit. One was finally discharged home on postoperative day 209 and was still alive seven months after his return home at the time the report was prepared for publication. Autopsy in four patients, including the three with fatal strokes, showed thrombus formation on the atrial struts. The device has since been modified in the hope of minimizing this complication.

6. CLINICAL TRIALS

One of the earliest reports involving a significant number of patients treated with an LVAD as BTT was published in 1999 and detailed the experience of 13 European centers implanting the Novacor N 100 PC wearable left ventricular assist device (World Heart Corp, Oakland, CA) in 118 patients.[25] There was no standardized protocol for patient selection or management, hence results varied considerably according to the implanting center. Overall survival was 64% with a median implant time of 115 days. One third of patients were discharged home with the device. Infection, bleeding, thromboembolism and stroke were common, with sepsis (37%), multi-organ failure (14%) and right HF (12%) accounting for the majority of deaths.

Frazier and co-workers reported in 2001 a study of 280 patients who received the HeartMate Vented Electric Left Ventricular Assist System (HeartMate VE LVAS) (Thoratec Corp, Pleasanton, CA) as BTT.[21] Median implant time was 112 days. Device-related infection, bleeding, thromboembolism and neurologic dysfunction occurred in 40%, 11%, 6% and 5% of patients, respectively. Comparison of this non-randomized patient group to a cohort of 48 historical control patients who did not receive a device showed improved outcomes with 67% of the HeartMate VE patients surviving to transplant *vs.* 33% of the control group (p < 0.001).

Miller and co-workers published in 2007 the first large series of patients implanted with a device using continuous-flow, rotary-pump technology, the HeartMate II Left Ventricular Assist System (Thoratec, Pleasanton, CA).[28] The theoretical advantages to such a device were smaller size, longer durability, and it was hoped, fewer complications. This observational study enrolled 133 patients with Class IV HF who were awaiting heart transplantation. Median duration of support was 126 days. At six and 12 months, overall survival to transplantation, recovery or continued circulatory support with transplant eligibility was 75% and 68%, respectively, which compares favorably to survival data from pulsatile flow device trials. Bleeding (mostly postoperative) was the most frequently encountered complication with 53% of patients receiving at least two units of packed red blood cell transfusion and 31% requiring re-operation. Device related infection occurred in 14% of patients, ischemic stroke in 6% and hemorrhagic stroke in 2%. When comparing events per patient year in this cohort to Frazier's study with the HeartMate VE device, rates of bleeding requiring surgery, stroke, other non-stroke neurologic events and right HF were considerably lower. However, there were considerable differences in baseline demographics favoring the later HeartMate II study (Table 1).[21,28] Similar to prior trials, sepsis, multi-organ failure and ischemic stroke were common terminal events. Extended follow-up of the 281 patients (including the 133 in this study) with the HeartMate II as BTT demonstrated a 72% survival at 18 months and significantly improved quality of life.[38] The early Euorpean experience with the HeartMate II has been similar.[39]

Table 1.   Comparison of Demographics: Historical Controls and HeartMate VE and HeartMate II in BTT studies.

	Historical Medical Controls[21]	HeartMateVE LVAD[21]	HeartMate II[28]
n	48	280	133
Mean Age (years)	50	55	50
% Male	87%	83%	79%
% Ischemic	46%	46%	37%
% Inotropic Support	1005	100%	87%
Systolic BP	86.1 ± 15.4	75.5 ± 9.7	95.8 ± 14.6
Cardiac Index (L/min/M2)	2.03 ± 0.7	1.67 ± 0.4	2.0 ± 0.6
Pulmonary Capillary Wedge Pressure (mm Hg)	27.8 ± 7.6	27.4 ± 6.6	26.1 ± 7.9
Serum Creatinine (mg/dL)	1.58 ± 0.5	1.7 ± 1.0	1.4 ± 0.5
Total Bilirubin (mg/dL)	1.69 ± 1.29	2.1 ± 2.9	1.2 ± 0.8
Hematocrit (%)	NA	32.4 ± 6.0	34.8 ± 5.2

A randomized, multi-center trial comparing the HeartMate II to the HeartMate XVE for DT was recently reported.[40] The primary end-point of survival free of disabling stroke and re-operation at 24 months was achieved in 46% (62 of 134) of HeartMate II subjects compared to 11% of Heartmate XVE subjects (p < 0.001) with an actuarial 24 months survival of 58% *vs.* 24% in HeartMate II *vs.* HeartMate XVE subjects, respectively (p = 0.008). Of the 59 patients implanted with a Heartmate XVE, 20 patients required 21 pump replacements (for bearing wear, valve malfunction, or infection), one required urgent transplantation and three required device explantation (for bearing wear, valve malfunction, or infection). Of the 133 subjects with a Heartmate II implantation 12 required 13 device replacements (ten for breakage of a percutaneous lead, two for pump thrombosis, and one for outflow elbow disconnection). Of some concern is the lack of a significant difference in disabling stroke with 14% and 17% of subjects suffering disabling strokes in two years in the

HeartMate XVE and HeartMate II groups, respectively, suggesting that non-pulsatile devices do not result in fewer strokes as had been hoped.

The Randomized Evaluation of Mechanical Assistance for the Treatment of Congestive HF (REMATCH) Trial was the first reported randomized, multi-center trial of left ventricular assist devices.[20] Unlike the trials discussed previously, patients in this study were ineligible for cardiac transplantation and were receiving the device for DT. One hundred and twenty-nine patients with advanced HF and left ventricular ejection fractions of ≤25% were entered into the study. Sixty-one patients were assigned to optimal medical therapy and 68 underwent implantation of the HeartMate Vented Electric (VE) device. Results showed a significant survival benefit to the device patients, with a 48% reduction in risk of death from any cause (the primary endpoint of the study). Median survival was 408 days in the device-treated patients *vs.* 150 days in medically treated patients. Kaplan-Meier estimated survival at one and two years was 52% and 25% *vs.* 23% and 8% in the device-treated *vs.* medically managed groups, respectively. Serious adverse events occurred 2.35 (95% confidence interval, 1.86 to 2.95) times more frequently in the device group compared to the medical-therapy group, with a predominance of infection, bleeding, and malfunction of the device. The quality of life was significantly improved at one year in the device group. There were no device failures at 12 months but 35% had device failure at 24 months. Ten devices were replaced and seven device failures resulted in death. As with the Novacor device, infection and bleeding were common with the HeartMate LVAD. Although the risk of neurologic events was high in the device group (4.35 times that of the medically treated group), only 10% of device patients suffered an ischemic stroke, perhaps attributable to the unique textured interior surface of the HeartMate VE.

The Clinical Utility Baseline Study (CUBS) was the first European observational study of DT using the LionHeart 2000 (LionHeart LVAD). Trancutaneous energy transfer is used to power this device and there is no need for externalized drive lines.[41] Infection rates did appear improved compared to REMATCH, with

30% developing sepsis, and 35% developing pump-pocket infections, although higher than anticipated with a totally implantable device. One- and two-year survival rates were 39% and 22% in the 23 patients enrolled; poorer than survival seen in REMATCH though the patients were more critically ill than those in REMATCH.

Rogers and co-workers reported the results of a non-randomized trial (INTREPID) comparing the Novacor LVAD. The study included 55 patients with New York Heart Association Class IV symptoms who had failed weaning from inotropic support and were not candidates for cardiac transplantation (WorldHeart, Inc., Oakland, CA).[24] All patients were offered support, 37 agreed and were compared to the 18 patients who declined. The two groups were relatively well matched for disease severity and demographics. The LVAD recipients had better survival at six months (46% *vs.* 26%, p = 0.03) and 12 months (27% *vs.* 11%). In the medically treated group, 17 of 18 patients died, all of cardiac dysfunction. In the LVAD group, 29 of 37 patients died — five of cardiac dysfunction (14%), seven of infection (18%) and ten of cerebrovascular accidents (27%). These results were inferior to the HeartMate XVE arm of the REMATCH trial but poorer survival may be explained by the high operative risk of INTREPID patients.

Ongoing clinical trials will be essential to the continuing evolution of ventricular assist devices and total artificial hearts. As devices become smaller and more durable, randomized trials to compare long-term outcomes with device therapy *vs.* medical therapy or transplantation in patients with NYHA Class III HF will be critical. In the meantime the Registry of Mechanically Assisted Circulatory Support (INTERMACS) is enrolling patients with approved devices at the time of implantation.[42] This registry has already provided considerable information about assessment of operative risk for VADs.[43,44]

7. USE AS AN ASSIST IN MYOCARDIAL RECOVERY

Although, the utility of temporary VADs in supporting patients with acute myocardial failure (as in post cardiotomy patients and patients with acute myocarditis) is reasonably well established, the role of these devices in promoting myocardial recovery in patients with long-standing myocardial dysfunction and HF is less well defined. Several reports have indicated that successful explantation of an LVAD with long-term freedom from HF or death occurred in less than 10% of patients with chronic non-ischemic cardiomyopathy.[44-47] More recent data have been somewhat more encouraging. Birks and co-workers reported that 11 of 15 patients with LVADs implanted for HF due to chronic non-ischemic cardiomyopathy had adequate myocardial recovery to allow explantation of the device.[48] One patient died within 24 hours and one died of cancer 27 months after explantation. Of the nine survivors in whom the LVAD was explanted after a mean of 320 ± 186 days of support, 100% and 89% survived free of HF for one and four years, respectively. Patients were treated with intense neurohormonal blockade with an ACE inhibitor, non-selective beta-blocker, aldosterone antagonist, and an angiotensin receptor blocker. Once myocardial recovery had occurred, the non-selective beta-blocker was switched to clenbuterol, a β-2 adrenergic receptor agonist, to promote physiologic myocardial hypertrophy. The Harefield Recovery Protocol Study for Patients With Refractory Chronic HF (HARPS) is currently studying this strategy in 40 subjects in a U.S. multi-center trial (www.clinicaltrials.gov; NCT00585546).

8. COMPLICATIONS

Major complications of device placement include bleeding, infections, CNS events, device failure, right HF, arrhythmias, and thromboembolic events. Intra- and perioperative bleeding are common with placement of mechanical hearts. As previously discussed, the large size of first generation devices mandated extensive surgical dissection, which increased the risk of perioperative bleeding. Moreover, many patients who require mechanical hearts have significant coagulopathies, exacerbating bleeding risk. Indeed, both prolongation of the protime and thrombocytopenia are significant predictors of mortality with mechanical hearts.[43] In the cohort of patients implanted with the HeartMate I for BTT, 48% of patients had bleeding significant enough to return to the operating room or cause death,[21] while in the

Heartmate II BTT study bleeding requiring surgery occurred in 31% of patients and 53% of patients required transfusion of ≥2 units of packed red blood cells.[28] Although this 64% reduction in bleeding may be attributable to the limited surgical dissection, it is important to remember that the earlier trials included sicker patients (Table 1). Late bleeding is exacerbated by the need for chronic anticoagulation in all devices except the HeartMate XVE, which requires only antiplatelet therapy. A recent report compared gastrointestinal bleeding at least 15 days postoperatively between first generation pulsatile flow devices and second generation axial flow devices.[14] There were 63 bleeding events/100 patient-years for non-pulsatile devices and 6.8 events/100 patient-years for pulsatile devices (p = 0.0004). Although mortality was similar between groups (15% non-pulsatile *vs.* 17% pulsatile, p = 0.6965), significant bleeding is likely to reduce quality of life, increase costs, and eventually affect mortality — major concerns, especially for DT patients who will have the device for much longer than BTT patients. The excess bleeding did not appear to be due only to the need for anticoagulation in the non-pulsatile devices as bleeding in patients with mechanical heart valves is similar to that seen with pulsatile flow devices and the INR is generally maintained at a higher level in valve recipients than in LVAD recipients.[49] Furthermore, other non-GI bleeding was not increased.

The use of red blood cell and platelet transfusions may lead to human leukocyte antigen (HLA) sensitization. Antibodies to HLA antigens may significantly prolong the time to finding a compatible donor heart and may also increase the risk of rejection post transplantation. There is controversy about whether some mechanical hearts are more likely to provoke HLA sensitization than others.[50]

All second generation and third continuous flow devices require anticoagulation with warfarin as well as aspirin. Intracranial bleeding appears to be decreased with second generation non-pulsatile devices but as discussed earlier, this may be due to higher-risk patients receiving the first-generation non-pulsatile devices.[21,28] The recently completed study comparing the HeartMate XVE to the HeartMate II for DT did not demonstrate a difference in disabling strokes over two years.[40] The significant rate of intraoperative bleeding has been reduced with the use of recombinant activated Factor VII but this has been associated with a significant and dose-related excess of thromboembolic events early postoperatively.[51] Newer third generation devices are smaller and some are placed directly in the pericardial space, obviating the need for the creation of a preperitoneal space for the device. This and the smaller associated drive lines are likely to further reduce the rate of intraoperative bleeding.

Device-related infections are of three types: drive line infections, pocket infections, and device or blood stream infections. Infections were a common source of morbidity and mortality with first generation pulsatile devices. Drive line and pocket infections are often treatable but device infections are associated with at least a 50% mortality.[52] Indeed, 41% of all deaths in REMATCH were due to sepsis.[20] Antibiotic protocols and immobilization of the drive line have decreased infections in the later enrolled REMATCH patients compared to those enrolled earlier. However, in a recent single center study of primarily first generation devices, device-related infections occurred in 42% of patients within 60 days of VAD implantation and did not decrease in the eras 1996–2000 to 2001–2006.[51] Infections do appear to be decreasing significantly with second generation continuous flow devices probably due to the need for less surgical dissection.[21,28] Devices without percutaneous drive lines, such as the LionHeart, also appear to be associated with fewer infections.[41]

Neurologic events such as cerebrovascular accident (CVA), transient ischemic attack (TIA), and encephalopathy are relatively common post VAD and occurred in 44% of patients in REMATCH and 57% of subjects in the CUBS LionHeart study. Twenty-one percent and 43%, respectively, suffered permanent neurologic deficit.[20,39] It has been suggested that strokes would occur less frequently with second generation devices, but the recent randomized trial already discussed comparing HeartMate XVE *vs.* Heartmate II did not demonstrate a difference.[21,28,40]

Device failure is a significant problem with the Thoratec HeartMate SVE. In the REMATCH trial, 35% of subjects had device failure at 18 months.[20] The second generation non-pulsatile devices have significantly improved durability with no device failure for the HeartMate II at 18 months and the expected durability of these devices is likely to be much longer.[38] Device thrombosis occurred in 1.4% of Heartmate II recipients with a mean follow-up of 155 days but may be a significant long-term problem if anticoagulation lapses.

Postoperative right ventricular failure complicates 10% to 35% of LVAD implantations.[53,54] Increased venous return that occurs after successful LVAD implantation may stress an already diseased right ventricle. In addition, increased cardiac output may worsen pulmonary hypertension, again burdening the right ventricle. Bowing of the interventricular septum to the left by suction from the LVAD may dysynchronize the right ventricle as well. The need for postoperative inotropic therapy to support the right ventricle is associated with a poor outcome.[55]

Arrhythmias occur frequently post LVAD implantation, with atrial arrhythmias in 21.5% and ventricular arrhythmias in 22% to 52%.[53,56] These can prolong hospitalization and may result in right ventricular failure or even death. Ventricular arrhythmias may be precipitated by "suction" events in the left ventricle but many episodes are unrelated to suction. Ventricular arrhythmias appear to be associated with higher mortality and thus an implantable defibrillator may be appropriate.[55] Gastrointestinal complications such as early satiety that were common with the large first generation devices have largely disappeared with smaller devices. Clinically significant hemolysis is rare and generally signals device thrombosis.

9. COSTS

The costs associated with chronic VADs or TAHs are substantial and vary considerably from patient to patient. Major costs are the device itself (for example the HeartMate I is about US$60,000) and the costs of hospitalization. Hospitalization costs vary with the number of complications, with total costs ranging from US$119,874 to US$869,199 per patient.[57] Recently Hernandez and co-workers analyzed costs for Medicare fee-for-service beneficiaries who received a ventricular assist device between February 2000 and June 2006 as primary therapy (n = 1476) or after cardiotomy (n = 1467). For patients in the 2000–2005 cohorts, mean one-year Medicare payments for inpatient care were US$178,714 (SD: US$142,549) in the primary device group and US$111,769 (SD: US$95,413) in the post cardiotomy group.[58] These were primarily first generation devices and hopefully complications and hospital stays will decline with second and third generation devices.

These costs significantly increase the cost per quality adjusted life year (QALY) of cardiac transplantation when the average time to transplantation is six months, but as device reliability improves and the devices provide several quality life years per patient, the cost per QALY will decrease significantly and may approach the €22,000 to €33,000 (US$33,000–US$49,000) per QALY threshold suggested by the United Kingdom's National Institute of Health and Clinical Excellence (NICE).[59]

10. FUTURE DIRECTIONS

Improvements in mechanical hearts are occurring rapidly. Devices and drive lines are now smaller, requiring less surgical dissection. They are quieter and significantly more durable, although long-term durability remains to be demonstrated. Infections remain a problem, in part due to the need for drive lines, but the development of transcutaneous energy sources has not yet eliminated infection and further study is needed to determine if there are immunologic defects created by the devices. While non-pulsatile flow appears to restore vital organ function as well as pulsatile flow, we are just beginning to understand that other problems, such as an excess of GI bleeding, may be associated with non-pulsatile flow. The need for chronic anticoagulation in all newer devices increases the risk of bleeding and limits quality of life. Significant progress has been made in determining perioperative risk for device implantation but we need significantly more information about who is likely to have a long-term survival and good quality of life if DT is to be cost effective.

REFERENCES

1. Ammar KA, Jacobsen SJ, Mahoney DW, Kors JA, et al. (2007) Prevalence and prognostic significance of heart failure stages: application of the American College of Cardiology/American Heart Association heart failure staging criteria in the community. *Circulation* 115: 1563–1570.

2. Spencer FC, Eiseman B, Trinkle JK, Rossi NP (1965) Assisted circulation for cardiac failure following intracardiac surgery with cardiopulmonary bypass. *J Thorac Cardiovasc Surg* 49: 56–73.

3. DeBakey ME (1971) Left ventricular bypass pump for cardiac assistance. Clinical experience. *Am J Cardiol* 27: 3–11.

4. Portner PM, Oyer PE, McGregor CGA, Baldwin JC, et al. (1985) First human use of an electrically powered implantable ventricular assist system. *Artificial Organs* 9(A): 36(Abstract).

5. Akutsu T, Kolff WJ (1958) Permanent substitutes for valves and hearts. *Trans Am Soc Artif Intern Organs* 4: 230–235.

6. Cooley DA, Liotta D, Hallman GL, Bloodwell RD, et al. (1969) Orthotopic cardiac prosthesis for two-staged cardiac replacement. *Am J Cardiol* 24: 723–730.

7. DeVries WC, Anderson JL, Joyce LD, Anderson FL, et al. (1984) Clinical use of the total artificial heart. *N Engl J Med* 310: 273–278.

8. DeVries WC (1988) The permanent artificial heart. Four case reports. *JAMA* 259: 849–859.

9. Lindstrom SJ, Pellegrino VA, Butt WW (2009) Extracorporeal membrane oxygenation. *Med J Aust* 191: 178–182.

10. Kamdar F, Boyle A, Liao K, Colvin-Adams M, et al. (2009) Effects of centrifugal, axial, and pulsatile left ventricular assist device support on end-organ function in heart failure patients. *J Heart Lung Transplant* 28: 352–359.

11. Garcia S, Kandar F, Boyle A, Colvin-Adams M, et al. (2008) Effects of pulsatile- and continuous-flow left ventricular assist devices on left ventricular unloading. *J Heart Lung Transplant* 27: 261–267.

12. Garatti A, Bruschi G, Colombo T, Russo C, et al. (2008) Clinical outcome and bridge to transplant rate of left ventricular assist device recipient patients: comparison between continuous-flow and

pulsatile-flow devices. *Eur J Cardiothorac Surg* 34: 275–280; discussion 280.

13. Amir O, Radovancevic B, Delgado RM, 3rd, Kar B, *et al.* (2006) Peripheral vascular reactivity in patients with pulsatile vs axial flow left ventricular assist device support. *J Heart Lung Transplant* 25: 391–394.

14. Crow S, John R, Boyle A, Shumway S, *et al.* (2009) Gastrointestinal bleeding rates in recipients of nonpulsatile and pulsatile left ventricular assist devices. *J Thorac Cardiovasc Surg* 137: 208–215.

15. Farrar DJ (2000) The thoratec ventricular assist device: a paracorporeal pump for treating acute and chronic heart failure. *Semin Thorac Cardiovasc Surg* 12: 243–250.

16. Farrar DJ, Holman WR, McBride LR, Kormos RL, *et al.* (2002) Long-term follow-up of Thoratec ventricular assist device bridge-to-recovery patients successfully removed from support after recovery of ventricular function. *J Heart Lung Transplant* 21: 516–521.

17. Hill JD, Farrar DJ, Naka Y, Chen JM, *et al.* (2006) Positive displacement ventricular assist devices. In: Kirklin J, Frazier OH (eds.) *Mechanical Circulatory Assistance.* Elsevier, Philadelphia, pp. 53–77.

18. Malaisrie SC, Pelletier MP, Yun JJ, Sharma K, *et al.* (2008) Pneumatic paracorporeal ventricular assist device in infants and children: initial Stanford experience. *J Heart Lung Transplant* 27: 173–177.

19. Dowling RD, Park SJ, Pagani FD, Tector AJ, *et al.* (2004) HeartMate VE LVAS design enhancements and its impact on device reliability. *Eur J Cardiothorac Surg* 25: 958–963.

20. Rose EA, Gelijns AC, Moskowitz AJ, Heitjan DF, *et al.* (2001) Long-term mechanical left ventricular assistance for end-stage heart failure. *N Engl J Med* 345: 1435–1443.

21. Frazier OH, Rose EA, Oz MC, Dembitsky W, *et al.* (2001) Multicenter clinical evaluation of the HeartMate vented electric left ventricular assist system in patients awaiting heart transplantation. *J Thorac Cardiovasc Surg* 122: 1186–1195.

22. Kalya AV, Tector AJ, Crouch JD, Downey FX, *et al.* (2005) Comparison of Novacor and HeartMate vented electric left ventricular assist devices in a single institution. *J Heart Lung Transplant* 24: 1973–1975.

23. El-Banayosy A, Arusoglu L, Kizner L, Tenderich G, *et al.* (2000) Novacor left ventricular assist system versus Heartmate vented electric left ventricular assist system as a long-term mechanical circulatory support device in bridging patients: a prospective study. *J Thorac Cardiovasc Surg* 119: 581–587.

24. Rogers JG, Butler J, Lansman SL, Gass A, *et al.* (2007) Chronic mechanical circulatory support for inotrope-dependent heart failure patients who are not transplant candidates: results of the INTrEPID Trial. *J Am Coll Cardiol* 50: 741–747.

25. El-Banayosy A, Deng M, Loisance DY, Vetter H, *et al.* (1999) The European experience of Novacor left ventricular assist (LVAS) therapy as a bridge to transplant: a retrospective multi-centre study. *Eur J Cardiothorac Surg* 15: 835–841.

26. El-Banayosy A, Arusoglu L, Kizner L, Morshuis M, *et al.* (2003) Preliminary experience with the LionHeart left ventricular assist device in patients with end-stage heart failure. *Ann Thorac Surg* 75: 1469–1475.

27. John R (2008) Current axial-flow devices — the HeartMate II and Jarvik 2000 left ventricular assist devices. *Semin Thorac Cardiovasc Surg* 20: 264–272.

28. Miller LW, Pagani FD, Russell SD, John R, *et al.* (2007) Use of a continuous-flow device in patients awaiting heart transplantation. *N Engl J Med* 357: 885–896.

29. Goldstein D, Zucker M, Pagain FD, Frazier OH (2006) Rotary ventricular assist devices. In: Frazier OH, Kirklin JK (eds.) *Mechanical Circulatory Support.* Elsevier, Philadelphia, pp. 77–104.

30. Haj-Yahia S, Birks EJ, Rogers P, Bowles C, *et al.* (2007) Midterm experience with the Jarvik 2000 axial flow left ventricular assist device. *J Thorac Cardiovasc Surg* 134: 199–203.

31. Westaby S, Siegenthaler M, Beyersdorf F, Massetti M, *et al.* Destination therapy with a rotary blood pump and novel power delivery. *Eur J Cardiothorac Surg* 37: 350–356.

32. Bruschi G, Ribera E, Lanfranconi M, Russo C, *et al.* (2006) Bridge to transplantation with the MicroMed DeBakey ventricular assist device axial pump: a single centre report. *J Cardiovasc Med* 7: 114–118.

33. Schmid C, Tjan TD, Etz C, Schmidt C, *et al.* (2005) First clinical experience with the Incor left ventricular assist device. *J Heart Lung Transplant* 24: 1188–1194.

34. Slaughter MS, Sobieski MA, 2nd, Tamez D, Horrell T, *et al.* (2009) HeartWare miniature axial-flow ventricular assist device: design and initial feasibility test. *Tex Heart Inst J* 36: 12–16.

35. Meyns B, Klotz S, Simon A, Droogne W, *et al.* (2009) Proof of concept: hemodynamic response to long-term partial ventricular support with the synergy pocket micro-pump. *J Am Coll Cardiol* 54: 79–86.

36. Copeland JG, Smith RG, Arabia FA, Nolan PE, *et al.* (2004) Cardiac replacement with a total artificial heart as a bridge to transplantation. *N Engl J Med* 351: 859–867.

37. Dowling RD, Gray LA, Jr., Etoch SW, Laks H, *et al.* (2004) Initial experience with the AbioCor implantable replacement heart system. *J Thorac Cardiovasc Surg* 127: 131–141.

38. Pagani FD, Miller LW, Russell SD, Aaronson KD, *et al.* (2009) Extended mechanical circulatory support with a continuous-flow rotary left ventricular assist device. *J Am Coll Cardiol* 54: 312–321.

39. Struber M, Sander K, Lahpor J, Ahn H, *et al.* (2008) HeartMate II left ventricular assist device; early European experience. *Eur J Cardiothorac Surg* 34: 289–294.

40. Slaughter MS, Rogers JG, Milano CA, Russell SD, *et al.* (2009) Advanced heart failure treated with continuous-flow left ventricular assist device. *N Engl J Med* 361: 2241–2251.

41. Pae WE, Connell JM, Adelowo A, Boehmer JP, *et al.* (2007) Does total implantability reduce infection with the use of a left ventricular assist device? The LionHeart experience in Europe. *J Heart Lung Transplant* 26: 219–229.

42. Kirklin JK, Naftel DC, Stevenson LW, Kormos RL, *et al.* (2008) INTERMACS database for durable devices for circulatory support: first annual report. *J Heart Lung Transplant* 27: 1065–1072.

43. Lietz K, Long JW, Kfoury AG, Slaughter MS, *et al.* (2007) Outcomes of left ventricular assist device implantation as destination therapy in the post-REMATCH era: implications for patient selection. *Circulation* 116: 497–505.

44. Holman WL, Kormos RL, Naftel DC, Miller MA, *et al.* (2009) Predictors of death and transplant in patients with a mechanical circulatory support device: a multi-institutional study. *J Heart Lung Transplant* 28: 44–50.

45. Mancini DM, Beniaminovitz A, Levin H, Catanese K, *et al.* (1998) Low incidence of myocardial recovery after left ventricular assist device implantation in patients with chronic heart failure. *Circulation* 98: 2383–2389.

46. Dandel M, Weng Y, Siniawski H, Potapov E, *et al.* (2005) Long-term results in patients with idiopathic dilated cardiomyopathy after weaning from left ventricular assist devices. *Circulation* 112: 137–45.

47. Maybaum S, Mancini D, Xydas S, Starling RC, *et al.* (2007) Cardiac improvement during mechanical circulatory support: a prospective multicenter study of the LVAD Working Group. *Circulation* 115: 2497–2505.

48. Birks EJ, Tansley PD, Hardy J, George RS, *et al.* (2006) Left ventricular assist device and drug therapy for the reversal of heart failure. *N Engl J Med* 355: 1873–1884.

49. Cannegieter SC, Rosendaal FR, Briet E (1994) Thromboembolic and bleeding complications in patients with mechanical heart valve prostheses. *Circulation* 89: 635–641.

50. Drakos SG, Kfoury AG, Kotter JR, Reid BB, *et al.* (2009) Prior human leukocyte antigen-allosensitization and left ventricular assist device type affect degree of post-implantation human leukocyte antigen-allosensitization. *J Heart Lung Transplant* 28: 838–842.

51. Bruckner BA, DiBardino DJ, Ning Q, Adeboygeun A, *et al.* (2009) High incidence of thromboembolic events in left ventricular assist device patients treated with recombinant activated factor VII. *J Heart Lung Transplant* 28: 785–790.

52. Gordon RJ, Quagliarello B, Lowy FD (2006) Ventricular assist device-related infections. *Lancet Infect Dis* 6: 426–437.

53. Genovese EA, Dew MA, Teuteberg JJ, Simon MA, *et al.* (2009) Incidence and patterns of adverse event onset during the first 60 days after ventricular assist device implantation. *Ann Thorac Surg* 88: 1162–1170.

54. Matthews JC, Koelling TM, Pagani FD, Aaronson KD (2008) The right ventricular failure risk score a pre-operative tool for assessing the

risk of right ventricular failure in left ventricular assist device candidates. *J Am Coll Cardiol* 51: 2163–2172.

55. Schenk S, McCarthy PM, Blackstone EH, Feng J, *et al.* (2006) Duration of inotropic support after left ventricular assist device implantation: risk factors and impact on outcome. *J Thorac Cardiovasc Surg* 131: 447–454.

56. Andersen M, Videbaek R, Boesgaard S, Sander K, *et al.* (2009) Incidence of ventricular arrhythmias in patients on long-term support with a continuous-flow assist device (HeartMate II). *J Heart Lung Transplant* 28: 733–735.

57. Christiansen S, Klocke A, Autschbach R (2008) Past, present, and future of long-term mechanical cardiac support in adults. *J Card Surg* 23: 664–676.

58. Hernandez AF, Shea AM, Milano CA, Rogers JG, *et al.* (2008) Long-term outcomes and costs of ventricular assist devices among Medicare beneficiaries. *JAMA* 300: 2398–2406.

59. Appleby J, Devlin N, Parkin D, Buxton M, *et al.* (2009) Searching for cost effectiveness thresholds in the NHS. *Health Policy* 91: 239–245.

VI. How to Create an Artificial Heart *Ex Vivo?*

B. Engineered Heart Tissue

Thomas Eschenhagen and Wolfram-Hubertus Zimmermann

1. INTRODUCTION

Therapeutic regeneration of heart muscle takes different roads and tissue engineering of heart muscle is one of it. The idea to create an organ or an organ patch *in vitro* and use it to substitute or to support the function of a diseased organ *in vivo* is principally straightforward and has attracted much attention over the past two decades. Yet, despite enormous progress, particularly in the stem cell field, we are still far away from the total bioartificial heart as announced more than ten years ago (review in Ref. 1). The less ambitious and likely more realistic goal of myocardial tissue engineering is the development of an engineered patch of myocardium that can be transplanted onto a diseased human heart to improve its function.

2. IS THERE A CLINICAL NEED FOR TISSUE ENGINEERED MYOCARDIAL PATCHES?

Compared to current treatment options (drugs, devices, transplantation, prostheses, cell therapy) cardiac repair with tissue engineered myocardial patches appears as quite a complicated, expensive and ambitious enterprise. Thus, it is important to ask whether there is a realistic clinical perspective for therapeutic tissue engineering in patients with heart disease. In our view, there are two major areas where tissue engineering could prove valuable. The first, and maybe most obvious, is cardiac malformations in childhood, the other, quantitatively much more important, is myocardial infarction and heart failure.

A variety of malformations could benefit from restoration using biological, autologous muscle patches. This includes several forms of ventricular dysplasia, such as the hypoplastic left heart syndrome or tricuspid atresia with right ventricular dysplasia. Right now most of these infants undergo palliative surgery by performing a univentricular correction ("Fontan-surgery"). Biologically active muscle patches may allow biventricular correction and true ventricular restoration. In addition, there are quite frequent indications for a right ventricular outflow tract patch plasty in patients with pulmonary atresia or "Fallot"-type malformations, which might also benefit from a biologic solution. Tissue engineered heart valves have been used for quite some time for pulmonary valve replacement in these patients.

In contrast to cardiac malformations, myocardial infarction and its consequences belong to the most common diseases with huge socio-economic impact. For the reasons mentioned above, only a minor fraction of patients surviving an acute myocardial infarction will be candidates for biological cardiac repair. Clinical studies employing bone marrow-derived cells focus on patients with relatively well-preserved LV function after acute myocardial infarction.[2–7] These patients are very common and easy to study, but several arguments indicate that they are not the best-suited population for tissue engineering strategies. First, if LV function is only mildly affected, the effect of an intervention aiming at providing the heart with new contractile myocardium is small. Second, current treatment with beta-blockers, statins, aspirin and ACE inhibitors is simple, safe and quite effective with a well-documented reduction of hard endpoints such as re-infarction, stroke and death by 50% to 60%.[8] Third, spontaneous recovery, e.g. by improved perfusion, occurs and would obscure potential effects of tissue patches. Finally, implanting tissue patches requires open-heart surgery which will not be justified in mild forms of cardiac disease.

More likely, therefore, implantation of tissue patches will be considered in patients with chronic heart failure resulting from a large myocardial infarction. The incidence of patients with chronic heart failure is rapidly increasing with 600,000 newly diagnosed cases per year in the United States. The WHO predicts congestive heart failure to be the leading cause of death in 2020. Particularly in the beginning, implantation of a tissue-engineered patch would be combined with standard surgical interventions recommended by current guidelines. Examples are surgical revascularization procedures, valve repair or replacement in patients with very poor LV function and ventricular restoration surgery ("Dor-plasty"). The place of the latter, however, is unclear at present, because the first randomized trial in 1000 patients failed to show an advantage in symptoms, exercise tolerance or outcome when surgical ventricular reconstruction was added to coronary artery bypass grafting alone.[9] It is possible that a limiting factor in this procedure is the non-biologically active Dacron patch currently used and that a viable, tissue engineered patch would improve the results.

Whereas it is hardly disputable that numerous patients would profit from a direct support of contractile function, the question is whether tissue engineered patches could play a significant role in this scenario when compared to current or foreseeable alternatives. First, mechanical LV assist devices (LVAD) acutely support the pumping activity of the LV and thereby allow the heart to recover, a process leading to "reverse remodeling".[10] Recent studies suggest that, at least for a selected fraction of patients, LVADs could serve as a long-term treatment ("destination therapy"[11,12]). Second, implantable defibrillators/cardioverters can protect against lethal arrhythmias and thereby prolong survival,[13] but in itself have no effect on contractile function and heart failure symptoms. Third, cardiac resynchronization therapy by biventricular pacing, in contrast, leads to improved heart function and has a documented effect on survival.[14] Yet, only a limited number of patients profit from this new therapy. Fourth, the ACORN device[15] may be valuable in a limited subset of patients with large ventricles, but finally it seems clear that heart transplantation is the only curative therapy of terminal heart failure. Yet, the number of candidates largely exceeds the number of suitable donor organs and this ratio did not improve over the years.[16]

When considering a tissue engineering strategy for patients with terminal heart failure two scenarios can be envisioned. The simple one is to add patches of tissue engineered heart muscle on the diseased heart, the more elaborated is to excise a piece of non-functioning ventricle and replace it with an engineered heart muscle patch. Obviously, the latter approach requires high mechanical stability of the patch to withstand the systolic pressure and this may be quite difficult to achieve. Thus, adding tissue engineered patches to the heart rather than replacing diseased parts is the more realistic intermediate perspective.

3. IDEAL PROPERTIES OF AN ENGINEERED MYOCARDIAL PATCH

The patch needs to come close to true myocardial tissue, i.e. it should consist of differentiated cardiomyocytes and all the other cells that make up normal heart tissue and it should exhibit a similar organization. These criteria may seem trivial, but is neither fulfilled yet nor is it undisputed. On the one hand, evidence exists that implantation of cell-free scaffolds[17] can have favorable effects on heart function, at least in animal models. On the other hand, cardiac reconstructions with Dacron patches or other non-biological materials are clinical reality already. Thus, the necessity to create nature-like patches of heart muscles in the culture dish with all the problems discussed can well be challenged. Arguments for it are, at this point, partly derived experimentally, partly theoretical. Important experimental evidence comes from the failure of the "skeletal myoplasty approach" in heart failure. It was based on the observation that chronic electrical stimulation of skeletal muscle induces the expression of slow, "cardiac-like" myosin isoforms which led to the hope that skeletal myocytes have the developmental plasticity to differentiate into cardiomyocytes or at least to electrically coupling cardiomyocyte-like cells. One procedure was the *latissimus dorsi plastic*, in which a vascularized piece of this muscle was surgically wrapped around the heart and electrically stimulated.[18] The other one was the injection of autologous skeletal myoblasts.[19] The first large randomized, placebo-controlled study with skeletal myoblasts was prematurely terminated in early 2006 due to a lack of efficacy and safety.[20] Though difficult to deduce from clinical studies it is very likely that the absence of cardiac differentiation and the paucity of electrical coupling between the implanted skeletal myocytes and the host cardiomyocytes[21] can induce re-entry[22] which may give rise to life-threatening arrhythmia reported from these trials. Thus, the available evidence suggests that skeletal myoblasts are not suitable for myocardial tissue engineering or cell therapy. Arguments for the heart-like patch are that only true cardiomyocytes can couple to host myocardium, develop force necessary to reverse the contractile deficit of a failing heart and rhythmically relax in diastole to allow unrestricted inflow of blood.

3.1. Contractile Force

Maximal forces of intact cardiac preparations are in the range of 50 mN/mm^2.[23] The force an engineered tissue patch develops can be taken as a good indicator of the cardiac tissue quality. It depends on several factors including the fraction of the tissue occupied by cardiomyocytes (approximately 80% in normal cardiac tissue), the fraction of cardiomyocytes occupied by sarcomeres (normally 45%), the maturation of the sarcomeres (normally a semi-crystal, highly organized structure), the isoforms of myofilament proteins (normally the fast α-myosin heavy chain is the adult isoform in rodents, the slow β-myosin heavy chain isoform the fetal) as well as the orientation of muscle strands with regard to the direction of force measurement (normally cardiac muscle fibers are aligned in parallel, albeit twisted in the entire heart). The ideal patch should simulate this physiological structure and function to be able to support normal heart contraction.

3.2. Form, Critical Size and Vascularization

Immature cells such as those from the embryonic chick and neonatal rat heart exhibit a striking degree of remodeling capacity and, under suitable conditions such as gyratory shaking, form beating, 3D heart-like tissues as already shown in the early 1950s.[24] Similarly, embryonic stem cells form embryoid bodies, which regularly contain beating cardiomyocyte aggregates.[25] Histologically, these structures have some resemblance with cardiac tissues. The usefulness of unstructured beating cardiomyocyte aggregates in cardiac repair appears, however, limited due to their small size (a few 10–100 μm)and a spherical/cystic structure with randomly oriented cardiomyocytes. Force development of these micro-cardioaggregates is naturally undirected. Conversely, cardiac tissue engineering, especially if developed for cardiac repair, requires custom-fabrications of 3D geometries and muscle strand orientations (for directed force development) according to the specific needs of an individual therapeutic application. In general, a longitudinal and parallel orientation of cardiac muscle strands will be desirable, because this is a major requisite for directed force development. Yet, the normal heart has a complex macro-architecture where different layers of muscle are interwoven and perform a wringing motion rather than a straight contraction.[26] Thus, it may ultimately be necessary to consider a similarly complex tissue engineering design.

The patch needs to be fabricated at a size large enough to realistically affect the function of an entire left ventricle or to be useful for cardiac reconstruction in case of malformations in childhood. It has been estimated that one gram of human myocardium contains 40 millions of cardiomyocytes, i.e. the entire human heart consists of five billion and the left ventricle of 2.5 billion cardiomyocytes.[27] If an infarction was critical when >30% of the LV mass was affected, then an ideal therapy would have to replace 750 million cardiomyocytes and, in the absence of any reliable experimental data, one may estimate that a minimum of 20% of these cells had to be replaced or added to affect cardiac function at all (150 millions). In case of cardiac malformation, commonly used patches range between 1 and 15 cm^2, which would correspond to 20 to 300 million cardiomyocytes (considering a thickness of 5 mm and a normal myocyte density).

A critical biological size of a tissue cannot be surpassed without perfusion of nutrients and oxygen, generally by vascularization. In the absence of capillarization and perfusion, tumors implanted into nude mice did not reach a size of more than 2 to 3 mm.[28] These values, however, depend both on the metabolic demand and the cellular density of the respective tissue. Beating cardiomyocytes obviously have a very high metabolic activity. Accordingly, the density of capillaries in the adult heart is very high, amounting to approximately 2400–3300/mm^2 in rat and human at different postnatal stages (intercapillary distance ~19–20 μm).[29,30] On the other hand, the early embryonic rat heart (until embryonic day 16) as well as the adult frog heart are avascular and nourished exclusively from the lumen by blood circulating within the trabecular system.[31,32] Thus, large heart muscles either develop through vascularization or through intense trabecularization with single strands smaller than 50–75 μm. Given that an implanted patch has to survive in a non-perfused environment (the surface of the heart), sufficiently large patches for the human heart appear only realistic if the patch contains a functioning vasculature which can be surgically connected to the host circulation.

3.3. Electrical Integration

Sudden cardiac death, most common on the basis of tachycardic arrhythmia, accounts for approximately 50% of cardiac death in patients with heart failure (1/3 of total death). Thus, the potential pro-arrhythmic risks of any tissue engineering approach in heart failure has to be rigorously evaluated. The ideal patch should have no pacemaker potential that may act as a focus of automaticity, except one aims at a "biological pacemaker".[33,34] Even more importantly, the ideal patch should robustly integrate into the electrical syncytium of the host myocardium, on the one hand to be paced by the normal sinus rhythm and develop synchronized force, on the other hand not to cause arrhythmia. Suboptimal coupling could be even worse than no coupling because delayed conduction inevitably favors re-entry circuits.

3.4. Immunological Responses

Organ transplantation represents a major progress in modern medicine. Its success justifies the need for life-long immunosuppression with its unavoidable side effects such as chronic renal failure and increased risk of malignancies. Acceptance of a non-autologous tissue engineered product would only be high if patients would experience a sustained and significant therapeutic benefit. The ultimate goal of tissue engineering-based myocardial repair is a complete biological cure and in this case, immunosuppression would be acceptable, similarly as in organ transplantation. Recent developments in stem cell biotechnology have also opened the door for the allocation of potentially autologous cells. Despite its appeal, however, it may eventually not be realistic to aim for autologous tissue engineering-based repair for two main reasons. On the one hand, autologous cells would likely carry disease-causing mutations. On the other hand, cell harvest, quality control, and processing for tissue engineering take time making an immediate application unrealistic. It may therefore be best to address the immunological challenge of tissue engineering-based myocardial repair either by refinement of immunosuppression or, and this is more likely, by establishing cell banks for immunological matching, similar to the practice in bone marrow transplantation.

3.5. Upscaling and GMP

If a tissue engineering strategy was to enter the clinical arena one day it had to be upscalable and compatible with good manufacturing practice procedures (GMP). This means, for example, that xenogenic compounds such as animal sera in the culture media have to be avoided, sterility needs to be assured and reproducibility must be below certain limits. This will make tissue engineering expensive, but will unlikely be the real limiting factor as these hurdles appear to be only of technical but not of biological nature.

4. PRINCIPAL STRATEGIES IN TISSUE ENGINEERING

Three-dimensional spontaneously beating cardiac spheres have been generated already 50 years ago,[24] but the systematic development of myocardial tissues for cardiac repair is a relatively young field and has gained a realistic perspective only with the recent advances in stem cell biology (see Chapter IV. B. to IV. C.). The principal strategies in tissue engineering of valves, vessels, and myocardium are quite similar and have been developed in parallel, partly independent from each other. They can be broadly categorized into four approaches: (1) cell seeding onto decellularized native tissues, (2) cell seeding onto biodegradable polymer scaffolds, (3) cell entrapment in naturally occurring biogels such as collagen I and fibrin, and (4) the cell sheet technique. Consistent evidence exists that tissue formation can be significantly improved by chronic mechanical loading, electrical stimulation and culture under non-static conditions in a bioreactor.

4.1. Decellularized Tissue

Decellularization of native tissues and repopulation with autologous cells is, in principle, the simplest technique and does not require any specific tissue engineering procedure. Conceptionally, the natural tissue provides the ideal 3D structure, and decellularization is expected to reduce immune reactions inherent to the cellular component of the tissue. Progress towards clinical application of this technique has been made in the field of heart valve replacements.[35] A recent study reported that whole rat and pig hearts can be decellularized with detergents and repopulated with neonatal rat cardiomyocytes, yielding spontaneously beating, force-developing engineered hearts.[36] Despite the apparent appeal of the idea, such a total bioartificial heart still has a long way to go, given that one would have to reconstitute not only the heart muscle, but also the valves and the vessels, and this on the scale of a human heart. Moreover, normal heart function requires billions of cardiomyocytes, endothelial cells and smooth muscle cells of the heart to be highly organized, i.e. to form a functional syncytium with capillaries associated with myocytes in a 1:1 ratio. In fact, repopulation of the intact heart via the coronary vasculature appears to be one of the biggest challenges, because cardiomyocytes will not migrate through a more or less intact vessel/capillary wall. Yet, the appeal of this approach remains, particularly since the extracellular matrix may itself instruct the (stem) cells to differentiate into the appropriate cell type as suggested by the authors. It will be interesting to see how this approach develops in the future.

4.2. Biodegradable Polymer Scaffolds

The second approach is to seed heart cells onto porous, biodegradable polymer scaffolds. Conceptionally, the polymer serves as a three-dimensional template for the desired organ, and the cells are expected to degrade the polymer and produce their own extracellular matrix. This "conventional" tissue-engineering approach has the great advantage that the scaffold can theoretically be manufactured at any desired size, macro-, or microstructure and with various chemical and biological properties. A principal problem of the polymer approach is that it is difficult to find the right balance between premature polymer degradation with the risk of dissolution and too little degradation preventing tissue formation. For example, the early experiments with cardiac myocytes seeded on gelatine,[37] alginate,[38] or PGA[39] showed survival of cells just on the surface of the matrix, suggesting that these scaffold materials prevented rather than promoted tissue formation. Much effort is directed toward newer matrix materials that may overcome the current limitations. Material sciences and nanotechnology are progressing rapidly, and it may be that in a collaboration between engineers, physicists, biotechnologists, biologists, and morphologists the ideal scaffold material for creating truly "artificial" heart tissues will be designed.[40,41] Recent examples are tubular scaffolds for the engineering of blood vessels that were generated by electrospinning of collagen and elastin[42] and collagen tubes for cardiac-tissue engineering.[43] Other innovative approach include the use of MRI-derived three-dimensional images of an organ as a blueprint for the fully automated fabrication of a synthetic copy[44] and the application of laser technologies to generate growth and structure supporting micropatterns in biomaterials such as polyglycerolsabacate.[45]

4.3. Biopolymers

The third approach is based on the entrapment of the desired cardiovascular cell type in biopolymers such as collagen I and fibrin. These polymers form a gel in which the cells are trapped in a spatially defined 3D environment. The main advantage of this approach is that both collagen I and fibrin are approved materials in clinical use and apparently promote rather than prevent tissue formation. With this technique Weinberg and Bell generated blood vessels more than 20 years ago,[46] and in the meantime many groups have adopted a similar approach using fibroblasts or smooth muscle cells to create blood vessels[47,48] or heart valves[49] and cardiomyocytes for myocardial patches.[50–53] A disadvantage of the biopolymer approach is that, until now, the tensile strength of the resulting tissues has been relatively low.

4.4. Cell Sheets

The fourth strategy differs from all others in that it goes without any scaffold material. In this approach, first described by Okano and co-workers,[54,55] vascular smooth muscle cells were grown on plastic culture dishes as normal monolayers until confluency and were then removed from the culture dish as a cellular sheet and wrapped around a porous mandrel to form the media, nourished from the inside through the mandrel. In the same manner, a second layer of fibroblast sheets was wrapped around the media to form the adventitia. In a similar strategy, cardiomyocyte sheets can be stapled to form a contracting myocardial tissue.[56] In the latter approach, thermo-sensitive culture dishes facilitate detachment of the monolayer.

5. HEART-LIKE TISSUE STRUCTURE

Numerous studies have been published claiming successful tissue engineering of myocardial tissues. However, the cardiac tissue structure is often so poor in terms of cardiomyocyte differentiation, orientation, organization and viability that their usefulness for either *in vitro* or *in vivo* applications is questionable. Early experiments with synthetic or natural scaffolds such as polyglycolic acid, alginate or preformed collagen sponges showed that cardiac cells survive on such 3D structures, but do not form synchronously beating heart-like structures,[37,38,57] likely because the scaffolds prevented intense cell-cell-contacts necessary for tissue formation. Yet, several techniques are now available that yield myocardial tissues with a characteristic, native cardiac tissue structure, both on the cellular and subcellular level[51,53,56,58,59] (Figure 1). A consistent conclusion from these studies is that immature cardiac cells have the intrinsic capacity to form

Bioengineering   Biological assembly   Cell sheet   De-/Recellularization

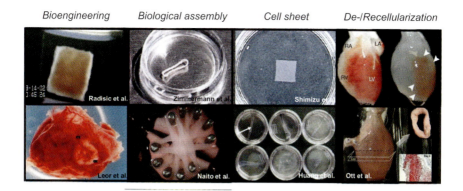

Figure 1.   Different tissue engineering technologies allow the generation of large *contracting bioartificial myocardium*. Cell seeding on preformed scaffolds.[27,42] Entrapment of cells in a defined cardiogenic environment.[36,57] Cell sheet technique.[19,46] Utilization of decellularized natural extracellular matrix.[39]

heart-like muscle structures[24,60] which can be directed and enhanced by mechanical load,[50,61] electrical stimulation,[59,62] oxygenation and perfusion,[63–65] growth factors[58] and a native heart cell mix.[66] Thus, on a small scale, true heart muscle patches can be engineered from primary heart cells.

5.1. Contractile Function

Such tissue engineered heart muscle patches beat spontaneously and develop contractile force. Reported values range between 0.05 and 2 mN/mm^2,[67] which is still considerably less than the biological limit of 50 mN/mm^2. Under optimized culture conditions including high oxygen, insulin and auxotonic stretch plus electrical stimulation, individual EHTs develop up to 5 mN/mm^2 (own unpublished data). The difference to native heart muscle is likely due to several factors. The occupancy of the EHT tissue by muscle tissue is relatively low (<30% *vs.* almost 100%), cardiomyocytes are thin compared to the thick brick-like adult cardiomyocytes of the ventricle, orientation may be not perfect and the developmental state of the myocytes remains immature. The importance of relative occupancy by muscle tissue is exemplified by the observation that replacement of collagen by fibrin leads to more extensive matrix degradation, thinner constructs and higher cell density, which is accompanied by relative forces of up to 12 mN/mm^2.[68]

It is questionable whether, for the purpose of cardiac repair, an "adult-like" degree of maturation would be helpful. Immature cells have a higher remodeling capacity and resistance to hypoxia (pO2 in the fetus 20–25 mmHg).[69] Both could be critical parameters for integration into the host myocardium. Moreover, myocytes can undergo further maturation after implantation.[52] The heart of a neonatal rat weighs approximately 30 mg and that of an adult rat 1000 mg. This increase is almost entirely due to an increase in size and not number, meaning that the volume of myocytes can increase 30-fold. Thus, it appears realistic to envision that the myocytes, initially implanted as a relatively loose network of thin cells, form a compact muscle tissue with adult contractile force over time. Indeed, this has been demonstrated.[65]

5.2. Directed Process of Fabrication

Numerous techniques exist to engineer tissue patches for cardiac repair, but few approaches have been developed towards a reproducible, robust procedure and have shown their usefulness *in vivo*, yet. These include the EHT multi-ring technique,[65] the cell sheet technique[70,71] and, more recently, a fibrin-based method which utilizes polydimethylsiloxane (PDMS) molds.[53] Chachques and co-workers generated non-contractile cell patches and investigated them in a first clinical trial (Magnum) with some success.[72] All of these techniques can be upscaled in terms of width and length, but, for reasons discussed above, not yet in thickness. Upscaling and "clinical" application may also require some automatization to become clinically feasible.[73] Others generate cardiospheres[60] or spheric microtissues and implant them,[74] but here cells are not under directed mechanical load and therefore not well oriented and cannot be handled as a tissue patch. Thus, it is not clear whether these spheres have an advantage compared to cells implanted in the heart as a suspension. Another recent approach developed by our group is a pouch-like, collagen-based construct that can be slipped over an entire (rat) heart.[75] Studies are needed to investigate whether such an approach offers advantages to others.

5.3. Critical Size and Vascularization

The usual size of currently developed engineered myocardial tissue is in the range of centimeters in width/length and 0.1–4 mm in total thickness (review in Ref. 67). While the width and length can be principally expanded indefinitely, the thickness of solid myocardial tissue strands is limited in all models to 50 to 100 μm. This value may be enhanced by perfusion and high oxygen to a maximal theoretical thickness of 200 μm,[64] but more seems unrealistic in the absence of vascularization and internal perfusion. A recent study demonstrated that multi-loop EHTs of an initial size of $15 \times 15 \times 4$ mm formed a compact new layer of myocardium at a mean thickness of 0.5 mm when implanted onto infarcted rat heart.[65] Similarly, implantation of stacked cell sheets gave rise to a thin layer of new myocardium. These data show, on the one hand, that engineered myocardial constructs survive when sutured/layered on the epicardial surface of the heart and that they are strongly vascularized. The time course of vascularization has not been studied in detail yet, but it must occur quite quickly to prevent implant degeneration. The high density of perfused blood vessels in the graft two to four weeks after implantation[58,65] most likely explains why the thickness of the newly formed myocardium exceeded the critical size reproducibly seen in cell culture. Yet, even a thickness of 0.5 mm is still far away from the thickness of the walls of the normal human left ventricle (mean: 9 mm) and is very likely that true vascularization is necessary to approach such values. One idea is to populate a piece of decellularized, but vascularized small intestine with cardiac cells.[76] Some hope is related to the observation that primitive vascular structures develop in EHTs,[58,66] as well as in stacked heart cell sheets[77] and in skeletal muscle constructs.[78] Recent experiments with fibrin-based EHTs showed extensive and well oriented vascular structures (Hansen and Eschenhagen, unpublished). Importantly, however, these primitive blood vessel-like structures are quite clearly not perfused and therefore likely non-functional in terms of improved oxygen and nutrient supply. Evidence suggests that they may find contact to the host circulation after implantation[58,65,77,78] and in this regard, they may reduce the time from implantation to perfusion. Details of this process need to be better worked out. A realistic, albeit complex, approach is to preimplant engineered myocardial tissues close to a well-accessible defined blood vessel such as the epigastric artery/vene and, after ingrowth of perfusing side branches, explant the entire patch and connect it to an epicardial coronary artery. A variation of this technique has recently been described.[79] Here, myocardial tissue formation is done in a chamber isolating a branch of the epigastric artery *in situ* simply by casting a cell-collagen I/matrigel mixture into the chamber. This procedure leads to excellent cardiac tissue formation and vascularization.

5.4. Electrical Integration

The few data available today suggest that engineered sandwich constructs[71,80] and EHTs[63,65] both find electrical contact to the host myocardium after implantation. This is, in principal, in accordance with conclusions from very careful studies investigating coupling of isolated cardiomyocytes injected into the myocardium.[81] However, studying the question of electrical coupling of engineered constructs implanted onto the heart surface is not trivial.[82] Optical imaging with voltage-sensitive dyes[71] and electrical mapping experiments[65,71] suggest undelayed impulse propagation through the graft, but it is not entirely clear how the electrical contact is actually made between graft and host. Several additional questions need to be answered: First, how does the epithelial surface cell lining of both graft and host remodel after implantation? How is the distance between graft and host myocytes bridged? Is it helpful to partially injure the heart surface to facilitate coupling?

Engineered heart constructs from neonatal rat cardiomyocytes depolarize spontaneously (and therefore beat) and this may induce arrhythmias after coupling to the host myocardium. In particular, imperfect, slow conduction between graft and host could represent an arrhythmogenic substrate. No evidence for pro-arrhythmic effects has been reported yet, but clearly the present experiments in rodents cannot rule out this risk. Size, heart rate and basic electrophysiological features make the rodent heart not well suited to study arrhythmias. Thus, more data are necessary to clarify this issue.

5.5. Autologous *vs.* Allogeneic Patch

Tissue engineering at present almost exclusively utilizes primary cardiac cells from neonatal rat, mouse or embryonic chick hearts, which are, by definition, allogeneic. Accordingly, implantation studies either used immune deficient models such as nude rats or mice or pharmacological immunosuppression; the impact of which on graft integration remains unknown. A number of recent reports described techniques to generate engineered myocardial constructs from mouse and human embryonic stem cells[83–85] (own unpublished work). This is important, because this work opens for the first time a realistic perspective to generate sufficient numbers of human cardiomyocytes and cardiac tissues.[86] However, the embryonic stem cell approach remains allogeneic and is ethically controversial. This is the reason why the discovery of means to create induced pluripotent stem (iPS) cells stirred enormous attention.[87] Alternative methods to generate autologous cardiomyocytes include a method to reprogram cardiac fibroblast-like cells with 5-azacytidine,[88] spermatogonial stem cells[89,90] and parthenogenetic stem cells[91] (own unpublished data). Critical unanswered questions are whether these formally autologous cells conserve their immunological profile during the numerous harsh manipulations necessary to induce pluripotency and differentiation in cell culture and therefore indeed escape the immune surveillance after implantation.

Another unresolved issue of ES and iPS cell-derived tissue engineering approaches is that, at present, the degree of cardiomyocyte differentiation and the quality of derived engineered myocardial tissue remains relatively poor. It is not clear at present whether this is simply a technical question or whether the differentiation and tissue-forming capacity of ES- and iPS-derived cells is biologically limited. This is a central question of the field that needs to be answered soon (for further details see Chapter IV. D.).

5.6. Upscaling and GMP

Clinical application of tissue engineered myocardial patches requires means to upscale the procedure under the strict regulations of good manufacturing practice. Although this question may appear far away from the current state of the field, the technical effort and thus the costs will be important for the clinical success of a given technique. The sandwich EHT and Bursac techniques can principally be automated and performed in GMP facilities. However, the regulations of GMP also prohibit the use of animal sera and growth supplements at any given point of the manufacturing process. Solving this issue will be not trivial, because cardiomyocytes in culture are very dependent on serum supplementation. Yet, much progress has been made in recent years, particularly with regard to defined stem cell media.

6. BOTTLENECKS

Myocardial tissue engineering is still far from clinical application for several reasons as described above. Some problems can likely be solved by technical means such as upscaling, GMP and serum-free culture condition and optimization of scaffolds. Electrical integration is subject to intense research and increasing evidence suggests that myocardial grafts integrate well into the host myocardium. Two questions, however, are still open and constitute major bottlenecks in this field: cell source and size/vascularization.

The most obvious problem is a suitable cell source. Since cardiomyocytes normally do not divide at a relevant rate, only few options can be envisioned: In principle, cell cycling could be reinforced in cardiomyocytes and several groups are working on deciphering the right protocols (see Chapter IV. A.). The alternative is to use adult stem cells or human embryonic stem cells. The main principle advantage of the former is its autologous character, i.e. that cells could be taken from biopsies, amplified *ex vivo*, differentiated into cardiomyocytes and used to engineer a myocardial tissue patch which would be recognized by the patient's immune system as its own. Despite enormous research effort over the past ten years, essential questions are still unanswered: Do such cells exist in the body at all or at a sufficient number? Can they be harvested and amplified? And do they generate true cardiomyocytes and engineered tissues with functional properties of the human heart? The cardiomyocyte progenitor cells described by the group of Goumans and Doevendans may come close to that cell type,[88] but it is not clear yet whether they represent a specific stem cell population or rather the product of reprogramming by 5-Aza or both. 5-Aza leads to demethylation and has a mutagenic potential, which may obviate clinical application of such cells. The main advantage of human ES cells is that they unambiguously generate true cardiomyocytes (and many other cell types). Thus, the question comes down to the efficacy and feasibility of this process and much work is directed towards optimizing it.[92] The main disadvantages, however, are that ES cell-derived engineered tissues are allogeneic, i.e. their implantation would require life-long immunosuppression. Moreover, undifferentiated ES cells can form teratomas. Finally, legislative restrictions exist in many countries that, at present, represent hurdles in this part of research. It is difficult to predict which cell type will finally work and therefore research should be as open as possible.

Much hope at present lies on iPS as discussed above, because they can be easily obtained from a hair or a skin biopsy, are ethically uncritical and principally autologous. Yet, issues related to long-term consequences of genetic reprogramming such as tumor formation have to be addressed. Moreover, timing of cell allocation will be an issue, especially if substantial myocardial reconstruction is anticipated. For this, large quantities of well-characterized cells will be necessary. Upscaling and quality control will realistically take one year before an autologous tissue construct could be applied. Given the immediate need for support, it appears unlikely that a true autologous rather than an immune matched allogenic approach will finally turn out to be clinically applicable.

Size matters and it is possible that a sufficient size of an engineered tissue patch turns out to be the second most critical bottleneck. Even if adult stem cells have the capacity to differentiate into cardiomyocytes, all available data suggest that this is a rare event. If the rate was 1% (presently an optimistic estimate), the 150 millions of cardiomyocytes needed (see above) had to be isolated from 15 billions cells. Even with human embryonic stem cells the published rates of cardiac differentiation were, for a long time, so low that the construction even of small engineered heart muscles was a major undertaking. However, recent growth factor-based protocols have increased the rate to up to 50% (Ref. 92), own unpublished experience), making the human embryonic stem cell-derived cardiac patch a much more realistic scenario. Establishment of clinically applicable protocols for differentiating and purifying cardiomyocytes or cardiovascular progenitor cells from stem cell culture will now be decisive for the success of cardiac tissue engineering. As outlined above, creating a myocardial patch of sufficient size will not be possible without vascularization. It is therefore one of the urgent goals in the field to develop suitable biotechnological and surgical techniques to vascularize and perfuse engineered myocardial patches before implanting them on the heart.

7. PERSPECTIVES

Cardiac repair with engineered myocardial tissues is an ambitious, but exciting goal of modern medicine. It could provide new therapeutic options for the correction of congenital heart defects and for terminal heart failure. Its main promises are the living growing patch and the provision of new contractile myocardium that helps the failing heart to meet its demands. We believe that both premises can only be fulfilled by true heart muscle surrogates and therefore require the use of stem cells with a robust cardiomyocyte differentiation potential. The recent developments in stem cell technologies (such as the availability of non-embryonic pluripotent stem cells, guided differentiation, and bioreactor technologies) make this a realistic perspective. Finally, vascularization of engineered heart tissues will be essential to reach the ultimate goal of myocardial reconstruction *in vivo*.

ACKNOWLEDGMENTS

The work of the authors has been financially supported by the German Research Foundation (FOR 604), German Ministry of Science and Education (Bundesministerium für Bildung und Forschung 01GN0520), the European Commission (FP6 (EUGeneHeart) and FP7 (Angioscaff)), the Novartis foundation, the Foundation LeDucq, a Hengstberger-Stipend, and the Forschungsförderungsfonds of the University Medical Center Hamburg-Eppendorf.

REFERENCES

1. Zandonella C (2003) Tissue engineering: the beat goes on. *Nature* 421: 884–886.
2. Assmus B, Schachinger V, Teupe C, Britten M, *et al.* (2002) Transplantation of progenitor cells and regeneration enhancement in acute myocardial infarction (TOPCARE-AMI). *Circulation* 106: 3009–3017.
3. Stamm C, Westphal B, Kleine HD, Petzsch M, *et al.* (2003) Autologous bone-marrow stem-cell transplantation for myocardial regeneration. *Lancet* 361: 45–46.
4. Strauer BE, Brehm M, Zeus T, Kostering M, *et al.* (2002) Repair of infarcted myocardium by autologous intracoronary mononuclear bone marrow cell transplantation in humans. *Circulation* 106: 1913–1918.
5. Wollert KC, Meyer GP, Lotz J, Ringes-Lichtenberg S, *et al.* (2004) Intracoronary autologous bone-marrow cell transfer after myocardial infarction: the BOOST randomised controlled clinical trial. *Lancet* 364: 141–148.
6. Janssens S, Theunissen K, Boogaerts M, Van de Werf F (2006) Bone marrow cell transfer in acute myocardial infarction. *Nat Clin Pract Cardiovasc Med* 3(Suppl 1): S69–S72.
7. Schachinger V, Erbs S, Elsasser A, Haberbosch W, *et al.* (2006) Intracoronary bone marrow-derived progenitor cells in acute myocardial infarction. *N Engl J Med* 355: 1210–1221.
8. Hippisley-Cox J, Coupland C (2005) Effect of combinations of drugs on all cause mortality in patients with ischaemic heart disease: nested case-control analysis. *BMJ* 330: 1059–1063.

9. Jones RH, Velazquez EJ, Michler RE, Sopko G, *et al.* (2009) Coronary bypass surgery with or without surgical ventricular reconstruction. *N Engl J Med* 360: 1705–1717.

10. Wohlschlaeger J, Schmitz KJ, Schmid C, Schmid KW, *et al.* (2005) Reverse remodeling following insertion of left ventricular assist devices (LVAD): a review of the morphological and molecular changes. *Cardiovasc Res* 68: 376–386.

11. Stevenson LW, Miller LW, Desvigne-Nickens P, Ascheim DD, *et al.* (2004) Left ventricular assist device as destination for patients undergoing intravenous inotropic therapy: a subset analysis from REMATCH (Randomized Evaluation of Mechanical Assistance in Treatment of Chronic Heart Failure). *Circulation* 110: 975–981.

12. Long JW, Healy AH, Rasmusson BY, Cowley CG, *et al.* (2008) Improving outcomes with long-term "destination" therapy using left ventricular assist devices. *J Thorac Cardiovasc Surg* 135: 1353–1360.

13. Bardy GH, Lee KL, Mark DB, Poole JE, *et al.* (2005) Amiodarone or an implantable cardioverter-defibrillator for congestive heart failure. *N Engl J Med* 352: 225–237.

14. Cleland JG, Daubert JC, Erdmann E, Freemantle N, *et al.* (2005) The effect of cardiac resynchronization on morbidity and mortality in heart failure. *N Engl J Med* 352: 1539–1549.

15. Starling RC, Jessup M (2004) Worldwide clinical experience with the CorCap Cardiac Support Device. *J Card Fail* 10: S225–S233.

16. Cai J, Terasaki PI (2004) Heart transplantation in the United States 2004. *Clin Transpl* 331–344.

17. Leor J, Tuvia S, Guetta V, Manczur F, *et al.* (2009) Intracoronary injection of *in situ* forming alginate hydrogel reverses left ventricular remodeling after myocardial infarction in Swine. *J Am Coll Cardiol* 54: 1014–1023.

18. Carpentier A, Chachques JC (1985) Myocardial substitution with a stimulated skeletal muscle: first successful clinical case. *Lancet* 1: 1267.

19. Menasche P, Hagege AA, Scorsin M, Pouzet B, *et al.* (2001) Myoblast transplantation for heart failure. *Lancet* 357: 279–280.

20. Menasche P, Alfieri O, Janssens S, McKenna W, *et al.* (2008) The myoblast autologous grafting in ischemic cardiomyopathy (MAGIC) trial: first randomized placebo-controlled study of myoblast transplantation. *Circulation* 117: 1189–1200.

21. Reinecke H, Poppa V, Murry CE (2002) Skeletal muscle stem cells do not transdifferentiate into cardiomyocytes after cardiac grafting. *J Mol Cell Cardiol* 34: 241–249.

22. Abraham MR, Henrikson CA, Tung L, Chang MG, *et al.* (2005) Antiarrhythmic engineering of skeletal myoblasts for cardiac transplantation. *Circ Res* 97: 159–167.

23. Hasenfuss G, Mulieri LA, Blanchard EM, Holubarsch C, *et al.* (1991) Energetics of isometric force development in control and volume-overload human myocardium. Comparison with animal species. *Circ Res* 68: 836–846.

24. Moscona A, Moscona H (1952) The dissociation and aggregation of cells from organ rudiments of the early chick embryo. *J Anat* 86: 287–301.

25. Martin GR, Evans MJ (1975) Differentiation of clonal lines of teratocarcinoma cells: formation of embryoid bodies *in vitro*. *Proc Natl Acad Sci USA* 72: 1441–1445.

26. Davis JS, Hassanzadeh S, Winitsky S, Lin H, *et al.* (2001) The overall pattern of cardiac contraction depends on a spatial gradient of myosin regulatory light chain phosphorylation. *Cell* 107: 631–641.

27. Beltrami CA, Finato N, Rocco M, Feruglio GA, *et al.* (1994) Structural basis of end-stage failure in ischemic cardiomyopathy in humans. *Circulation* 89: 151–163.

28. Folkman J (1971) Tumor angiogenesis: therapeutic implications. *N Engl J Med* 285: 1182–1186.

29. Korecky B, Hai CM, Rakusan K (1982) Functional capillary density in normal and transplanted rat hearts. *Can J Physiol Pharmacol* 60: 23–32.

30. Rakusan K, Flanagan MF, Geva T, Southern J, *et al.* (1992) Morphometry of human coronary capillaries during normal growth and the effect of age in left ventricular pressure-overload hypertrophy. *Circulation* 86: 38–46.

31. Ratajska A, Ciszek B, Sowinska A (2003) Embryonic development of coronary vasculature in rats: corrosion casting studies. *Anat Rec A Discov Mol Cell Evol Biol* 270: 109–116.

32. Sys SU, Pellegrino D, Mazza R, Gattuso A, *et al.* (1997) Endocardial endothelium in the avascular heart of the frog: morphology and role of nitric oxide. *J Exp Biol* 200: 3109–3118.

33. Kehat I, Khimovich L, Caspi O, Gepstein A, *et al.* (2004) Electro-mechanical integration of cardiomyocytes derived from human embryonic stem cells. *Nat Biotechnol* 22: 1282–1289.

34. Plotnikov AN, Shlapakova I, Szabolcs MJ, Danilo P, Jr, *et al.* (2007) Xenografted adult human mesenchymal stem cells provide a platform for sustained biological pacemaker function in canine heart. *Circulation* 116: 706–713.

35. Cebotari S, Lichtenberg A, Tudorache I, Hilfiker A, *et al.* (2006) Clinical application of tissue engineered human heart valves using autologous progenitor cells. *Circulation* 114: I132–I137.

36. Ott HC, Matthiesen TS, Goh SK, Black LD, *et al.* (2008) Perfusion-decellularized matrix: using nature's platform to engineer a bioartificial heart. *Nat Med* 14: 213–221.

37. Li RK, Jia ZQ, Weisel RD, Mickle DA, *et al.* (1999) Survival and function of bioengineered cardiac grafts. *Circulation* 100: II63–II69.

38. Leor J, Aboulafia-Etzion S, Dar A, Shapiro L, *et al.* (2000) Bioengineered cardiac grafts: a new approach to repair the infarcted myocardium? *Circulation* 102: III56–III61.

39. Carrier RL, Papadaki M, Rupnick M, Schoen FJ, *et al.* (1999) Cardiac tissue engineering: cell seeding, cultivation parameters, and tissue construct characterization. *Biotechnol Bioeng* 64: 580–589.

40. Buxton DB, Lee SC, Wickline SA, Ferrari M (2003) Recommendations of the national heart, lung, and blood institute nanotechnology working group. *Circulation* 108: 2737–2742.

41. Davis ME, Hsieh PC, Grodzinsky AJ, Lee RT (2005) Custom design of the cardiac microenvironment with biomaterials. *Circ Res* 97: 8–15.

42. Boland ED, Matthews JA, Pawlowski KJ, Simpson DG, *et al.* (2004) Electrospinning collagen and elastin: preliminary vascular tissue engineering. *Front Biosci* 9: 1422–1432.

43. Yost MJ, Baicu CF, Stonerock CE, Goodwin RL, *et al.* (2004) A novel tubular scaffold for cardiovascular tissue engineering. *Tissue Eng* 10: 273–284.

44. Sodian R, Fu P, Lueders C, Szymanski D, *et al.* (2005) Tissue engineering of vascular conduits: fabrication of custom-made scaffolds using rapid prototyping techniques. *Thorac Cardiovasc Surg* 53: 144–149.

45. Engelmayr GC, Jr, Cheng M, Bettinger CJ, Borenstein JT, *et al.* (2008) Accordion-like honeycombs for tissue engineering of cardiac anisotropy. *Nat Mater* 7: 1003–1010.

46. Weinberg CB, Bell E (1986) A blood vessel model constructed from collagen and cultured vascular cells. *Science* 231: 397–400.

47. L'Heureux N, Germain L, Labbe R, Auger FA (1993) *In vitro* construction of a human blood vessel from cultured vascular cells: a morphologic study. *J Vasc Surg* 17: 499–509.

48. Niklason LE, Gao J, Abbott WM, Hirschi KK, *et al.* (1999) Functional arteries grown *in vitro*. *Science* 284: 489–493.

49. Vesely I (2005) Heart valve tissue engineering. *Circ Res* 97: 743–755.

50. Eschenhagen T, Fink C, Remmers U, Scholz H, *et al.* (1997) Three-dimensional reconstitution of embryonic cardiomyocytes in a collagen matrix: a new heart muscle model system. *FASEB J* 11: 683–694.

51. van Luyn MJ, Tio RA, Gallego y van Seijen XJ, Plantinga JA, *et al.* (2002) Cardiac tissue engineering: characteristics of in unison contracting two- and three-dimensional neonatal rat ventricle cell (co)-cultures. *Biomaterials* 23: 4793–4801.

52. Zimmermann WH, Schneiderbanger K, Schubert P, Didie M, *et al.* (2002) Tissue engineering of a differentiated cardiac muscle construct. *Circ Res* 90: 223–230.

53. Bian W, Liau B, Badie N, Bursac N (2009) Mesoscopic hydrogel molding to control the 3D geometry of bioartificial muscle tissues. *Nat Protoc* 4: 1522–1534.

54. Okano T, Yamada N, Okuhara M, Sakai H, *et al.* (1995) Mechanism of cell detachment from temperature-modulated, hydrophilic-hydrophobic polymer surfaces. *Biomaterials* 16: 297–303.

55. L'Heureux N, Paquet S, Labbe R, Germain L, *et al.* (1998) A completely biological tissue-engineered human blood vessel. *FASEB J* 12: 47–56.

56. Shimizu T, Yamato M, Isoi Y, Akutsu T, *et al.* (2002) Fabrication of pulsatile cardiac tissue grafts using a novel 3-dimensional cell sheet manipulation technique and temperature-responsive cell culture surfaces. *Circ Res* 90: e40.

57. Bursac N, Papadaki M, Cohen RJ, Schoen FJ, *et al.* (1999) Cardiac muscle tissue engineering: toward an *in vitro* model for electrophysiological studies. *Am J Physiol* 277: H433–H444.

58. Zimmermann WH, Didie M, Wasmeier GH, Nixdorff U, *et al.* (2002) Cardiac grafting of engineered heart tissue in syngenic rats. *Circulation* 106: I151–I157.

59. Radisic M, Park H, Shing H, Consi T, *et al.* (2004) Functional assembly of engineered myocardium by electrical stimulation of cardiac myocytes cultured on scaffolds. *Proc Natl Acad Sci USA* 101: 18129–18134.

60. Akins RE, Boyce RA, Madonna ML, Schroedl NA, *et al.* (1999) Cardiac organogenesis *in vitro*: reestablishment of three-dimensional tissue architecture by dissociated neonatal rat ventricular cells. *Tissue Eng* 5: 103–118.

61. Fink C, Ergun S, Kralisch D, Remmers U, *et al.* (2000) Chronic stretch of engineered heart tissue induces hypertrophy and functional improvement. *FASEB J* 14: 669–679.

62. Tandon N, Cannizzaro C, Chao PH, Maidhof R, *et al.* (2009) Electrical stimulation systems for cardiac tissue engineering. *Nat Protoc* 4: 155–173.

63. Zimmermann WH, Didie M, Doker S, Melnychenko I, *et al.* (2006) Heart muscle engineering: an update on cardiac muscle replacement therapy. *Cardiovasc Res* 71: 419–429.

64. Radisic M, Deen W, Langer R, Vunjak-Novakovic G (2005) Mathematical model of oxygen distribution in engineered cardiac tissue with parallel channel array perfused with culture medium containing oxygen carriers. *Am J Physiol Heart Circ Physiol* 288: H1278–H1289.

65. Zimmermann WH, Melnychenko I, Wasmeier G, Didie M, *et al.* (2006) Engineered heart tissue grafts improve systolic and diastolic function in infarcted rat hearts. *Nat Med* 12: 452–458.

66. Naito H, Melnychenko I, Didie M, Schneiderbanger K, *et al.* (2006) Optimizing engineered heart tissue for therapeutic applications as surrogate heart muscle. *Circulation* 114: I72–I78.

67. Eschenhagen T, Zimmermann WH (2005) Engineering myocardial tissue. *Circ Res* 97: 1220–1231.

68. Hansen A, Eder A, Bonstrup M, Flato M, *et al.* (2010) Development of a drug screening platform based on engineered heart tissue. *Circ Res* 107: 35–44.

69. Hollenberg M, Honbo N, Samorodin AJ (1976) Effects of hypoxia on cardiac growth in neonatal rat. *Am J Physiol* 231: 1445–1450.

70. Miyahara Y, Nagaya N, Kataoka M, Yanagawa B, *et al.* (2006) Monolayered mesenchymal stem cells repair scarred myocardium after myocardial infarction. *Nat Med* 12: 459–465.

71. Furuta A, Miyoshi S, Itabashi Y, Shimizu T, *et al.* (2006) Pulsatile cardiac tissue grafts using a novel three-dimensional cell sheet manipulation technique functionally integrates with the host heart, *in vivo. Circ Res* 98: 705–712.

72. Chachques JC, Trainini JC, Lago N, Masoli OH, *et al.* (2007) Myocardial assistance by grafting a new bioartificial upgraded myocardium (MAGNUM clinical trial): one year follow-up. *Cell Transplant* 16: 927–934.

73. Shimizu T, Sekine H, Yamato M, Okano T (2009) Cell sheet-based myocardial tissue engineering: new hope for damaged heart rescue. *Curr Pharm Des* 15: 2807–2814.

74. Kelm JM, Djonov V, Ittner LM, Fluri D, *et al.* (2006) Design of custom-shaped vascularized tissues using microtissue spheroids as minimal building units. *Tissue Eng* 2: 2151–2160.

75. Yildirim Y, Naito H, Didie M, Karikkineth BC, *et al.* (2007) Development of a biological ventricular assist device: preliminary data from a small animal model. *Circulation* 116: I16–I23.

76. Bar A, Dorfman SE, Fischer P, Hilfiker-Kleiner D, *et al.* (2009) The pro-angiogenic factor CCN1 enhances the re-endothelialization of biological vascularized matrices *in vitro. Cardiovasc Res* 85: 806–813.

77. Sekine H, Shimizu T, Hobo K, Sekiya S, *et al.* (2008) Endothelial cell coculture within tissue-engineered cardiomyocyte sheets enhances neovascularization and improves cardiac function of ischemic hearts. *Circulation* 118: S145–S152.

78. Levenberg S, Rouwkema J, Macdonald M, Garfein ES, *et al.* (2005) Engineering vascularized skeletal muscle tissue. *Nat Biotechnol* 23: 879–884.

79. Morritt AN, Bortolotto SK, Dilley RJ, Han X, *et al.* (2007) Cardiac tissue engineering in an *in vivo* vascularized chamber. *Circulation* 115: 353–360.

80. Haraguchi Y, Shimizu T, Yamato M, Kikuchi A, *et al.* (2006) Electrical coupling of cardiomyocyte sheets occurs rapidly via functional gap junction formation. *Biomaterials* 27: 4765–4774.

81. Rubart M, Pasumarthi KB, Nakajima H, Soonpaa MH, *et al.* (2003) Physiological coupling of donor and host cardiomyocytes after cellular transplantation. *Circ Res* 92: 1217–1224.

82. Eschenhagen T, Zimmermann WH, Kleber AG (2006) Electrical coupling of cardiac myocyte cell sheets to the heart. *Circ Res* 98: 573–575.

83. Caspi O, Lesman A, Basevitch Y, Gepstein A, *et al.* (2007) Tissue engineering of vascularized cardiac muscle from human embryonic stem cells. *Circ Res* 100: 263–272.

84. Stevens KR, Pabon L, Muskheli V, Murry CE (2009) Scaffold-free human cardiac tissue patch created from embryonic stem cells. *Tissue Eng Part A* 15: 1211–1222.

85. Tulloch NL, Muskheli V, Razumova MV, Korte FS, *et al.* (2011) Growth of engineered human myocardium with mechanical loading and vascular coculture. *Circ Res* 109: 42–59.

86. Zimmermann WH, Eschenhagen T (2007) Embryonic stem cells for cardiac muscle engineering. *Trends Cardiovasc Med* 17: 134–140.

87. Takahashi K, Okita K, Nakagawa M, Yamanaka S (2007) Induction of pluripotent stem cells from fibroblast cultures. *Nat Protoc* 2: 3081–3089.

88. Smits AM, van Vliet P, Metz CH, Korfage T, *et al.* (2009) Human cardiomyocyte progenitor cells differentiate into functional mature cardiomyocytes: an *in vitro* model for studying human cardiac physiology and pathophysiology. *Nat Protoc* 4: 232–243.

89. Guan K, Nayernia K, Maier LS, Wagner S, *et al.* (2006) Pluripotency of spermatogonial stem cells from adult mouse testis. *Nature* 440: 1199–1203.

90. Conrad S, Renninger M, Hennenlotter J, Wiesner T, *et al.* (2008) Generation of pluripotent stem cells from adult human testis. *Nature* 456: 344–349.

91. Cibelli JB, Grant KA, Chapman KB, Cunniff K, *et al.* (2002) Parthenogenetic stem cells in nonhuman primates. *Science* 295: 819.

92. Yang L, Soonpaa MH, Adler ED, Roepke TK, *et al.* (2008) Human cardiovascular progenitor cells develop from a KDR+ embryonic-stem-cell-derived population. *Nature* 453: 524–528.

VII. The Complexity of Organ Regeneration

Tatyana Novoyatleva and Felix B. Engel

1. INTRODUCTION

Thomas Morgan defined the term "Regeneration" in 1901 as "the re-placement of missing structures following injury." In the online lexicon Wikipedia is stated: "In biology, an organism is said to regenerate a lost or damaged part if the part re-grows so that the original function is restored." In the field of regenerative medicine we focus on restoring the function of an organ whereas it is not necessarily important that the original morphology is restored as well. The function of the liver, for example, is restored after resection of a liver lobe by increasing the size of the remaining lobes and not by re-growing the resected lobe.[1]

Regeneration occurs in many different organisms, e.g. the fresh-water planarian *Schmidtea mediterranea*, the cnidarian *Hydra vulgaris*, the starfish *Allostichaster capensis*, the zebrafish *Danio rerio*, the snail *Helix lucorum*, the cricket *Gryllus bimaculatus*, the newt *Notophthalmus viridescens* and humans. It appears that the ability to regenerate is randomly distributed in the animal kingdom and varies greatly in terms of efficiency among species.[2] For example, mammals can recover from blood loss and damage to the skeletal muscle, the peripheral nervous system or the liver but they lack the ability of amphibians and fish to re-grow appendages, heart, lens, retina, and central nervous system.[3]

Naturally occurring organ regeneration is a highly complex event and is characterized by several phases: detection of injury, wound healing and initiation of regeneration, remodeling of the tissue and termination of regeneration. On the cellular level there are three basic mechanisms underlying regeneration. First, missing cells or structures are replaced by utilizing stem or progenitor cells. For example, planarians can be cut in over 200 pieces and each piece will re-grow all missing body parts. This is possible as planarians contain a stem cell population called neoblasts.[4,5] Second, adult cells at the site of injury undergo dedifferentiation and form a blastema containing a pool of progenitor cells. The best studied examples for this type of regeneration is the salamander *Ambystoma mexicanum* (the axolotl) and the zebrafish *Danio rerio*.[6,7] Third, regeneration can be achieved by proliferation of functional cells in the remaining tissue, like during liver regeneration.[8]

One goal of regenerative medicine is to elucidate the mechanisms that govern regeneration in model organisms and to apply this knowledge to repair human organs after injury. In order to choose the optimal strategy for a specific organ it is important to know the composition of the organ and the changes that occur after injury like scarring.

2. THE COMPLEXITY OF THE HEART

At first glance the heart is a simple pump that provides the organism with oxygen-enriched blood and nutrients. At a closer look the heart is a truly spectacular organ. The heart is the first organ that forms during mammalian development and pumps constantly blood until death. The human heart pumps blood with a frequency of around 80 beats per minute at rest. Thus, the heart of a 60-year-old person has contracted over 2.5 billion times. If one considers that this heart consists still of over 50% of the cells that made up this heart at birth[9] its efficiency is simply astonishing.

The heart is a complex organ consisting of four chambers (two atria and two ventricle) and four valves. In order to provide the body with oxygen-enriched blood the contraction of these four chambers is highly regulated by an electrical system, the cardiac conduction system. The main constituents of this system are the sinoatrial (SA) node, the atrioventricular (AV) node and the His-Purkinje system (Figure 1). The heartbeat is initiated by an electrical

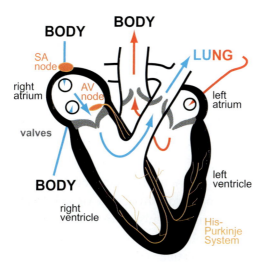

Figure 1.   The heart. This schematic drawing illustrates the morphology of the heart as well as the blood flow. Red arrows indicate the flow of oxygen-enriched blood from the lung into the left atrium to the left ventricle and from there into the body. The blue arrows indicate the flow of blood carrying carbon dioxide and cellular waste from the body into the right atrium to the right ventricle into the lung. The heartbeat is initiated in the sinoatrial (SA) node and subsequently spreads through the atrial cells, the atrioventricular (AV) node, the His-Purkinje system and the ventricular cells to enable the concerted contraction of the chambers.

signal from the SA node spreading across the cells of the atria causing their contraction to push blood into the ventricle. The signal
moves to the AV node where it slows for an instant to allow the
right and left ventricles to fill with blood. Subsequently, the signal
is released, travels along the His-Purkinje system, spreads across the
cells of the ventricle walls, and both ventricles contract. The signal
passes and the ventricle relax. It is important to realize, however,
that the blood is not simply ejected from the ventricles by a simple
concentric contraction. Torrent-Guasp's model of the helical heart
illustrates that the heart is a pump of non-uniform behavior requiring a highly coordinated function and explains the different
position-dependent orientation of cardiomyocytes.[10,11] Sarnoff and
Mitchell declared "The muscular ventricular walls squeeze down on
the contained blood much as one would milk a cow or squeeze a
lemon in a clenched fist."[12] In order to perform these coordinated
contractions the heart requires a stress-tolerant, viscoelastic extracellular matrix.[13]

Investigations of the cellular composition of the heart revealed
that cardiomyocytes account for around 75% of the heart volume.
However, 50% to 70% of the cells in the heart are non-myocytes.
The composition of the cardiac tissue varies depending on the
regional functional requirements. The main cell types in the
heart are: cardiomyocytes, fibroblasts, endothelial cells, and vascular
smooth muscle cells.[14,15] Other less abundant but still important
cell types are pericytes, neurons, perineurial cells, and Schwann
cells.[16–18] In addition, cardiomyocytes of different compartments
differ in the expression of transcription factors, their contractility
and their electrophysiological properties.[19–21] Thus, it is of great
importance for regenerative therapies to introduce the proper
myocardial cell type at the right place in the heart. For instance,
myocardial conduction cells are needed to repair the sinus node but
when placed in the working myocardium they might lead to life-
threatening arrhythmias.

Taken together, the heart is a highly complex structured organ
consisting of several different cell types and many subtypes of
cardiomyocytes.

3. REGENERATION OF THE HEART

Natural occurring wound healing and regeneration are highly ordered and well-coordinated processes that include several cell types. Injury to the skin, for example, initiates a cascade of events including inflammation, cell proliferation, matrix deposition, and tissue remodeling and requires cell-to-cell as well as cell-to-matrix interactions. The re-epithelialization of the wound exemplifies the involvement of different cell types: Upon injury keratinocyte growth factor is secreted by dermal fibroblasts and $\gamma\delta$ T-cells acting on the FGFR2IIIb receptor of keratinocytes. Subsequently, keratinocytes proliferate and migrate affecting wound re-epithelialization.[22]

Another good example is liver regeneration.[8] After partial hepatectomy, most of the remaining hepatocytes enter the cell cycle. However, liver regeneration is much more than just induction of hepatocyte proliferation.[3,8] Other cell types including Kupffer cells, biliary epithelial cells, and endothelial cells lining the blood vessels start to proliferate at a later time point. In addition, the proliferative phase of liver regeneration is followed by a remodeling phase to re-establish normal organization.[8,23] Interestingly, liver regeneration depends on several cytokines and growth factors that are not all delivered by liver cells. After injury by partial hepatectomy several signals are initiated simultaneously. Molecules of the innate immunity system activate Kupffer cells, resident macrophages, which then release growth factors and cytokines that induce hepatocyte proliferation. In addition, factors are released from the pancreas (insulin), duodenum or salivary gland (epidermal growth factor; EGF), adrenal gland (norepinepherine), thyroid gland (triodothronine; T3) and stellate cells (hepatocyte growth factor; HGF). After liver mass and function has been restored another set of growth factors is released to stop liver regeneration. The signals that mediate this regulation are poorly understood. If hepatocyte proliferation is compromised the liver can also regenerate based on progenitor cells.[24,25]

Taken together, it appears unlikely given the complexity of the heart that the generation of new cardiomyocytes, the predominant functional cells in the heart, will be sufficient to restore heart function.

Regeneration of the infarct area through cardiomyocyte proliferation without increasing capillary density for example appears impossible.[22] For example, tumors implanted into nude mice did not reach a size of more than 2 to 3 mm in the absence of capillarization and perfusion.[26] Due to the high need of oxygen of cardiac tissue the maximal thickness of heart tissue in the absence of vascularization is probably significantly smaller (see Chapter VI. B.). Furthermore, blocking epithelial cell proliferation and neo-vascularization impairs wound-healing. Similarly, liver regeneration relies on the expansion of the sinusoidal and portal vascular systems.[27]

3.1. Induction of Cardiomyocyte Proliferation

The idea to regenerate the heart by inducing cardiomyocyte proliferation is based on the observation that cardiac regeneration in newt and zebrafish is based on cardiomyocyte proliferation and that the mammalian heart during prenatal development increase in size due to proliferation of already contracting cardiomyocytes (see Chapters III. A. and III. B.). The hope is to induce partial dedifferentiation of adult cardiomyocytes to allow proliferation and to maintain at the same time the ability to contract to exhibit their function. In addition, it is likely that the right subtype of cardiomyocytes will be generated as one can assume that the surviving neighboring cardiomyocytes are of the same type as the ones that got lost due to injury. However, dedifferentiation harbors also risks. For example, the cells might transdifferentiate in an inappropriate cell type or overproliferate causing cancer. Moreover, cardiomyocyte cell division requires the disassembly of the contractile apparatus at least during cytokinesis (see Chapter IV. A.). Thus, they will stop beating for at least a short time. Whereas this appears to be tolerated in newt and zebrafish hearts as well as in fetal mammalian hearts it might cause problems in an adult mammalian heart with a much higher workload.

It is unclear how this approach can generate other cell types lost during injury like fibroblasts, endothelial cells, vascular smooth muscle cells, pericytes, neurons, perineurial cells, and Schwann cells or to restore the extracellular matrix. However, the fact that transgenic mice overexpressing cyclin D2 under the control of the cardiomyo-cyte-specific alpha-MHC promoter induce increased cardiomyocyte proliferation after cardiac injury forming new cardiac tissue (see Chapter IV. A. for more details) suggests that proliferating cardiomyocytes themselves can induce proliferation, migration and differentiation of other cell types to form new cardiac tissue.[28] To our knowledge there is so far no study available investigating a possible link between induction of cardiomyocyte proliferation and the behavior of other cell types. However, it has been shown that cardiomyocytes during early development have the potential to attract blood vessels by the secretion of pro-angiogenic cytokines like VEGF-A or angiopoietin-1.[29–32] In addition, it has been shown that GATA-4 is a critical stress responsive factor in the murine heart that is able to activate a pro-angiogenic gene profile in cardiomyocytes *in vivo*.[33] It will be important in future studies to investigate how induction of cardiomyocyte proliferation affects other cell types with regards to tissue regeneration. Another possibility is that mechanisms responsible for tissue homeostasis are responsible for re-innervation, neo-vascularization and tissue formation once new cardiomyocytes are formed.

3.2. Stem Cell-Based Regeneration

Rumyantsev wrote in 1977: "Unlike skeletal muscle, myocardium is very rarely subject to neoplastic transformation, which is strongly suggestive of a stable inhibition of DNA synthesis and mitosis of the myonuclei and of the absence of a pool of undifferentiated myogenic cells." In addition, he summarizes all the observations that suggest that cardiomyocytes "possess an intrinsic capacity to dedifferentiate and multiply at the borders of necroses."[34] Not surprisingly, the field of cardiac regeneration concentrated subsequently on ways to promote cardiomyocyte proliferation. This changed dramatically when Orlic and co-workers published their study suggesting that bone marrow stem cells can regenerate the heart.[35] From then on the majority of labs concentrated on utilizing stem cells for cardiac regeneration and on identifying cardiac progenitor cells (see Chapters IV. B. to IV. D.).

The "stem cell" approach has the advantage that stem cells or progenitor cells can, in principle, differentiate in all the cell types that have to be replaced during regeneration. Mammalian cardiogenesis requires the generation of diverse muscle and non-muscle cell lineages: atrial/ventricular cardiomyocytes, conduction system cells, smooth muscle/endothelial cells of the coronary arteries and veins, endocardial cells, valvular components and connective tissue. As described in Chapter IV. B. to IV. D., a wide variety of cells from embryonic stem cells to resident cardiac stem cells have been investigated. Unfortunately, the so far utilized stem cells and application forms rarely result in the generation of cardiomyocytes even though the stem cell approach can improve heart function. In contrast, the injected cells can regenerate other components of the heart like the vasculature, which is known to improve heart function.[36,37] Moreover, stem cell approaches harbor the risk of cancer. Thus, it will be important in the future to better understand how to guide differentiation of stem cells into the required cells; especially differentiation into cardiomyocytes.

Taken together, we are far from being able to regenerate the heart after an injury. However, the generated data for cardiomyocyte proliferation as well as stem cell approaches are very promising and give hope that we will resolve the remaining issues and establish a regenerative therapy for heart disease. If we integrate our knowledge from the analysis of the development of the heart with the data derived from stem cell and proliferation studies it might become possible to regenerate the heart by manipulating cardiomyocytes and non-myocardial cells (endogenous stem cells or fibroblasts) to proliferate and/or to differentiate into the required cell types present at the location where they are needed.

4. FROM REGENERATION MODEL SYSTEM TO THERAPY

In order to establish regenerative therapies it is important to understand how the organ of interest grows during development and the basic principles of regeneration. Newt and zebrafish can regenerate heart but mammals cannot. This raises the question how and why these hearts are different? In what way do the cardiomyocytes differ? The group of Ken Poss identified in 2008 one very important difference. They showed that heart growth and maintenance in adult zebrafish is maintained by cardiomyocyte proliferation and by adding epicardial cells.[38] This homeostatic turnover of cardiac tissue is also known to occur in humans. However, the cell source is unknown and the turnover is less than 50% in 50 years.[9] Thus, it is important to determine why zebrafish and also newt cardiomyocytes maintain their proliferation potential. It is possible that these cells get arrested in the cell cycle at a certain time during development and dedifferentiate upon injury to regenerate the heart. In that case, the study of these systems might help us to find a way to promote cardiomyocyte proliferation. However, it is also possible that these cardiomyocytes never get arrested in the cell cycle.

It is also possible that the type of injury influences the regenerative capacity. Cardiac regeneration in lower animals has been studied after resection of the apex. This kind of injury is very different from injuries that patients sustain or that are used in mammalian injury models. Thus it might be useful to determine whether these differences matter. For example, it would be important to determine the response of zebrafish to a cardiac injury caused by ischemia (e.g. pinching off the apex) or by necrosis (e.g. heat). Similar experiments have recently been performed in newts (see Chapter III. A.)

Induction of regeneration in mammals after injury requires more than inducing the mechanisms that occur during natural regeneration. Injury to the myocardium causes secondary changes to the heart like scarring and remodeling. Thus, the approach to induce regeneration has to prevent these changes or even has to revert them. Previously, it has been shown for other tissues that scar tissue is hostile and inhibits regeneration. Chondroitin sulfate proteoglycans for example are the principal inhibitory component of glial scars, which form after damage to the adult central nervous system and act as a barrier to regenerating axons. Interestingly, administration of chondroitinase ABC, an enzyme that selectively cleaves GAG chains from the chondroitin sulfate proteoglycans, promotes the regeneration of sensory and corticospinal tract axons after lesion.[39] Nevertheless, the work from Loren Field's group has clearly shown that generation of new cardiomyocytes can improve heart function significantly without reducing scar size.[28] Moreover, recent data indicate that scar formation does not impair heart regeneration in zebrafish.[40] However, formation of isolated cardiac tissue islands electrically isolated by scar from the host myocardium can cause severe arrythmia. This problem became very apparent during studies attempting to regenerate the heart using skeletal myoblasts.[41]

5. THE EXTRACELLULAR MATRIX

The microenvironment appears to be a very important issue in regeneration and thus it is important to also consider the role of the extracellular matrix (ECM). Historically, the ECM was primarily viewed as a supporting inert scaffold. It is a heterogeneous mixture of water, saccharides, and various protein components traditionally classified into four categories: collagens, proteoglycans, noncollagenous glycoproteins, and elastins. However, in recent years it has been shown that the ECM is a complex, dynamic and critical component that regulates, besides cell adhesion, also cell survival, proliferation, migration, and differentiation for example by modulating the activity, bioavailability, or presentation of growth factors to cell surface receptors.[42,43] For example, during limb regeneration matrix metalloproteinases (MMPs) are upregulated very early after amputation and are required for regeneration. It has been postulated that they play a role in ECM degradation, contributing to formation of the wound epidermis.[44,45] During liver regeneration cytokines appear to activate growth factors via activation of MMPs. The cytokine TNF can activate an MMP called TGF-alpha-converting enzyme (TACE), resulting in release of TGF-alpha and activation of hepatocyte proliferation *in vitro*.[46] Mice lacking a specific inhibitor of TACE show elevated levels of TNF protein and earlier entry into S phase after hepatectomy.[47] In addition, EGF induced hepatocyte proliferation depends on TNF priming which is mediated by activation of MMPs.[48]

The ECM regulates proliferation in tissue by regulating the bioavailability of growth factors. In addition, ECM proteins can directly affect cells and inhibit their proliferation. It has been shown that ECM proteins can tame aggressive carcinoma cells to form normal tissues or can be reverted to a normal phenotype.[49-51] The cardiac ECM might also have antiproliferative properties. Interestingly, application of fibroblast growth factor 1, which is present in the heart in high amounts, can induce cardiomyocyte mitosis, indicating that the inhibitory effect of the ECM is dose dependent.[52] Also alteration of the ECM by injecting the ECM protein Periostin has been shown to result in cardiomyocyte proliferation.[53]

The ECM plays also an important role in stem cell-based therapies. The success of stem cell-based therapies depends on the efficient differentiation of stem cells into cardiomyocytes. This problem is best exemplified by the behavior of embryonic stem cells (ES). Undifferentiated ES cells, when implanted into their normal microenvironment, the embryonic blastocysts, will participate in normal development and will not develop tumors. Outside this environment (in the heart or at other ectopic locations), however, they will form teratomas that are characterized by uncontrolled proliferation and differentiation.[54] Recently, it has also been shown that injection of mesenchymal stem cells into the heart differentiate in bone-like cells.[55] In fact, none of the applied stem cell therapies has yet unambiguously shown that injected stem cells of any type differentiate in a significant amount into cardiomyocytes.

It remains a challenge to understand precisely how the combination of tissue repair mechanisms with reactivation of embryonic programs in urodele and zebrafish can generate growth, pattern formation, and morphogenesis in an adult animal. Whether such a "perfect" regeneration can be achieved in mammals is difficult to estimate. However, full regeneration might not be necessary. As mentioned above the liver has extraordinary regenerative capabilities, however, it is rather compensatory growth than regeneration. After excision of one of the five liver lobes the liver does not restore the original morphology. Instead, the remaining four lobes grow to compensate for the loss in liver mass.[56] Thus, functional regeneration to improve cardiac output might suffice in most clinical settings to restore a satisfactory functional performance.

6. SUMMARY AND CONCLUDING REMARKS

We understand the cellular and molecular mechanisms of naturally occurring regeneration in some details but we do not understand why some tissues and organs regenerate and others not. Moreover, the regenerative capacity of a distinct organ is species-dependent. Why can the heart of newt and zebrafish regenerate but not the human heart?

It is commonly believed that regenerative capacity is inversely correlated with the complexity of an organ or species. However, a more thorough review of the regenerative potential across the animal kingdom revealed that also some "primitive" forms do not regenerate, while close relatives do. For example, some annelids can regenerate the whole body from small fragments, while similarly complex nematodes do not regenerate at all.[2] Another surprising finding is that newts are able to regenerate their lenses, while axolotls, the champions of regeneration, are incapable of doing so.[57]

Considerable effort has been invested to determine whether it is possible to induce regeneration in non-regenerating species. It has been shown that cellular behavior can be altered without genetic manipulation. For example, terminally differentiated mouse muscle cells dedifferentiate when treated with extracts from salamander regenerating limbs.[58] Terminally differentiated adult rat cardiomyocytes divide upon stimulation with FGF1/p38 inhibitor, Periostin or Neuregulin (see Chapter IV. A.). Moreover, some spectacular experiments indicated that even regeneration could be promoted. The diminished regenerative potential of skeletal muscle in old mice was reversed by joining their circulatory systems with those of young mice.[59] FGF2-soaked beads can stimulate embryonic chick limb buds to regenerate digit-like structures.[60] Skeletal muscle regeneration could be improved by targeted overexpression of IGF.[61]

Taken together, these studies suggest that the basic machinery necessary for regeneration still exists in mammalian organs and that providing the triggers of regeneration might be sufficient to induce at least partial regeneration. In our opinion, it is worthwhile to invest in the field of regenerative medicine to establish in the future novel strategies to improve human health. Independent of the target organ there are several alternative strategies that all should be further investigated: (1) Stimulation of intrinsic stem/progenitor cells utilizing cytokines, growth factors or small molecules. (2) Activation of dedifferentiation and proliferation of remaining functional cells. (3) Expansion of progenitor and stem cells *in vitro* and differentiating them into specific cell types. (4) Growing organs and tissues *in vitro*.

ACKNOWLEDGMENTS

This work was supported by a grant from the Alexander von Humboldt Foundation (Sofja Kovalevskaja Award to F. B. E.), the Excellence Cluster Cardio-Pulmonary System (DFG) and the International Research Training Group: Protecting the Heart from Ischemia (PROMISE, DFG).

REFERENCES

1. Michalopoulos GK, DeFrances MC (1997) Liver regeneration. *Science* 276: 60–66.
2. Alvarado AS, Tsonis PA (2006) Bridging the regeneration gap: genetic insights from diverse animal models. *Nat Rev Genet* 7: 873–884.
3. Stoick-Cooper CL, Moon RT, Weidinger G (2007) Advances in signaling in vertebrate regeneration as a prelude to regenerative medicine. *Genes Dev* 21: 1292–1315.
4. Newmark PA, Sanchez Alvarado A (2000) Bromodeoxyuridine specifically labels the regenerative stem cells of planarians. *Dev Biol* 220: 142–153.
5. Forsthoefel DJ, Newmark PA (2009) Emerging patterns in planarian regeneration. *Curr Opin Genet Dev* 19: 412–420.
6. Poss KD, Keating MT, Nechiporuk A (2003) Tales of regeneration in zebrafish. *Dev Dyn* 226: 202–210.
7. Kragl M, Knapp D, Nacu E, Khattak S, *et al.* (2009) Cells keep a memory of their tissue origin during axolotl limb regeneration. *Nature* 460: 60–65.

8. Taub R (2004) Liver regeneration: from myth to mechanism. *Nat Rev Mol Cell Biol* 5: 836–847.

9. Bergmann O, Bhardwaj RD, Bernard S, Zdunek S, *et al.* (2009) Evidence for cardiomyocyte renewal in humans. *Science* 324: 98–102.

10. Coghlan HC, Coghlan AR, Buckberg GD, Gharib M, *et al.* (2001) The structure and function of the helical heart and its buttress wrapping. III. The electric spiral of the heart: the hypothesis of the anisotropic conducting matrix. *Semin Thorac Cardiovasc Surg* 13: 333–341.

11. Torrent-Guasp F, Buckberg GD, Clemente C, Cox JL, *et al.* (2001) The structure and function of the helical heart and its buttress wrapping. I. The normal macroscopic structure of the heart. *Semin Thorac Cardiovasc Surg* 13: 301–319.

12. Sarnoff SJ, Mitchell JH (1962) Section 2. Circulation. In: Hamilton DF, Dow P (eds.) *Handbook of Physiology*. Williams & Wilkins, Baltimore, pp. 489–532.

13. Fomovsky GM, Thomopoulos S, Holmes JW (2009) Contribution of extracellular matrix to the mechanical properties of the heart. *J Mol Cell Cardiol* 48: 490–496.

14. Vliegen HW, van der Laarse A, Cornelisse CJ, Eulderink F (1991) Myocardial changes in pressure overload-induced left ventricular hypertrophy. A study on tissue composition, polyploidization and multinucleation. *Eur Heart J* 12: 488–494.

15. Banerjee I, Fuseler JW, Price RL, Borg TK, *et al.* (2007) Determination of cell types and numbers during cardiac development in the neonatal and adult rat and mouse. *Am J Physiol Heart Circ Physiol* 293: H1883–H1891.

16. Pauziene N, Pauza DH, Stropus R (2000) Morphology of human intracardiac nerves: an electron microscope study. *J Anat* 197 Pt 3: 437–459.

17. Chottova Dvorakova M, Wiegand S, Pesta M, Slavikova J, *et al.* (2008) Expression of neuropeptide Y and its receptors Y1 and Y2 in the rat heart and its supplying autonomic and spinal sensory ganglia in experimentally induced diabetes. *Neuroscience* 151: 1016–1028.

18. Higuchi K, Hashizume H, Aizawa Y, Ushiki T (2000) Scanning electron microscopic studies of the vascular smooth muscle cells and pericytes in the rat heart. *Arch Histol Cytol* 63: 115–126.

19. Boukens BJ, Christoffels VM, Coronel R, Moorman AF (2009) Developmental basis for electrophysiological heterogeneity in the ventricular and outflow tract myocardium as a substrate for life-threatening ventricular arrhythmias. *Circ Res* 104: 19–31.

20. Moorman AF, Christoffels VM (2003) Cardiac chamber formation: development, genes, and evolution. *Physiol Rev* 83: 1223–1267.

21. Hoogaars WM, Barnett P, Moorman AF, Christoffels VM (2007) T-box factors determine cardiac design. *Cell Mol Life Sci* 64: 646–660.

22. Grose R, Werner S (2004) Wound-healing studies in transgenic and knockout mice. *Mol Biotechnol* 28: 147–166.

23. Fausto N, Campbell JS, Riehle KJ (2006) Liver regeneration. *Hepatology* 43: S45–S53.

24. Fausto N (2004) Liver regeneration and repair: hepatocytes, progenitor cells, and stem cells. *Hepatology* 39: 1477–1487.

25. Santoni-Rugiu E, Jelnes P, Thorgeirsson SS, Bisgaard HC (2005) Progenitor cells in liver regeneration: molecular responses controlling their activation and expansion. *APMIS* 113: 876–902.

26. Folkman J (1971) Tumor angiogenesis: therapeutic implications. *N Engl J Med* 285: 1182–1186.

27. Abshagen K, Eipel C, Menger MD, Vollmar B (2006) Comprehensive analysis of the regenerating mouse liver: an *in vivo* fluorescence microscopic and immunohistological study. *J Surg Res* 134: 354–362.

28. Pasumarthi KBS, Nakajima H, Nakajima HO, Soonpaa MH, *et al.* (2005) Targeted expression of cyclin D2 results in cardiomyocyte DNA synthesis and infarct regression in transgenic mice. *Circ Res* 96: 110–118.

29. Giordano FJ, Gerber HP, Williams SP, VanBruggen N, *et al.* (2001) A cardiac myocyte vascular endothelial growth factor paracrine pathway is required to maintain cardiac function. *Proc Natl Acad Sci USA* 98: 5780–5785.

30. Hsieh PC, Davis ME, Lisowski LK, Lee RT (2006) Endothelial-cardiomyocyte interactions in cardiac development and repair. *Annu Rev Physiol* 68: 51–66.

31. Suri C, Jones PF, Patan S, Bartunkova S, *et al.* (1996) Requisite role of angiopoietin-1, a ligand for the TIE2 receptor, during embryonic angiogenesis. *Cell* 87: 1171–1180.

32. Kelly BD, Hackett SF, Hirota K, Oshima Y, *et al.* (2003) Cell type-specific regulation of angiogenic growth factor gene expression and induction of angiogenesis in nonischemic tissue by a constitutively active form of hypoxia-inducible factor 1. *Circ Res* 93: 1074–1081.

33. Heineke J, Auger-Messier M, Xu J, Oka T, *et al.* (2007) Cardiomyocyte GATA4 functions as a stress-responsive regulator of angiogenesis in the murine heart. *J Clin Invest* 117: 3198–3210.

34. Rumyantsev PP (1977) Interrelations of the proliferation and differentiation processes during cardiact myogenesis and regeneration. *Int Rev Cytol* 51: 186–273.

35. Orlic D, Kajstura J, Chimenti S, Bodine DM, *et al.* (2001) Transplanted adult bone marrow cells repair myocardial infarcts in mice. *Ann N Y Acad Sci* 938: 221–229.

36. Smart N, Risebro CA, Melville AA, Moses K, *et al.* (2007) Thymosin beta4 induces adult epicardial progenitor mobilization and neovascularization. *Nature* 445: 177–182.

37. Lavine KJ, White AC, Park C, Smith CS, *et al.* (2006) Fibroblast growth factor signals regulate a wave of Hedgehog activation that is essential for coronary vascular development. *Genes Dev* 20: 1651–1666.

38. Wills AA, Holdway JE, Major RJ, Poss KD (2008) Regulated addition of new myocardial and epicardial cells fosters homeostatic cardiac growth and maintenance in adult zebrafish. *Development* 135: 183–192.

39. Busch SA, Silver J (2007) The role of extracellular matrix in CNS regeneration. *Curr Opin Neurobiol* 17: 120–127.

40. Kikuchi K, Holdway JE, Werdich AA, Anderson RM, *et al.* (2010) Primary contribution to zebrafish heart regeneration by gata4(+) cardiomyocytes. *Nature* 464: 601–605.

41. Menasche P (2004) Skeletal myoblast transplantation for cardiac repair. *Expert Rev Cardiovasc Ther* 2: 21–28.

42. Aszodi A, Legate KR, Nakchbandi I, Fassler R (2006) What mouse mutants teach us about extracellular matrix function. *Annu Rev Cell Dev Biol* 22: 591–621.

43. Daley WP, Peters SB, Larsen M (2008) Extracellular matrix dynamics in development and regenerative medicine. *J Cell Sci* 121: 255–264.

44. Call MK, Tsonis PA (2005) Vertebrate limb regeneration. *Adv Biochem Eng Biotechnol* 93: 67–81.

45. Vinarsky V, Atkinson DL, Stevenson TJ, Keating MT, *et al.* (2005) Normal newt limb regeneration requires matrix metalloproteinase function. *Dev Biol* 279: 86–98.

46. Argast GM, Campbell JS, Brooling JT, Fausto N (2004) Epidermal growth factor receptor transactivation mediates tumor necrosis factor-induced hepatocyte replication. *J Biol Chem* 279: 34530–34536.

47. Mohammed FF, Smookler DS, Taylor SE, Fingleton B, *et al.* (2004) Abnormal TNF activity in Timp3-/- mice leads to chronic hepatic inflammation and failure of liver regeneration. *Nat Genet* 36: 969–977.

48. Serandour AL, Loyer P, Garnier D, Courselaud B, *et al.* (2005) TNFalpha-mediated extracellular matrix remodeling is required for multiple division cycles in rat hepatocytes. *Hepatology* 41: 478–486.

49. Mintz B, Illmensee K (1975) Normal genetically mosaic mice produced from malignant teratocarcinoma cells. *Proc Natl Acad Sci USA* 72: 3585–3589.

50. Weaver VM, Howlett AR, Langton-Webster B, Petersen OW, *et al.* (1995) The development of a functionally relevant cell culture model of progressive human breast cancer. *Semin Cancer Biol* 6: 175–184.

51. Weaver VM, Petersen OW, Wang F, Larabell CA, *et al.* (1997) Reversion of the malignant phenotype of human breast cells in three-dimensional culture and *in vivo* by integrin blocking antibodies. *J Cell Biol* 137: 231–245.

52. Engel FB, Hsieh PC, Lee RT, Keating MT (2006) FGF1/p38 MAP kinase inhibitor therapy induces cardiomyocyte mitosis, reduces scarring, and rescues function after myocardial infarction. *Proc Natl Acad Sci USA* 103: 15546–15551.

53. Kuhn B, Del Monte F, Hajjar RJ, Chang YS, *et al.* (2007) Periostin induces proliferation of differentiated cardiomyocytes and promotes cardiac repair. *Nat Med* 13: 962–969.

54. Nussbaum J, Minami E, Laflamme MA, Virag JA, *et al.* (2007) Transplantation of undifferentiated murine embryonic stem cells in the heart: teratoma formation and immune response. *FASEB J* 21: 1345–1357.

55. Breitbach M, Bostani T, Roell W, Xia Y, *et al.* (2007) Potential risks of bone marrow cell transplantation into infarcted hearts. *Blood* 110: 1362–1369.
56. Hata S, Namae M, Nishina H (2007) Liver development and regeneration: from laboratory study to clinical therapy. *Dev Growth Differ* 49: 163–170.
57. Godwin JW, Brockes JP (2006) Regeneration, tissue injury and the immune response. *J Anat* 209: 423–432.
58. McGann CJ, Odelberg SJ, Keating MT (2001) Mammalian myotube dedifferentiation induced by newt regeneration extract. *Proc Natl Acad Sci USA* 98: 13699–13704.
59. Conboy IM, Conboy MJ, Wagers AJ, Girma ER, *et al.* (2005) Rejuvenation of aged progenitor cells by exposure to a young systemic environment. *Nature* 433: 760–764.
60. Taylor GP, Anderson R, Reginelli AD, Muneoka K (1994) FGF-2 induces regeneration of the chick limb bud. *Dev Biol* 163: 282–284.
61. Musaro A, McCullagh K, Paul A, Houghton L, *et al.* (2001) Localized Igf-1 transgene expression sustains hypertrophy and regeneration in senescent skeletal muscle. *Nat Genet* 27: 195–200.

VIII. Cell Therapy Trials in Patients with Coronary Heart Disease

Kai C. Wollert

1. INTRODUCTION

Modern reperfusion strategies and advances in pharmacological management have resulted in an increasing proportion of patients surviving after an acute myocardial infarction (AMI). Unfortunately, this decrease in early mortality has resulted in an increased incidence of chronic heart failure (HF) in the survivors.[1] Because none of our current therapies addresses the central problem in ischemic HF, i.e. the massive loss of cardiomyocytes, vascular cells, and interstitial cells, these patients continue to experience frequent hospitalizations and premature death.[2]

Cell transplantation has been conceptualized more than ten years ago as a means to augment myocyte numbers and to improve cardiac function after AMI.[3,4] Regeneration of the infarcted heart is a daunting task, however, considering (i) that the myocyte deficit in infarction-induced human HF may be on the order of one billion cardiac myocytes,[5] (ii) that not only cardiomyocytes but also supporting (e.g. vascular) cells have to be supplied, and (iii) that the environmental cues that are required to guide transplanted cells into multicellular 3D heart structures may be lacking in infarcted and scarred myocardium.[6] In many of the early experimental studies, fetal, neonatal, and adult cardiomyocytes have been transplanted and were shown to form stable grafts in injured hearts.[7] Because of their limited availability, differentiated cardiomyocytes are not a realistic cell source for large-scale clinical applications, however.

Based on their capacities for self-renewal, infinite *ex vivo* proliferation, and differentiation into mature specialized cells, stem cells have now emerged as the prime cell source for regenerative therapies. Pluripotent stem cells can differentiate into cells derived from all three germ layers, a typical example being embryonic stem (ES) cells that can be isolated from the inner cell mass of blastocysts and that can give rise to all cardiac cell types. Adult stem cells, by contrast, are multipotent and restricted in their differentiation potential to cell lineages of the organ in which they are located [e.g. hematopoietic stem cells (HSC) giving rise to mature hematopoietic cells or mesenchymal stem cells (MSC) giving rise to osteoblasts, chondrocytes, and adipocytes]. Progenitor cells are even more restricted in their differentiation potential and have a limited capacity to self-renew [e.g. endothelial progenitor cells (EPC) or skeletal myoblasts]. Experiments conducted at the beginning of this millennium appeared to challenge the traditional concept that adult stem and progenitor cells are lineage restricted and suggested instead that these cells can transdifferentiate into cell types outside their original lineage (e.g. HSC differentiating into neurons or cardiomyocytes).[8,9] This new concept of adult stem cell "plasticity," combined with a large body of animal data indicating that adult stem and progenitor cell transplantation can improve contractile function after MI, provided the scientific rationale to treat cardiac patients with adult stem and progenitor cells.[10]

The field has made rapid progress. Some clinical trials have provided a signal that cell therapy can improve cardiac function after AMI. Parallel investigations of the involved mechanisms, however, have shattered the concept of adult stem cell plasticity and have highlighted instead the possible importance of paracrine effects. This scientific roller coaster ride has generated excitement and confusion. Confusion is caused also by the lack of a universally accepted nomenclature for stem and progenitor cells, by the occasional imprecise use of terminology (e.g. "stem cell therapy" instead of bone marrow cell therapy), by the large number of different cell types that are undergoing clinical testing, and by a lack of standardization in cell isolation protocols that would facilitate a comparison of clinical trial results from different institutions. This chapter uses the cell nomenclature that has been used by the respective investigators; the reader is encouraged to go back to the original publications to learn more about the cell isolations protocols that were used in the cited references.

2. WHAT HAVE WE LEARNT FROM THE FIRST CLINICAL TRIALS?

We will start our discussion with a brief review of the largest randomized-controlled clinical trials that have explored the prospects of cell therapy in patients with coronary heart disease (Table 1). Most of these studies used unfractionated bone marrow cells (BMC) as an easily accessible source of HSC, MSC, and EPC.

2.1. Acute Myocardial Infarction

In the BOOST trial, 60 patients were randomized to intracoronary nucleated BMC (nBMC) transfer, on average 4.8 days after acute percutaneous coronary intervention (PCI), or to a control group in which neither bone aspiration nor placebo injection were performed. Cardiac function was assessed in a blinded fashion using magnetic

Table 1.   Randomized Cell Therapy Trials in Patients with Acute Myocardial Infarction or Ischemic Heart Failure.

Study	Patient	Cell Types	Dose	Route of Delivery	Timing of Delivery	Primary Endpoint	Comments
Acute myocardial infarction							
BOOST	60	nBMC	128 mL	i.c.	Day 6±1	LVEF ↑	Effect diminished after 18 and 61 months
REPAIR-AMI	187	mnBMC	50 mL	i.c.	Day 3–6	LVEF ↑	
Leuven-AMI	66	mnBMC	130 mL	i.c.	Day 1	LVEF ↔	Regional contractility ↑ Infarct size ↓
ASTAMI	97	mnBMC	50 mL	i.c.	Day 6±1	LVEF ↔	
FINCELL	77	mnBMC	80 mL	i.c.	Day 3	LVEF ↑	
REGENT	117	mnBMC (unselected vs. CD34+/CXCR4+)	50–70 mL (unselected) 100–120 mL (selected)	i.c.	Day 3–12	LVEF ↑ with both cell types	
HEBE	189	mnBMC vs. mnPBC		i.c.	Day 3–8	Regional contractility ↔	
Ischemic heart failure							
TOPCARE-CHD	58	mnBMC vs. CPC	50 mL	i.c.	Month 81±72	LVEF ↑ (mnBMC) LVEF ↔ (CPC)	
MAGIC	97	SkM	400 or 800 × 10^6	i.m.	>Week 4	LVEF ↔	LVEDV ↓ LVESV ↓

Only patients with complete imaging studies are considered here. Dose refers to the average amount of bone marrow that was harvested, or the number of transplanted skeletal myoblasts. nBMC, nucleated bone marrow cells; mnBMC, mononucleated bone marrow cells; mnPBC, mononucleated peripheral blood cells; CPC, circulating blood-derived progenitor cells; SkM, skeletal myoblasts; i.c., intracoronary; i.m., intramuscular; LVEF, left ventricular ejection fraction; LVEDV, left ventricular resonance imaging (MRI) at serial time points before cell transfer, and after six, 18, and 61 months. As compared to the control group, the nBMC group showed an enhanced left ventricular (LV) contractility in the infarct border zone, and a significant improvement of global LV ejection fraction (LVEF) by 6.0 percentage points after six months.[11] In the overall study cohort, the differences in LVEF between the nBMC and control groups were diminished and no longer statistically significant after 18 months (2.8 percentage points) and 61 months (0.8 percentage points).[12,13] In a post hoc analysis, patients with greater infarcts and an infarct transmurality greater than the median appeared to benefit from nBMC transfer with a sustained improvement of LVEF also at the later time points.[12,13] After six and 18 months, the control group had developed echocardiographic signs of mild diastolic dysfunction, which were significantly attenuated in the nBMC group.[14]

The Leuven AMI trial randomized 67 patients to intracoronary mononucleated BMC (mnBMC) or placebo infusion within 24 hours after acute PCI.[15] MRI assessment of LVEF at three to four days and four months after PCI did not demonstrate a significant impact of mnBMC therapy on LVEF recovery, the primary endpoint of the trial. Notably, however, the reduction of infarct volume after four months, as measured by serial contrast-enhanced MRI, was greater in mnBMC-treated patients than in controls.[15] Moreover, a significant improvement in regional contractility was observed by MRI in the mnBMC group with the greatest infarct transmurality at baseline.[15] Echocardiographic strain rate imaging confirmed that mnBMC infusions improved the recuperation of myocardial function in the infarct region, suggesting that quantitative assessment of regional systolic function may be more sensitive than measuring global LVEF for the evaluation of cell therapy after myocardial infarction.[16]

In the REPAIR-AMI trial, 204 patients were randomized to an intracoronary infusion of mnBMC or placebo, on average 4.4 days after acute PCI. LV function was assessed by contrast angiography. mnBMC infusion promoted an increase in LVEF of 2.5 percentage points after four months as compared to the control group.[17] In a subgroup of 54 patients who underwent serial MRI investigations, the treatment effect of mnBMC infusion on LVEF amounted to 2.8 percentage points (P = 0.26) at 12 months,[18] similar to what has been observed in the BOOST trial after 18 months.[13]

In the ASTAMI trial, 100 patients with anterior MI were randomized to intracoronary infusion of mnBMC or a control group in which neither bone aspiration nor placebo injection was performed. Cells were infused, on average, six days after acute PCI. After six and 12 months no significant effects of mnBMC therapy on LVEF, LV volumes, or infarct size were observed by single-photon-emission computed tomography, echocardiography, and MRI.[19,20] Fewer mnBMC were infused in the ASTAMI trial as compared to the REPAIR-AMI trial (median, 68×10^6 *vs.* 198×10^6 cells).[17,19] It has also been proposed that the cell isolation protocol used in the ASTAMI trial may have recovered a mnBMC population with impaired functionality, as assessed by *in vitro* migratory and colony forming capacities, as well as *in vivo* capacity to promote blood flow recovery in a mouse model of hind-limb ischemia.[21]

The HEBE trial randomized 200 patients with AMI to an intracoronary infusion of mononucleated cells isolated from bone marrow (mnBMC) or peripheral blood (mnPBC) or to a control group that did not undergo bone marrow aspiration or placebo injection.[22] The final results of the HEBE trial were presented at the American Heart Association Meeting in 2008,[23] and showed that intracoronary infusion of mnBMC or mnPBC did not improve global or regional LV systolic function at four months as assessed by MRI. Further discussion of these data will have to await publication of the HEBE trial.

In the FINCELL trial, 80 patients with myocardial infarction treated with thrombolytic therapy followed by PCI were randomized to intracoronary mnBMC or placebo infusions. Cells were infused immediately after PCI, which was performed two to six days after thrombolysis. As shown by LV contrast angiography before and six months after cell transfer, mnBMC infusions improved LVEF recovery by 5.0 percentage points as compared to the control group. Paired echocardiographic investigations yielded similar results.[24]

In the REGENT trial, 200 patients with anterior MI were randomized to an intracoronary infusion of unselected mnBMC or

CXCR4[+] CD34[+] mnBMC, on average seven days after acute PCI, or to a control group in which neither bone aspiration nor placebo injection was performed.[25] Paired MRI images to assess LVEF at baseline and after six months were available in 117 patients. Significant improvements in LVEF from baseline to follow-up were noted within the unselected and selected mnBMC groups (3.0 percentage points, each), but not in the control group (no change in LVEF). However, changes in LVEF between the groups, the primary endpoint of the trial, were not significantly different between the mnBMC and control groups. This trial is limited by imbalances in baseline LVEF and incomplete follow-up. Nevertheless, the REGENT trial indicates that a specific BMC population expressing progenitor cell surface markers may carry much of the treatment effect observed after mnBMC transfer in post AMI patients.

So far, no safety concerns related to intracoronary BMC infusions have emerged. An increased risk of in-stent restenosis has been observed in a small non-randomized study after intracoronary infusion of CD133[+] mnBMC.[26] In the placebo-controlled FINCELL trial, no increased risk of in-stent restenosis was observed by intravascular ultrasound after six months.[24] In two meta-analyses, the risks of target vessel restenosis or repeat revascularization were not increased in cell-treated patients.[27,28] Moreover, none of the clinical trials reported an increased incidence of symptomatic arrhythmias after intracoronary BMC transfer. An electrophysiological study performed in the BOOST trial,[11] and a careful assessment of microvolt T-wave alternans and signal-averaged ECG measures in the FINCELL trial,[24] provide further assurance in this respect.

2.2. Ischemic Heart Failure

Skeletal myoblasts were the first cell type undergoing clinical testing in HF patients.[29] Skeletal myoblasts are lineage-restricted progenitor cells that can be isolated from skeletal muscle biopsies and expanded *in vitro*. When transplanted into an infarct scar, myoblasts differentiate into myotubes that, for the most part, remain electromechanically isolated from their host cardiomyocytes.

Early non-randomized clinical studies confirmed the feasibility of transplanting autologous skeletal myoblasts either transepicardially during cardiac surgery or transendocardially using an injection needle catheter advanced across the aortic valve and positioned against the endocardial surface.[30] The recently published MAGIC trial was the first randomized, placebo-controlled trial in the field.[31] Ninety-seven patients with ischemic HF received transepicardial injections of culture-expanded, autologous skeletal myoblasts or placebo in and around akinetic segments during bypass surgery. After six months, myoblast transplantation did not promote an incremental improvement in the primary efficacy endpoints, i.e. regional and global LV function, over that provided by bypass surgery alone. Notably, significant reductions in left ventricular end-diastolic and end-systolic volumes were noted after myoblast therapy.[31] No significant difference in major adverse cardiac events between the placebo and cell-treated groups was observed. However, a trend towards a greater incidence of arrhythmias was noted in the myoblast-treated patients, thus confirming a safety concern that had already been raised by earlier non-randomized trials.[32] Further investigations of skeletal myoblast therapy need to establish that the potential benefits outweigh the risk of increased arrhythmogenicity.[32]

In the TOPCARE-CHD trial, 75 patients with ischemic HF were randomized to receive no cell infusion or intracoronary infusions of mnBMC or circulating blood-derived progenitor cells into the patent coronary artery supplying the most dyskinetic left ventricular area. To obtain circulating blood-derived progenitor cells, mononucleated cells were isolated from the peripheral blood by Ficoll density gradient centrifugation and cultured *ex vivo* in medium containing vascular endothelial growth factor, a statin, and autologous serum. After 3 months, the absolute change in LV ejection fraction, as assessed by contrast angiography, was significantly greater among patients receiving mnBMC (+2.9 percentage points) than among those receiving circulating blood-derived progenitor cells (−0.4 percentage points) or no cell infusion (−1.2 percentage points).[33] No side effects were reported.

2.3. Chronic Myocardial Ischemia

An increasing number of patients with coronary artery disease have chronic myocardial ischemia and experience angina that is refractory to medical therapy and not amenable to revascularization. Chronic myocardial ischemia can be associated with a regional impairment of contractile function that is partially reversible when tissue perfusion is restored (hibernating myocardium). There is a need for new therapeutic strategies aimed at delivering oxygenated blood to the myocardium in these patients. Three randomized, double-blind, placebo-controlled cell therapy trials have been completed in this setting.[34-36] Cells were injected transendocardially into ischemic myocardial areas that were identified by nuclear perfusion imaging and electromechanical mapping in these trials. Except for one episode of ventricular tachycardia occurring during the mapping procedure,[34] and one pericardial infusion that was treated with cardiocentesis,[36] transendocardial injections were safe. No safety concerns related to the cell injections, especially no evidence for an increased risk of arrhythmias, emerged from these studies.

Losordo and co-workers explored the therapeutic potential of CD34+ cells in 24 patients.[34] CD34+ cells were collected from the peripheral blood after five daily injections of granulocyte colony-stimulating factor (G-CSF). Angina frequency and exercise time showed trends in favor of CD34+ cell therapy, but perfusion imaging at three and six months did not provide clear-cut evidence for a greater reduction in myocardial ischemia in the cell-treated group. A larger phase IIb study is now underway.[34]

In the PROTECT-CAD trial that included 28 patients, mnBMC injections were associated with improvements in NYHA class, exercise time, LVEF, and regional wall thickening over the target regions after six months. Angina class decreased similarly in cell and placebo-treated patients. On myocardial perfusion imaging, stress-induced perfusion defects tended to decrease more in the mnBMC treated patients.[35]

Van Ramshorst and co-worker randomized 50 patients to intramyocardial injections of mnBMC or placebo. A two- to five-fold higher number of mnBMC were administered in this study as compared to the PROTECT-CAD trial. At three months, the cell-injected group showed significantly greater improvements in LVEF and myocardial perfusion, and a more pronounced improvement in angina class, exercise capacity, and quality of life,[36] thus providing the strongest evidence, so far, that cell therapy can improve myocardial perfusion and anginal symptoms in patients with chronic myocardial ischemia. Considering that previous adjunctive therapies have failed to improve myocardial perfusion in patients with chronic myocardial ischemia,[37,38] these data are encouraging. It remains to be investigated if these treatment effects are sustained over time and associated with reduced morbidity and mortality (although symptomatic benefit may be the primary goal in these patients).

3. WHAT CAN WE EXPECT FROM ONGOING CLINICAL TRIALS?

Considering the heterogeneity in cell isolation protocols, trial design, and the methods used for outcome measurement, and given the fact that autologous cell preparations represent a medical product whose complexity far exceeds that of any drug that is currently prescribed to patients with coronary heart disease, it may not be surprising that mixed results have emerged from the first clinical trials in this emerging field. In the setting of AMI, the fairly large database provides encouraging evidence that intracoronary BMC transfer after AMI has the potential to enhance the recovery of LV systolic function beyond what can be achieved by current interventional and medical therapies. This conclusion is supported by two meta-analyses of randomized and non-randomized controlled clinical trials of intracoronary BMC transfer in patients with AMI.[27,28] In both of these analyses, cell transfer was associated with a significant improvement in global LVEF by 3.0 percentage points that was accompanied by a significant reduction in LV end-systolic volume and a non-significant reduction in LV end-diastolic volume, suggesting that cell transfer exerts a beneficial effect on LV systolic function and remodeling.[27,28] It has been pointed out that the magnitude of these effects is quite comparable to what is achieved by established therapies after AMI, including acute PCI and ACE inhibitor or beta-blocker therapy.[39] The effects of BMC transfer on LV function and remodeling beyond an observation period of approximately four to six months remain poorly characterized at this time, emphasizing the need to obtain long-term follow-up data in clinical trials.

Based on the favorable safety profile and the promising efficacy data, several clinical trials are underway to further explore the prospects of cell therapy in patients with various manifestations of coronary heart disease. As discussed below, important issues are addressed in these trials in an attempt to maximize patient benefit (Table 2). Considering that previous clinical trials have used apparently similar cell isolation protocols yet have observed largely different outcomes,[17,19,23,24] it will be absolutely critical to establish assays to assess cell functionality and the quality of the cell product. This however requires a better understanding which cellular functions determine clinical benefit.

3.1. Patient Selection

For safety reasons, initial studies in patients with AMI have included mostly individuals with moderately depressed baseline LVEF. These patients, however, have a favorable prognosis and may not be in need of cell therapy.[40] Moreover, it appears that patients with more extensive infarct damage benefit more from intracoronary BMC transfer in terms of LVEF improvement; these patients have been identified by more severely depressed baseline LVEF or stroke volumes,[17,25,33] or a greater transmural extent of the infarct.[12,13,15] Many of the ongoing trials focus on these higher-risk patients. Conversely, the presence of microvascular obstruction in the reperfused infarct territory, as identified by late enhancement MRI, may identify a patient subpopulation that does not respond to intracoronary BMC therapy (Stefan Janssens, personal communication).[15]

3.2. Procedural Details

Procedural details such as the timing of cell transfer, cell dose, cell type and cell isolation protocols, and the mode of cell delivery need

Table 2.   Ongoing Cell Therapy Trials in Patients with Coronary Heart Disease.

Study Identifier	Principal Investigator(s)	Acronym	Patients	Cells	Primary Endpoint	Route of Cell Delivery
Non-ST-elevation acute coronary syndrome						
ClinicalTrials NCT00711542	Andreas M Zeiher	REPAIR-ACS	100	Bone marrow-derived progenitor cells	Coronary flow reserve	Intracoronary
Acute myocardial infarction						
Controlled Trials ISRCTN17457407	Helmut Drexer Kai C. Wollert	BOOST 2	200	Bone marrow cells Low vs. high cell number Non-irradiated vs. irradiated cells	LVEF	Intracoronary
ClinicalTrials NCT00355186	Roberto Corti	SWISS-AMI	150	Bone marrow-derived stem cells	LVEF	Intracoronary Day 5–7 vs. day 21–28
ClinicalTrials NCT00684021	Carl Pepine *et al.*	TIME	120	Bone marrow mononuclear cells	LVEF	Intracoronary Day 3 vs. day 7 post MI
ClinicalTrials NCT00684060	Carl Pepine *et al.*	Late TIME	87	Bone marrow mononuclear cells	LVEF	Intracoronary 2–3 weeks post MI

(Continued)

Table 2. (*Continued*)

Study Identifier	Principal Investigator(s)	Acronym	Patients	Cells	Primary Endpoint	Route of Cell Delivery
ClinicalTrials NCT00501917	Hyo-Soo Kim	MAGIC Cell-5	116	Peripheral blood stem cells mobilized with G-CSF vs. G-CSF and darbepoetin	LVEF	Intracoronary
ClinicalTrials NCT00877903	Anthony DeMaria *et al.*		220	Allogeneic mesenchymal stem cells	LVESV	Intravenous
ClinicalTrials NCT00677222	Marc Penn Warren Sherman		28	Allogeneic mesenchymal stem cells	Safety	Perivascular
ClinicalTrials NCT00442806	Patrick Serruys	APOLLO	48	Adipose tissue-derived stem cells	Safety	
Ischemic heart failure						
ClinicalTrials NCT00526253	Warren Sherman *et al.*	MARVEL	390	Skeletal myoblasts	6-min walk test QOL, LVEF	Transendocardial
ClinicalTrials NCT00824005	Carl Pepine *et al.*	FOCUS	87	Bone marrow mononuclear cells	MVO$_2$, LVESV Ischemic area	Transendocardial
ClinicalTrials NCT00747708	Anthony Mathur	REGENERATE-IHD	165	G-CSF-stimulated bone marrow-derived stem/ progenitor cells	LVEF	Transendocardial vs. intracoronary

(*Continued*)

Table 2. (*Continued*)

Study Identifier	Principal Investigator(s)	Acronym	Patients	Cells	Primary Endpoint	Route of Cell Delivery
ClinicalTrials NCT00326989	Andreas M Zeiher	Cellwave	100	Bone marrow mononuclear cells	LVEF	Extracorporal shock wave, then intracoronary cell therapy
ClinicalTrials NCT00285454	Eric W Alton		60	Bone marrow mononuclear cells	Safety, perfusion Systolic function	Retrograde coronary venous delivery
ClinicalTrials NCT00462774	Boris Nasseri Christof Stamm	Cardio133	60	CD133$^+$ bone marrow cells	LVEF	Transepicardial during CABG
ClinicalTrials NCT00810238	Jozef Bartunek André Terzic	C-Cure	240	Bone marrow-derived cardiopoietic cells	LVEF	Transendocardial
ClinicalTrials NCT00768066	Joshua M Hare *et al.*	TAC-HFT	60	Bone marrow cells vs. mesenchymal stem cells	Safety	Transendocardial
ClinicalTrials NCT00644410	Jens Kastrup		60	Mesenchymal stem cells	LVEF	Transendocardial
ClinicalTrials NCT00587990	Joshua M. Hare *et al.*	PROMETHEUS	45	Mesenchymal stem cells	Safety	Transepicardial during CABG
ClinicalTrials NCT00721045	Nabil Dib *et al.*		60	Allogeneic mesenchymal precursor cells	Safety	Transendocardial
ClinicalTrials NCT00474461	Roberto Bolli		40	Cardiac stem cells harvested from right atrial appendage	Safety	Intracoronary

Unless otherwise stated, autologous cell sources are used. LVEF, left ventricular ejection fraction; LVESV, left ventricular end-systolic volume; QOL, quality of life; MVO$_2$, maximal oxygen consumption; CABG, coronary artery bypass grafting.

to be tailored to the specific disease setting resulting in hundreds of possible permutations thus highlighting the complexity of optimizing cell therapy protocols for patients with cardiovascular disease.

The success of BMC therapy after AMI may critically depend on the timing of cell transfer and cell dose. As indicated by a subgroup analysis of the REPAIR-AMI trial and one meta-analysis, intracoronary BMC transfer in the first days after AMI may be associated with less improvement in LVEF as compared with later cell delivery.[17,28] The same meta-analysis also suggested that the improvement in LVEF may correlate with the administered BMC dose.[28] Results from a small randomized-controlled clinical trial comparing the effects of low-dose and high dose mnBMC transfer after AMI support the hypothesis that there is a dose-response relationship. This hypothesis is currently addressed in the BOOST 2 trial.

Head-to-head comparisons of cell delivery strategies (REGENERATE-IHD trial) and cell types (TAC-HF trial), exploration of alternative cell delivery methods (e.g. transcoronary venous infusion or transcoronary arterial injection into the perivascular space), and improvements in intramyocardial injection needle design (e.g. needles which limit immediate washout and promote cell dispersion), may help to optimize existing cell therapy protocols (Table 2).[41–44] Progress can also be expected from the use of more comprehensive imaging techniques that help to characterize the target tissue (especially in patients with advanced and chronic disease) and to facilitate delivery of cells to tissue sites on the basis of their physiological characteristics and anatomic location.[41,45] As recently shown in a swine model of myocardial infarction of different ages, fusion of 3D magnetic resonance images with 2D fluoroscopic images can be used to precisely target transendocardial cell injections to the remote, peri-infarct, and infarct locations. This technique can be applied without the need for a combined X-ray/MRI suite and may be combined with electroanatomic mapping, if desired.[45]

State-of-the art imaging techniques and endpoint evaluation by external core labs are required to unequivocally demonstrate moderate functional effects of cell therapy. LV dimensions and systolic function, for example, should be evaluated by MRI rather than echocardiography or angiography.

3.3. Clinical Endpoints

Previous trials have not been powered to assess the impact of cell therapy on mortality and other clinical endpoints. In the REPAIR-AMI trial, the cumulative endpoint of death, recurrent myocardial infarction, or necessity for revascularization was significantly reduced in the mnBMC group as compared with placebo after 12 months. Likewise, the combined endpoint of death, AMI, and HF hospitalizations was significantly reduced after mnBMC transfer.[46] In a recently completed study in AMI patients, intracoronary mnBMC transfer was associated with a significant reduction of all-cause mortality after five years.[47] It should be noted that patients who denied cell treatment served as controls in this non-randomized study.[47] Trends in favor of BMC therapy with regard to the endpoints of death, risk of recurrent AMI, and HF hospitalizations, have also emerged from the meta-analyses.[27,28] Ultimately, outcome trials will have to be conducted.

4. WHERE DO WE GO FROM HERE?

Cell therapy has been shown in randomized-controlled clinical trials to improve the recovery of systolic function in patients with myocardial infarction. These first-generation trials highlight, for the first time, an exciting therapeutic avenue to enhance healing and functional regeneration of the infarcted heart. Trial results are somewhat inconsistent, however, and uncertainty persists regarding patient selection and cell isolation protocols. Obvious procedural issues are currently being investigated in second-generation trials. These aim to refine the protocols and identify patient populations who benefit most from cell therapy. Third-generation trials using pre-conditioning strategies are expected to address the current limitations of cell therapy, i.e. low transplanted cell retention and survival and impaired cell functionality in advanced cardiovascular disease. It is increasingly being recognized that the transplanted cells secrete factors that exert a paracrine effect. Identification of these factors by secretome analyses and bioinformatic approaches could advance protein-based therapies to promote healing and regeneration of the infarcted heart.[48]

ACKNOWLEDGMENTS

K. C. W. is supported by the Deutsche Forschungsgemeinschaft (KFO 136).

REFERENCES

1. Velagaleti RS, Pencina MJ, Murabito JM, Wang TJ, Parikh NI, D'Agostino RB, Levy D, Kannel WB, Vasan RS (2008) Long-term trends in the incidence of heart failure after myocardial infarction. *Circulation* 118: 2057–2062.
2. McMurray JJ, Pfeffer MA (2005) Heart failure. *Lancet* 365: 1877–1889.
3. Soonpaa MH, Koh GY, Klug MG, Field LJ (1994) Formation of nascent intercalated disks between grafted fetal cardiomyocytes and host myocardium. *Science* 264: 98–101.
4. Taylor DA, Atkins BZ, Hungspreugs P, Jones TR, Reedy MC, Hutcheson KA, Glower DD, Kraus WE (1998) Regenerating functional myocardium: improved performance after skeletal myoblast transplantation. *Nat Med* 4: 929–933.
5. Robey TE, Saiget MK, Reinecke H, Murry CE (2008) Systems approaches to preventing transplanted cell death in cardiac repair. *J Mol Cell Cardiol* 45: 567–581.
6. Chien KR, Domian IJ, Parker KK (2008) Cardiogenesis and the complex biology of regenerative cardiovascular medicine. *Science* 322: 1494–1497.
7. Dowell JD, Rubart M, Pasumarthi KB, Soonpaa MH, Field LJ (2003) Myocyte and myogenic stem cell transplantation in the heart. *Cardiovasc Res* 58: 336–350.
8. Blau HM, Brazelton TR, Weimann JM (2001) The evolving concept of a stem cell: entity or function? *Cell* 105: 829–841.
9. Wagers AJ, Weissman IL (2004) Plasticity of adult stem cells. *Cell* 116: 639–648.
10. Wollert KC, Drexler H (2005) Clinical applications of stem cells for the heart. *Circ Res* 96: 151–163.
11. Wollert KC, Meyer GP, Lotz J, Ringes-Lichtenberg S, Lippolt P, Breidenbach C, Fichtner S, Korte T, Hornig B, Messinger D, Arseniev L,

Hertenstein B, Ganser A, Drexler H (2004) Intracoronary autologous bone-marrow cell transfer after myocardial infarction: the BOOST randomised controlled clinical trial. *Lancet* 364: 141–148.

12. Meyer GP, Wollert KC, Lotz J, Pirr J, Rager U, Lippolt P, Hahn A, Fichtner S, Schaefer A, Arseniev L, Ganser A, Drexler H (2009) Intracoronary bone marrow cell transfer after myocardial infarction: 5-year follow-up from the randomized-controlled BOOST trial. *Eur Heart J* 30: 2978–2984.

13. Meyer GP, Wollert KC, Lotz J, Steffens J, Lippolt P, Fichtner S, Hecker H, Schaefer A, Arseniev L, Hertenstein B, Ganser A, Drexler H (2006) Intracoronary bone marrow cell transfer after myocardial infarction: eighteen months' follow-up data from the randomized, controlled BOOST (BOne marrOw transfer to enhance ST-elevation infarct regeneration) trial. *Circulation* 113: 1287–1294.

14. Schaefer A, Meyer GP, Fuchs M, Klein G, Kaplan M, Wollert KC, Drexler H (2006) Impact of intracoronary bone marrow cell transfer on diastolic function in patients after acute myocardial infarction: results from the BOOST trial. *Eur Heart J* 27: 929–935.

15. Janssens S, Dubois C, Bogaert J, Theunissen K, Deroose C, Desmet W, Kalantzi M, Herbots L, Sinnaeve P, Dens J, Maertens J, Rademakers F, Dymarkowski S, Gheysens O, Van Cleemput J, Bormans G, Nuyts J, Belmans A, Mortelmans L, Boogaerts M, Van de Werf F (2006) Autologous bone marrow-derived stem-cell transfer in patients with ST-segment elevation myocardial infarction: double-blind, randomised controlled trial. *Lancet* 367: 113–121.

16. Herbots L, D'Hooge J, Eroglu E, Thijs D, Ganame J, Claus P, Dubois C, Theunissen K, Bogaert J, Dens J, Kalantzi M, Dymarkowski S, Bijnens B, Belmans A, Boogaerts M, Sutherland G, Van de Werf F, Rademakers F, Janssens S (2009) Improved regional function after autologous bone marrow-derived stem cell transfer in patients with acute myocardial infarction: a randomized, double-blind strain rate imaging study. *Eur Heart J* 30: 662–670.

17. Schachinger V, Erbs S, Elsasser A, Haberbosch W, Hambrecht R, Holschermann H, Yu J, Corti R, Mathey DG, Hamm CW, Suselbeck T, Assmus B, Tonn T, Dimmeler S, Zeiher AM (2006) Intracoronary bone

marrow-derived progenitor cells in acute myocardial infarction. *N Engl J Med* 355: 1210–1221.

18. Dill T, Schachinger V, Rolf A, Mollmann S, Thiele H, Tillmanns H, Assmus B, Dimmeler S, Zeiher AM, Hamm C (2009) Intracoronary administration of bone marrow-derived progenitor cells improves left ventricular function in patients at risk for adverse remodeling after acute ST-segment elevation myocardial infarction: results of the Reinfusion of Enriched Progenitor cells And Infarct Remodeling in Acute Myocardial Infarction study (REPAIR-AMI) cardiac magnetic resonance imaging substudy. *Am Heart J* 157: 541–547.

19. Lunde K, Solheim S, Aakhus S, Arnesen H, Abdelnoor M, Egeland T, Endresen K, Ilebekk A, Mangschau A, Fjeld JG, Smith HJ, Taraldsrud E, Grogaard HK, Bjornerheim R, Brekke M, Muller C, Hopp E, Ragnarsson A, Brinchmann JE, Forfang K (2006) Intracoronary injection of mononuclear bone marrow cells in acute myocardial infarction. *N Engl J Med* 355: 1199–1209.

20. Lunde K, Solheim S, Forfang K, Arnesen H, Brinch L, Bjornerheim R, Ragnarsson A, Egeland T, Endresen K, Ilebekk A, Mangschau A, Aakhus S (2008) Anterior myocardial infarction with acute percutaneous coronary intervention and intracoronary injection of autologous mononuclear bone marrow cells: safety, clinical outcome, and serial changes in left ventricular function during 12–months' follow-up. *J Am Coll Cardiol* 51: 674–676.

21. Seeger FH, Tonn T, Krzossok N, Zeiher AM, Dimmeler S (2007) Cell isolation procedures matter: a comparison of different isolation protocols of bone marrow mononuclear cells used for cell therapy in patients with acute myocardial infarction. *Eur Heart J* 28: 766–772.

22. van der Laan A, Hirsch A, Nijveldt R, van der Vleuten PA, van der Giessen WJ, Doevendans PA, Waltenberger J, Ten Berg JM, Aengevaeren WR, Zwaginga JJ, Biemond BJ, van Rossum AC, Tijssen JG, Zijlstra F, Piek JJ (2008) Bone marrow cell therapy after acute myocardial infarction: the HEBE trial in perspective, first results. *Neth Heart J* 16: 436–439.

23. Hirsch A, Nijveldt R, van der Vleuten PA, Tio RA, van der Giessen WJ, Marques KM, Doevendans PA, Waltenberger J, Ten Berg JM, Aengevaeren WR, Biemond BJ, Tijssen JG, van Rossum AC, Piek JJ,

Zijlstra F (2008) Intracoronary infusion of autologous mononuclear bone marrow cells in patients with acute myocardial infarction treated with primary PCI: Pilot study of the multicenter HEBE trial. *Catheter Cardiovasc Interv* 71: 273–281.

24. Huikuri HV, Kervinen K, Niemela M, Ylitalo K, Saily M, Koistinen P, Savolainen ER, Ukkonen H, Pietila M, Airaksinen JK, Knuuti J, Makikallio TH (2008) Effects of intracoronary injection of mononuclear bone marrow cells on left ventricular function, arrhythmia risk profile, and restenosis after thrombolytic therapy of acute myocardial infarction. *Eur Heart J* 29: 2723–2732.

25. Tendera M, Wojakowski W, Ruzyllo W, Chojnowska L, Kepka C, Tracz W, Musialek P, Piwowarska W, Nessler J, Buszman P, Grajek S, Breborowicz P, Majka M, Ratajczak MZ (2009) Intracoronary infusion of bone marrow-derived selected CD34+CXCR4+ cells and non-selected mononuclear cells in patients with acute STEMI and reduced left ventricular ejection fraction: results of randomized, multicentre Myocardial Regeneration by Intracoronary Infusion of Selected Population of Stem Cells in Acute Myocardial Infarction (REGENT) Trial. *Eur Heart J* 30: 1313–1321.

26. Mansour S, Vanderheyden M, De Bruyne B, Vandekerckhove B, Delrue L, Van Haute I, Heyndrickx G, Carlier S, Rodriguez-Granillo G, Wijns W, Bartunek J (2006) Intracoronary delivery of hematopoietic bone marrow stem cells and luminal loss of the infarct-related artery in patients with recent myocardial infarction. *J Am Coll Cardiol* 47: 1727–1730.

27. Lipinski MJ, Biondi-Zoccai GG, Abbate A, Khianey R, Sheiban I, Bartunek J, Vanderheyden M, Kim HS, Kang HJ, Strauer BE, Vetrovec GW (2007) Impact of intracoronary cell therapy on left ventricular function in the setting of acute myocardial infarction: a collaborative systematic review and meta-analysis of controlled clinical trials. *J Am Coll Cardiol* 50: 1761–1767.

28. Martin-Rendon E, Brunskill SJ, Hyde CJ, Stanworth SJ, Mathur A, Watt SM (2008) Autologous bone marrow stem cells to treat acute myocardial infarction: a systematic review. *Eur Heart J* 29: 1807–1818.

29. Menasche P, Hagege AA, Scorsin M, Pouzet B, Desnos M, Duboc D, Schwartz K, Vilquin JT, Marolleau JP (2001) Myoblast transplantation for heart failure. *Lancet* 357: 279–280.

30. Menasche P (2007) Skeletal myoblasts as a therapeutic agent. *Prog Cardiovasc Dis* 50: 7–17.

31. Menasche P, Alfieri O, Janssens S, McKenna W, Reichenspurner H, Trinquart L, Vilquin JT, Marolleau JP, Seymour B, Larghero J, Lake S, Chatellier G, Solomon S, Desnos M, Hagege AA (2008) The Myoblast Autologous Grafting in Ischemic Cardiomyopathy (MAGIC) trial: first randomized placebo-controlled study of myoblast transplantation. *Circulation* 117: 1189–1200.

32. Menasche P (2009) Stem cell therapy for heart failure: are arrhythmias a real safety concern? *Circulation* 119: 2735–2740.

33. Assmus B, Honold J, Schachinger V, Britten MB, Fischer-Rasokat U, Lehmann R, Teupe C, Pistorius K, Martin H, Abolmaali ND, Tonn T, Dimmeler S, Zeiher AM (2006) Transcoronary transplantation of progenitor cells after myocardial infarction. *N Engl J Med* 355: 1222–1232.

34. Losordo DW, Schatz RA, White CJ, Udelson JE, Veereshwarayya V, Durgin M, Poh KK, Weinstein R, Kearney M, Chaudhry M, Burg A, Eaton L, Heyd L, Thorne T, Shturman L, Hoffmeister P, Story K, Zak V, Dowling D, Traverse JH, Olson RE, Flanagan J, Sodano D, Murayama T, Kawamoto A, Kusano KF, Wollins J, Welt F, Shah P, Soukas P, Asahara T, Henry TD (2007) Intramyocardial transplantation of autologous CD34+ stem cells for intractable angina: a phase I/IIa double-blind, randomized controlled trial. *Circulation* 115: 3165–3172.

35. Tse HF, Thambar S, Kwong YL, Rowlings P, Bellamy G, McCrohon J, Thomas P, Bastian B, Chan JK, Lo G, Ho CL, Chan WS, Kwong RY, Parker A, Hauser TH, Chan J, Fong DY, Lau CP (2007) Prospective randomized trial of direct endomyocardial implantation of bone marrow cells for treatment of severe coronary artery diseases (PROTECT-CAD trial). *Eur Heart J* 28: 2998–3005.

36. van Ramshorst J, Bax JJ, Beeres SL, Dibbets-Schneider P, Roes SD, Stokkel MP, de Roos A, Fibbe WE, Zwaginga JJ, Boersma E, Schalij MJ, Atsma DE (2009) Intramyocardial bone marrow cell injection for chronic myocardial ischemia: a randomized controlled trial. *JAMA* 301: 1997–2004.

37. Arora RR, Chou TM, Jain D, Fleishman B, Crawford L, McKiernan T, Nesto RW (1999) The multicenter study of enhanced external counterpulsation (MUST-EECP): effect of EECP on exercise-induced

myocardial ischemia and anginal episodes. *J Am Coll Cardiol* 33: 1833–1840.

38. Leon MB, Kornowski R, Downey WE, Weisz G, Baim DS, Bonow RO, Hendel RC, Cohen DJ, Gervino E, Laham R, Lembo NJ, Moses JW, Kuntz RE (2005) A blinded, randomized, placebo-controlled trial of percutaneous laser myocardial revascularization to improve angina symptoms in patients with severe coronary disease. *J Am Coll Cardiol* 46: 1812–1819.

39. Reffelmann T, Konemann S, Kloner RA (2009) Promise of blood- and bone marrow-derived stem cell transplantation for functional cardiac repair: putting it in perspective with existing therapy. *J Am Coll Cardiol* 53: 305–308.

40. Moller JE, Hillis GS, Oh JK, Reeder GS, Gersh BJ, Pellikka PA (2006) Wall motion score index and ejection fraction for risk stratification after acute myocardial infarction. *Am Heart J* 151: 419–425.

41. Bartunek J, Sherman W, Vanderheyden M, Fernandez-Aviles F, Wijns W, Terzic A (2009) Delivery of biologics in cardiovascular regenerative medicine. *Clin Pharmacol Ther* 85: 548–552.

42. Perin EC, Lopez J (2006) Methods of stem cell delivery in cardiac diseases. *Nat Clin Pract Cardiovasc Med* 3(Suppl 1): S110–S113.

43. Silva SA, Sousa AL, Haddad AF, Azevedo JC, Soares VE, Peixoto CM, Soares AJ, Issa AF, Felipe LR, Branco RV, Addad JA, Moreira RC, Tuche FA, Mesquita CT, Drumond CC, Junior AO, Rochitte CE, Luz JH, Rabischoffisky A, Nogueira FB, Vieira RB, Junior HS, Borojevic R, Dohmann HF (2009) Autologous bone-marrow mononuclear cell transplantation after acute myocardial infarction: comparison of two delivery techniques. *Cell Transplant* 18: 343–352.

44. Yokoyama S, Fukuda N, Li Y, Hagikura K, Takayama T, Kunimoto S, Honye J, Saito S, Wada M, Satomi A, Kato M, Mugishima H, Kusumi Y, Mitsumata M, Murohara T (2006) A strategy of retrograde injection of bone marrow mononuclear cells into the myocardium for the treatment of ischemic heart disease. *J Mol Cell Cardiol* 40: 24–34.

45. de Silva R, Gutierrez LF, Raval AN, McVeigh ER, Ozturk C, Lederman RJ (2006) X-ray fused with magnetic resonance imaging (XFM) to target endomyocardial injections: validation in a swine model of myocardial infarction. *Circulation* 114: 2342–2350.

46. Schachinger V, Erbs S, Elsasser A, Haberbosch W, Hambrecht R, Holschermann H, Yu J, Corti R, Mathey DG, Hamm CW, Suselbeck T, Werner N, Haase J, Neuzner J, Germing A, Mark B, Assmus B, Tonn T, Dimmeler S, Zeiher AM (2006) Improved clinical outcome after intracoronary administration of bone-marrow-derived progenitor cells in acute myocardial infarction: final 1-year results of the REPAIR-AMI trial. *Eur Heart J* 27: 2775–2783.

47. Yousef M, Schannwell CM, Kostering M, Zeus T, Brehm M, Strauer BE (2009) The BALANCE Study: clinical benefit and long-term outcome after intracoronary autologous bone marrow cell transplantation in patients with acute myocardial infarction. *J Am Coll Cardiol* 53: 2262–2269.

48. Wollert KC, Drexler H (2010) Cell therapy for the treatment of coronary heart disease: a critical appraisal. *Nat Rev Cardiol* 7: 204–215.

Index